The Event Safety Guide

A Guide to Health, Safety and Welfare
at Live Entertainment Events
in the United States

event
safety
alliance

Skyhorse Publishing

Skyhorse Publishing books may be purchased in bulk at special discounts for sales promotion, corporate gifts, fund-raising, or educational purposes. Special editions can also be created to specifications. For details, contact the Special Sales Department, Skyhorse Publishing, 307 West 36th Street, 11th Floor, New York, NY 10018 or info@skyhorsepublishing.com.

Skyhorse® and Skyhorse Publishing® are registered trademarks of Skyhorse Publishing, Inc.®, a Delaware corporation.

Visit our website at www.skyhorsepublishing.com.

Further information on the Event Safety Alliance can be found at www.eventsafetyalliance.org.

10 9 8 7 6 5 4 3

Library of Congress Cataloging-in-Publication Data is available on file.

ISBN: 978-1-62914-761-1

Printed in the United States of America

Contents

Contents (continued)

Foreword

The Event Safety Alliance was born at a time of crisis. Several high-profile incidents threatened the reputation of the live event industry, which had previously maintained a generally positive safety record. It became apparent that systemic issues affecting safety had to be identified and addressed in order to avoid further tragedies.

As a group of experienced event professionals began to explore these issues, we concluded that the greatest threat facing productions came not from an external source such as unpredictable weather, nor from carelessness or indifference by event organizers or staff. Rather, the common thread was limited knowledge of and planning for life safety. Although reference materials pertaining to safety had been available for years, few addressed the unique conditions facing most events in a manner that was straightforward and accessible for people like us.

The first actions for the Event Safety Alliance became obvious. To reduce the risk of future incidents, we would provide our peers with a safety resource written specifically for our industry, in a format we could use. For that, we needed to ensure that the information was field tested and proven effective. We needed not only the insight of safety experts, but also the experience of people working in the trenches.

From the moment we began developing this *Event Safety Guide*, the response has been overwhelming. Hundreds of professionals from across the event industry have contributed the ideas, practices, and suggestions which, along with essential information from existing safety codes and standards, form the core of this document.

Now the real work begins.

To realize the *Event Safety Guide's* full potential, we must now take it into the field. As you apply these best practices to your daily operations, we hope you will continue to give us the benefit of your experience, whether you are a fresh-faced intern or an industry veteran. We intend the *Event Safety Guide* to be a "living" document, which we will revise and improve as new approaches and technologies emerge.

And we will continue to build new tools to help you run safer events. Next on the horizon are establishing a trade association dedicated to event safety, and creating an annual conference where innovators in safety products and ideas can be heard and new visions for life safety can come to life.

Now go forth and spread the gospel of life safety. The lives of our peers and our guests, and the very future of our industry, are in your hands.

~   Jim Digby, Founding Member and
    President of the Event Safety Alliance

1. Introduction

1.1 Our Agenda: Life Safety First

We are people who have made our careers in live entertainment, who have experience and expertise, who take our jobs seriously so that other people may safely have fun.

In just the last few years, we have seen lives shattered as outdoor stages have collapsed in Alberta and Ottawa, in Tulsa and Indianapolis, in Belgium, in Toronto. Other outdoor structures have been no less affected. More people died when a bar's party tent blew over in St. Louis, Missouri, when lightning struck a crowd evacuating a racetrack in Pennsylvania, when a temporary advertising scaffold fell outside a Cape Town, South Africa concert venue. For each fatality, many times that number were hurt, property was destroyed, and lawsuits often followed.

Particularly after the August 2011 collapse of the temporary stage roof structure at the Indiana State Fair, industry professionals began talking about why these tragedies happened and what could be done to prevent them. Starting in January 2012 at Tour Link, then Pollstar Live!, then the International Association of Venue Managers' Academy for Venue Safety & Security and Severe Weather Planning and Preparedness course, a group of us decided to take matters into our own hands. The Event Safety Alliance includes tour managers, event producers, engineers, riggers, equipment lessors, roadies, safety specialists, and many more. We are people of action. It is not in our nature to sit idly by when there is work to be done.

Our conversations focused on operational best practices and decision-making within the live entertainment industry. We realized that the sort of event safety manual we were talking about had been relied upon in the United Kingdom since 1999. Once we concluded that we did not have to invent this wheel ourselves, we pooled our collective knowledge and experience to update and supplement the U.K.'s "Purple Guide."

Our mission is to promote life safety first -- to set forth in easily understood language the best operational practices currently available in the live event industry, and to make the awareness and application of life safety the highest priority of industry professionals. This *Event Safety Guide* is our first collective effort toward that goal.

We intend for this *Guide* to help industry professionals know what safe workplace practices might be, to heighten their understanding of the importance of safety in everything we do, and to engage in these best practices in their daily work. Doing the right thing is the best risk management we know.

We do not intend for the *Event Safety Guide* to be a roadmap for lawyers seeking to assign liability for tragedies, but we are aware that it can serve that purpose. The fear of litigation can be a strong motivator, and if it serves the cause of safety in this instance, then we are comfortable with that ancillary effect, too.

The frequency of disasters in our industry has not numbed us to their impact. To the contrary, we are increasingly shocked and saddened with each incident. If we appear to seek the unattainable, we do so in an effort to avoid the unimaginable.

1.2 How to Use a "Best Practices" Guide

1.2.1 This *Event Safety Guide* is intended to provide the people who create and organize live events with operational best practices to help the events run safely. On a variety of topics within our collective expertise, we state what we feel is required and why it is necessary or sensible to do so.

1.2.2 The "why" part of a situation is often the key. In some matters, there is an absolute right answer, a single best and most correct way to do something. We have emphasized those few rules we consider unbreakable. In a great majority of situations, however, there is more than one safe way to do something. For those, we have tried to identify important issues for you to consider as you seek to apply a general safety standard or principle to the particular factual circumstances you actually face. In other words, we try to teach you to think about safety for yourself, not just to follow rules that may apply to you to varying degrees, or not at all.

1.2.3 The *Event Safety Guide* is based on widely accepted principles of safety and risk assessment that apply to events that take place at a variety of venues such as purpose-built arenas, sites not designed for public entertainment, and open-air venues, among others. These principles expressly acknowledge that each event will be different and will require a particular configuration of elements, management, services and provisions.

1.3 How the Guide is Arranged

1.3.1 Good planning and management are fundamental to the success of any music event. The first chapter of the guide (after the Introduction) gives event organizers essential points to consider in these areas as well as general advice on legal duties.

1.3.2 Subsequent chapters provide advice on specific arrangements for the health and safety of those involved in events, including the provision of services and facilities. There are also chapters which give some specific guidance for different types of events. These chapters should not, however, be read in isolation of all other chapters.

1.3.3 Where other guidance is available, event organizers are recommended to refer to this. Technical details contained in ANSI E1.21, NFPA 1, NFPA 101, and The (ICC) International Fire and Building Codes, among others, will be important to include.

1.3.4 All event organizers are recommended to use the chapter headings as a checklist for planning the requirements for their event. By applying a risk assessment approach to the type and size of event, it should be straightforward to decide which elements from each chapter are relevant and to assess the level and type of provisions needed at a particular event.

1.4 Notice and Legal Disclaimer

1.4.1 **Development Process.** The Event Safety Alliance has created this *Event Safety Guide* through a consensus best practices development process. This process brings together volunteers representing various viewpoints and interests to achieve consensus on safety issues related to live entertainment events. While the Event Safety Alliance administers the process and establishes rules to promote fairness in the development of consensus, it does not independently test, evaluate, or verify the accuracy of any information or the soundness of any judgment contained in the *Event Safety Guide*.

1.4.2 **Disclaimer of Liability.** The Event Safety Alliance of USA, Inc., disclaims liability for any personal injury, property or other damages of any kind which may directly or indirectly result from the use of or reliance on this *Event Safety Guide*.

1.4.3 **Use of Independent Judgment.** The Event Safety Alliance of USA, Inc., is not undertaking to render professional or other services for or on behalf of any person or entity. Anyone using this *Event Safety Guide* should rely on his or her own independent judgment, or, as appropriate, seek the advice of a competent professional in determining the exercise of reasonable care in any given circumstance. Reviewing the *Event Safety Guide* also does not replace the need for event organizers to consult with local authorities and emergency services, or to follow all applicable laws and regulations.

1.4.4 **Enforcement.** The Event Safety Alliance of USA, Inc., cannot, and does not undertake to, enforce compliance with the contents of the *Event Safety Guide*. Nor does the Event Safety Alliance of USA, Inc., list, certify, test, or inspect products, designs, or installations for compliance with this document or any others.

1.4.5 **Updating.** The *Event Safety Guide* is intended to be a "living" document. That is, the particular text of this document may be supplemented, modified, or superseded at any time by the issuance of a new edition. To determine whether a given document is the most current version, visit the Event Safety Alliance website at www.eventsafetyalliance.org, or contact the Event Safety Alliance at the address listed below.

1.4.6 **Copyright.** The *Event Safety Guide* is copyrighted by the Event Safety Alliance of USA, Inc. It is made available for use by reference in laws and regulations, and for private self-regulations and the promotion of safe practices and methods. By making this document available for use and adoption by public authorities and private users, the Event Safety Alliance of USA, Inc., does not waive any copyrights in this document.

This *Guide* is based in large measure on HSG195, the U.K.'s Event Safety Guide published by the Health and Safety Executive. United Kingdom Crown Copyright is duly acknowledged for passages which remain unchanged from the original. For practitioners wishing to research current guidance on planning safe events in the United Kingdom, please refer to the Health and Safety Executive's event web site (http://www.hse.gov.uk/event-safety/index.htm).

1.4.7 **Further Information.** All communications regarding safety practices recommended or discussed in the *Event Safety Guide*, or suggestions for revisions or supplements, should be sent to the attention of Administrator, Event Safety Alliance, 8776 E. Shea Boulevard, Suite 106-510, Scottsdale, AZ 85260. For more information about the Event Safety Alliance, visit the website at www.eventsafetyalliance.org.

2. Planning and Management

2.0.1 The chapters in this guide offer advice and recommendations for organizers in planning a safe, successful event. It explains the principles that emphasize good health and safety management and sets out a basic approach that event organizers can adopt.

2.0.2 Events, even comparable events, can differ significantly. Even events that occur annually in the same location must adapt to the current elements influencing the event and not rely on implementing a "copy and paste" strategy. Smooth operating, well-executed events are the result of meticulous planning and preparation by the organizers.

2.0.3 Consciously thinking about hazards, risks, mediation and event safety reminds us that we all continually face and mediate multiple levels of risk every day. We take a raincoat when stepping out into the rain, check for traffic before crossing the street, hold a handrail while descending stairs—the list of things we do every day to identify hazards and mediate risk is seemingly endless.

2.0.4 As with our personal daily routine, the more experience a person has identifying and addressing workplace hazard, the more "second nature" those tasks become. Filling out a risk assessment form for the first time can be daunting and even cause "paralysis by analysis," meaning the process can frustrate a person to the point of discontinuing the process. An organizer just starting to incorporate safety systems into their events should endeavor to complete the task. Risk assessments will get easier with experience, and they will also become more involved and detailed as the organizer's awareness evolves.

2.0.5 Typically, the workplace is the location where our exposure to risk is greatest. It is impossible to make products and processes 100 percent safe. That fact does not provide an excuse for careless thinking, poor planning, hazardous conditions or working in an unsafe manner.

2.0.6 Successful safety policies in the workplace are most often the result of appropriate choices being made continuously by the individuals performing the work. So, an essential safety device on a job site is an alert, well-trained, well-equipped and engaged staff.

> **An essential safety device on a job site is an alert, well-trained, well-equipped and engaged staff.**

2.0.7 The Federal Emergency Management Agency (FEMA) produced an Independent Study online course titled "IS-15: Special Events Contingency Planning for Public Safety Agencies." It was updated in 2010 and includes a manual titled *Special Events Contingency Planning Job Aids Manual*. This manual defines a "special event" as, "...a non-routine activity within a community that brings together a large number of people" (p. 1-1). The manual further defines a "mass gathering" as "...a subset of a special event" (p. 1-1).

2.0.8 The event organizer is the individual or organization who promotes and manages an event. This role comes with many obligations and liabilities and it is the organizer's burden to discover what those obligations are and the organizer's exposure to liability, keeping in mind those obligations and liabilities can change event to event, even if an event is replicated at a later date in the same location.

2.0.9 Organizers should consider obtaining legal and insurance advice early in the planning stage. Items that warrant consideration include:
- Liability for injuries;
- Liability for acts or omissions;
- Liability for financial obligations incurred in responding to major emergencies occasioned by the event; and
- Potential liability for the resultant effects of the event on normal emergency operations.

2.0.10 Planning an event can be complicated. Planning for the reasonably foreseeable risks and hazards associated *with* an event is even more difficult and essential to the event's success. Before scheduling the event, the organizer should consider the scope of the event, the risks to spectators and participants, community impact, and the emergency support required (personnel and logistics).

2.0.11 To protect the health, safety and welfare of people attending an event, as well as the event staff, contractors and subcontractors working at the event, health and safety has to be managed. It is important to plan for effective health and safety management beginning at the same time as the planning for all other aspects of the proposed event.

2.0.12 Some form of legislation usually governs or restricts public events or aspects of them. Some events, particularly extremely large or high-impact events, may require special state or local legislation. On occasion, when an event requires a lot of interaction with the local government or when multiple local governments are affected by the proposed event, organizers employ firms that specialize in navigating the workings of "City Hall" to facilitate the permit process. While not cheap, this method can be effective.

2.0.13 Organizers should assume the event will need a permit regardless of the size, location or timing. Investigate this inevitability early—many months before the event—to assure compliance. The permitting process is always intended to enhance safety and should be viewed as such.

2.0.14 Site inspections may be required by several authorities having jurisdiction including the fire, building, electrical, and health departments. These inspections and other public services will likely have a cost to the organizer. The organizer should always assume the local government's policy is "User Pays" and budget accordingly. Organizers should always do their homework before committing to an event and know their obligations and liabilities as well as their associated costs to avoid future issues.

2.1 Initial Planning Considerations

2.1.1 One important consideration often overlooked by event organizers is the increased strain their event will place on public service agencies such as emergency management, law enforcement, fire and rescue, public works/utilities, public health, and medical facilities.

2.1.2 The first concern of these agencies and facilities will likely be the timing and location of the event so they can verify they will have the resources available to service the event. Financially challenged local jurisdictions simply may not have the resources or contingency to accommodate some larger events. In some cases, those communities may have a resource sharing arrangement with other neighboring communities.

2.1.3 If the community's public service agencies cannot acquire the necessary resources for the proposed event's day or time, the organizer may need to reschedule or relocate the event to accommodate the availability of the necessary resources or acquire them from contractor sources. At the very least, the event organizer should consider the effect that the availability, or unavailability of public resources may have on the organizer's ability to address reasonably foreseeable issues that may arise during the event.

2.1.4 It is recommended the organizer not promote or go on sale with an event before confirming the date and time with the community's public service agencies.

2.1.5 Early in the event planning process, a lead agency should be determined and the contact for that agency identified and introduced to the organizer. The reasoning behind a lead agency is because, on large events with many agencies involved, there is an obvious risk of confusion in matters of leadership. Often, the lead agency is the community's emergency management agency.

2.1.6 On occasion, the work load on public service agencies can delay the decision making process. If the organizer is unable to determine which agency is the lead agency for their event, they may have to be more assertive in the discovery process, especially if time is of the essence.

2.1.7 Many communities have existing planning protocols and systems in place. If the community has an existing plan that has proven to be successful, the organizer should consider using that plan and adjusting their event plan where necessary. The organizer's event plan is doubtless more nimble and capable of modification than trying to change the established systems of a multi-agency operations plan.

2.1.8 In addition to ensuring event operations run smoothly, the event organizer is (generally) also responsible for making a profit. An organizer who has an unyielding schedule and for whatever reason, is not involved in the above planning process may not be familiar with the laws or regulations of the community and can therefore unintentionally jeopardize public safety. This is why it is recommended busy organizers employ an experienced safety coordinator (see section 2.13) and assemble an event safety team (the collection of individuals addressing safety issues at an event) to counsel the organizer. This group should be empowered to address the issues of event safety regardless of whether the organizer is involved in the above planning process.

2.1.9 To assist organizers during the planning process there is a series of excellent checklists in "Appendix A" of the FEMA IS-15 *Special Events Contingency Planning Job Aids Manual* (2010, p. A-1).

2.2 Health and Safety Management

2.2.1 The key elements of successful health and safety management include:
- Creating a health and safety policy;
- Developing an event-specific health and safety plan to ensure the policy is put into practice;
- Organizing an effective management structure and distribution of the policy to include the responsible person for monitoring health and safety implementation; and
- Analyzing and reviewing performance.

2.3 Health and Safety Policy

2.3.1 A health and safety policy is a document that demonstrates the organizer's commitment to health and safety. The policy should contain details and show how the event organizer will put it into practice. It should also describe the roles and responsibilities of those people who have been given safety duties, such as the event safety coordinator. Even though the policy may delegate the authority to do certain things, the ultimate responsibility remains with the organizer.

2.3.2 The organization section of the safety policy should also contain the event's informative elements, e.g., organizational diagrams, maps, procedures and checklists. Organizational charts with relevant contact information should be posted (at least) in the event office showing the delegation of safety duties and the identification of people with the authority and competence to monitor safety and the resources that are available for health and safety.

2.3.3 The policy should address items including the maintenance of a safe place of work, safe working methods, safe access, provision of information, training and consultation with employees.

2.3.4 The health and safety policy may relate to a series of events if these are to be organized by the same event organizer. An event health and safety policy prepared for a series of events will need to be reviewed for each particular event for the organization and arrangements for health and safety.

2.3.5 It is important that the health and safety policy defines the hierarchy of health and safety responsibility for the duration of the event and shows who is responsible for recording these details in the safety policy. (For the purposes of this chapter, the "duration of the event" includes the entire period the event occupies the site, i.e. the beginning of load-in through the completion of load-out.)

2.3.6 In some states, the organizer or their agent acting as the general contractor is responsible not only for their own violations of federal labor law but also for those violations committed by their contractors and subcontractors. According to the US Department of Labor (DOL), by outsourcing some or all aspects of the execution of an event, organizers do not relieve themselves of their legal obligations. Prior to planning an event, it would be wise for organizers to become familiar with the risks and liabilities for which they are legally responsible.

2.3.7 If you or your company has been hired to promote and manage an event on behalf of another company or organization (e.g., a charity, club or a corporate client), your company may not actually be an employer or have any employees. However, it will still be necessary to establish who has the overall responsibility for compliance with local laws to ensure that the responsible parties are noted. Although most state laws and federal regulations can now be accessed on the Internet, local laws, codes and ordinances are not always found online and may need to be requested of local authorities having jurisdiction.

2.3.8 In some instances, events are organized by people or organizations where there is no actual producer or employer (e.g., various community events), so there may be no legal requirement to produce a health and safety policy. However, there is still a responsibility for the management of the public, staff, contractors and subcontractors, etc., on site. Producing such a policy in these circumstances is still recommended as it demonstrates diligence and provides a framework around which you can manage health and safety at the event.

2.3.9 If an event is to be staged in an existing venue such as an auditorium, rental outdoor event space, arena or a sports stadium, the event organizer will need to coordinate with the venue management regarding the existing arrangements for health and safety. In this instance, the event-specific safety policy would supplement the venue's existing policy.

2.4 Planning for Safety

2.4.1 Effective planning is concerned with hazard identification and the mitigation or elimination of those hazards to reduce or eliminate risks. The amount of time needed for planning will depend upon the event's size, type, location and duration. For some large events, as much as 6 to 12 months lead-time is required to plan the event properly. Smaller events can be prepared in several weeks. If the organizer is not experienced with event planning, it is recommended they allow more time so they are not rushed—a hazard in itself.

2.5 The Phases of an Event

2.5.1 The process for planning an event can be considered in separate parts. Some people find it easier to view the planning process as a progression of the event by the various phases involved. With time, most organizers develop a multiphase process they use to divide the tasks of their events into segments. For the purposes of this chapter we will use four phases: Planning, Pre-Production, Production and Post-Production. The below is only an example of the elements that may be involved in each phase and should only be used only as a reference.

2.5.2 The Planning Phase is the first order of business for the organizer. This is when RFPs are received and replies sent, client relationships are developed, budgets are created and approved, funding is acquired, and the design process occurs—often including the venue selection and permit process. Depending on the organizer's method, the operations or production team may be omitted from this phase allowing the sales or account teams to guide this part of the process. There are arguments for and against inclusion of the operations or production team during this phase; it really comes down to the organizer's preference.

2.5.3 The Pre-Production Phase is the period of operations and production preparation before the event begins and includes: final design development and engineering; vendor selection; health and safety planning, planning for logistics, crowd management, signage, waste management, the load out; and, development of strategies for dealing with fire, first aid and major incidents.

2.5.4 The Production Phase is the operational period of the event. It begins when the event first occupies the venue and continues until all the event's elements are removed from that venue and the final walkthrough is performed. The three basic subparts of this phase may be referred to as: the 'install' or 'load-in,' which is when the event's operational elements are delivered, installed and checked; the "show" or "event," which generally refers to the event or performance, and includes the period before and after a performance when the public or attendees occupy the front-of-house (FOH) areas of the venue; and the "load-out," "strike" or "dismantle," which is when all of he installed elements are removed, this time period is often fast-paced and therefore increases risk to working staff due to issues like fatigue, the quantity of staff and equipment in operation, severe drops in temperature, etc.

2.5.5 The Post-Production Phase is the period after the event when the operational elements used during the Production Phase are returned, final accounting is completed, recaps are written and in some cases assets are stored and managed. This phase may also include the execution of a site restoration plan to return the venue to its original state before the event began.

2.6 Planning for the Pre-Production Phase

2.6.1 To minimize risks during the load-in, you want to ensure, to the extent that it is reasonably possible under the circumstances, that the venue is designed for safety (see Chapter 8, *Venue and Site Design*). It is also necessary that someone ensure the event's infrastructure (i.e., stages, seating, tents, stages or other structures) be safely erected and structurally sound and monitored during operation (see Chapter 19, *Structures*).

2.6.2 Prepare diagrams showing the location of items such as delivery truck routes, stages, barriers, front-of-house towers, delay towers, cable routes, artist transportation routes, entries and exit points, emergency routes, first-aid and triage areas with ambulance parking locations, positioning of toilets, and merchandising stalls. In order to do this, the event organizer may have to obtain existing venue plans from the venue's owner or manager. Copies of diagrams may need to be given to the contractors delivering and building the infrastructure to ensure safe ingress as well as correct placement of the various structures to be used at the event. Plan the arrivals of all contractors and ensure their activities on site are coordinated with others.

2.6.3 To the extent that it is possible during the Planning Phase of an event, the event organizer may ask contractors and subcontractors to provide copies of their own health and safety policies in order to help identify any hazards and risks associated with their work *before* the load-in begins. Engineering documents and calculations may also need to be obtained in relation to the stages, seating or other temporary structures. These diagrams, plans, documents, and calculations will likely be needed during pre-production meetings when discussing your event with inspectors, local authorities and emergency services. Organizers should verify the laws in their state regarding their responsibilities for compliance by event contractors and subcontractors.

2.6.4 Plan for the provision of first-aid facilities for the people who will be working on site during the entire production phase; ensure that they are sufficient and will be available from the time that work starts until load-out is completed.

2.6.5 It is good practice to draw up a set of site safety rules and communicate these rules to the contractors during the vendor selection process, again before they arrive, and again when they arrive on site to begin work. Signage printed with these rules should be posted at venue entrances, in event offices and other pertinent areas. This practice on behalf of the organizer to inform all staff of safe working practices expected of them demonstrates the organizer's commitment to safety and will encourage compliance.

2.7 Planning for the Production Phase

2.7.1 Once the venue's infrastructure is built, other 'top layer' equipment and services will need to be brought to the site and installed in or on those structures (e.g., the loading of the performers' equipment onto the stage, which typically requires manual handling by staff) and the delivery of equipment to be used in the front-of-house concession and merchandise areas. The logistics of these operations will require careful planning as there are often multiple elements competing for the same space at the same time.

2.7.2 Planning for the show requires preparing strategies for crowd management, transportation management, fire, first aid, major incident, contingency planning and more. More specific details about planning these aspects can be found in their respective chapters in this *Event Safety Guide*. Successful planning for the show requires a team approach in which information is shared among law enforcement, fire, the health authority, other local authorities, venue management and security contractors.

2.7.3 Organizers are encouraged to create an event safety management team to coordinate planning the safety aspects of the event. The event safety management team should include members of the lead agency or local authority having jurisdiction over the event as well as emergency services providers. It is advisable to set up at least one or a series of safety planning meetings between the parties and to ensure the relevant agencies are aware of the planning process.

2.7.4 For large and complex events, "tabletop" emergency planning exercises may also be useful to test the viability of the emergency plans under low stress conditions. This not only reveals weaknesses in the plans without putting real people at risk, but encourages a collaborative relationship among the team members.

2.7.5 An event safety management plan. The elements of this plan should include at least the following elements:

- The event safety policy statement detailing the organization chart and levels of safety responsibility;
- The event risk assessment (see the "Event Risk Assessment" section below);
- Basic details of the event including venue layout, structures, audience profile, demographic, venue capacity, duration, food, toilets, trash, water, fire precautions, first aid, special effects, access and exits;
- The site safety plan detailing the site safety rules, site managers and safety coordinator, structural safety calculations and drawings;
- The crowd management plan detailing the numbers and types of staffing, methods of working, chains of command and organizational charts;
- The transportation management plan detailing the parking arrangements, traffic management plans, public transportation arrangements, and a description of site vehicles and vehicular routes inside the venue perimeter;
- The emergency action plan (EAP) detailing action to be taken by designated people if there is a major incident;
- The first-aid plan detailing procedures for administering first-aid on site and arrangements with local hospitals.

2.7.6 The event safety management plan should be reviewed and updated regularly as new information is received before and during the event. It is only necessary to distribute this plan to the key members of the event safety team. Take the steps necessary to ensure diligent document control so redundant or outdated documents are not mistaken for the final version. Examples of document control include prominent color-coding and date and time stamps.

2.7.7 Event safety planning meetings are an ideal way to ensure that the event safety management team members are updated on the content of the plan, as well as providing a mechanism for ensuring a flow of safety information on a regular basis. These meetings can be arranged in the weeks or days leading up to the event. If the event is to take place over a few days, e.g., citywide events or multi-day festivals, meetings should take place at least once each day of the event to ensure that all of that day's team members have been briefed.

2.8 Planning for the Post-Production Phase

2.8.1 During post-production, there is a natural reduction in the number of event staff as well as the physical footprint the event occupies. Administrative staff numbers typically reduce down to the staff required to manage the event's financial wrap up, manage the demobilization of administrative assets and prepare the event's recap. The operations staff numbers are reduced to those necessary to demobilize equipment and vendors, as well as restore the site and perform the final walkthrough with venue management.

2.8.2 The frequencies of risks are likely to be reduced during this phase. While the pace of work and stress levels are more relaxed for those persons involved in the post-production phase, they should remain diligent when it comes to hazard identification and risk management. The daily safety meetings should continue and log entries maintained and stored in the event binder.

2.9 The Event Hazard/Risk Assessment

2.9.1 A first critical step in developing a comprehensive safety and health program is to identify physical and health hazards at the work site. This process is known as a "hazard/risk assessment." Potential hazards and risks may be physical or health-related and a comprehensive assessment should identify hazards in both categories. Examples of physical hazards include moving objects, fluctuating temperatures, high intensity lighting, rolling or pinching objects, electrical connections and sharp edges. Examples of health hazards include overexposure to harmful dusts, chemicals or radiation.

2.9.2 The authorities having jurisdiction (AHJ) over your event may not require a comprehensive risk assessment. It is nonetheless recommended that organizers produce a risk assessment, as the process of considering reasonably foreseeable risks tends to cause event organizers to take steps to minimize the likelihood of such risks leading to actual problems. In other words, thinking in advance about problems should lead to doing something to try to prevent them from occurring.

2.9.3 The purpose of a hazard/risk assessment is to identify hazards that could cause harm, assess the risks which may arise from those hazards and decide on suitable measures to eliminate or control those risks. A comprehensive hazard/risk assessment for the load-in, show and strike, can only be fully completed once information has been received from the various contractors, vendors and event staff who will be working on site. It is a good practice for the person preparing the event risk assessment to be personally familiar with the venue as configured for the event.

2.9.4 A hazard is anything with the potential to harm people, structures, and/or facilities. This could be an item or a dangerous property of an item or a substance, a condition, a situation or an activity.

2.9.5 Risk is the likelihood that the harm from a hazard is realized and the extent of it. In a risk assessment, risk should reflect both the likelihood that harm will occur and the probably severity of that harm.

2.9.6 Hazards associated with mass gatherings vary according to the nature of the event. The previous history of the performers and the audience that they attract can provide valuable information. The overall event risk assessment will then indicate areas where risks need to be mediated.

2.9.7 The hazard/risk assessment should begin with a walk-through survey of the facility to develop a list of potential hazards in the following basic hazard categories:
- Impact;
- Penetration;
- Compression (roll-over);
- Chemical;
- Heat/cold;
- Harmful dust;
- Light (optical) radiation; and

- Biologic.

2.9.8 To assess the risk associated with staging the event:
- Identify the hazards associated with the event's activities and where the activities must be carried out and how the activities will be undertaken;
- Identify those people who may be harmed and how;
- Identify existing precautions, e.g., venue design, operational procedures or existing operational measures employed to mediate those hazards;
- Evaluate the risks;
- Determine what further actions may be required to mediate those hazards and risks, e.g., improvement in venue design, safe systems of work such as personal protection equipment (PPE), additional staff and/or staff training.

2.9.9 In addition to noting the basic layout of the work site and reviewing any history of occupational illnesses or injuries, things to look for during the walk-through survey include:
- Sources of electricity;
- Sources of motion such as machines or processes where movement may exist that could result in an impact between personnel and equipment;
- Sources of high temperatures that could result in burns, eye injuries or fire;
- Types of chemicals used in the workplace;
- Sources of harmful dusts;
- Sources of light radiation, such as welding, brazing, cutting, furnaces, heat treating, high intensity lights;
- The potential for falling or dropping objects;
- Sharp objects that could poke, cut, stab or puncture;
- Biologic hazards such as blood or other potentially infected material.

2.9.10 When the walk-through is complete, the employer should organize and analyze the data so that they may be efficiently used to determine the proper types of personal protective equipment (PPE) needed at the work site. The employer should become aware of the different types of PPE available and the levels of protection offered. It is recommended that employers select PPE that will provide a level of protection greater than the minimum required to protect employees from hazards. See Chapter 16, *Personal Protective Equipment*, for more information.

2.9.11 The hazard/risk assessment findings will need to be recorded and a system developed to ensure the document is reviewed and, if necessary, revised as plans are modified.

2.9.12 Documentation of the hazard/risk assessment should be required through a written certification that includes the following information:
- Identification of the workplace evaluated;
- Name of the person conducting the assessment;
- Date of the assessment; and
- Identification of the document certifying completion of the hazard assessment.

2.9.13 Persons creating a hazard/risk assessment form can find a variety of examples online. In addition, the Occupational Safety & Health Administration (OSHA) has developed a helpful Job Hazard Analysis (2002 Revised), which can be found online as OSHA Publication 3071 at http://www.osha.gov/Publications/osha3071.pdf.

2.10 Planning for the Load Out

2.10.1 When the event has ended, the organizer's responsibilities toward health and safety remain in place. Ensure that you have considered how the equipment and services will be removed from the venue and the stages, tents and roof structures after the event's performance or period of public access ends.

2.10.2 The same rules apply to the load out as were applied to the install. Ensure that site safety procedures are in place during this phase of the event.

2.10.3 A point mentioned above deserves to be reiterated: the load out period can still be a dangerous time for an event. People are usually in a hurry to leave and get their equipment off site as soon as possible and their haste can increase risk to all persons in the area. When considering the load out, consider issues such as available lighting, hunger, dehydration, fatigue, staff numbers, contractor workspaces, heavy equipment required, and weather.

> **The load out period can still be a dangerous time for an event.**

2.11 Organizing for Safety

2.11.1 It is important to clearly organize the health and safety policy statement in order to clearly identify which party has authority and responsibility for the myriad issues that go into the creation of a live event.

2.11.2 An effectively organized health and safety policy will highlight competence, control, cooperation and communication.

2.11.2.1 Competence is about ensuring that all producers, event staff, contractors, vendors and subcontractors working on the site have the necessary training, experience, expertise and resources to carry out their work safely. Competence is also about ensuring the right level of expertise is available, particularly about specialist advice.

2.11.2.2 Ensure that the vendors, contractors or subcontractors you intend to hire are competent in managing their own health and safety when working on site. Vendor health and safety policies and state and federal incident databases may be checked for information about the vendor's existing operational practice. If there are no policies in place, this may simply mean the vendor is in the stage of developing their policies, or worse, they have not considered a health and safety policy. Depending on the law of the state in which the event is being held, the general contractor may be legally responsible for contractor and subcontractor compliance with regulations and may also be legally responsible for their safety and well-being.

2.11.2.3 Controlling and enforcing the event's safety policies are central to maintaining a disciplined site. Control starts with producing a health and safety review, which details specific vendor health and safety responsibilities. Control also ensures that the contractors and subcontractors understand they will be held accountable for safety on site. Make sure contractors understand how health and safety will be controlled, monitored and enforced before they begin work on site.

2.11.3 Effective cooperation relies on the involvement of all parties. Active involvement in monitoring the site makes everyone part of the solution and contributes to the greater good. This collaboration and the exchange of information enable the risks to be suitably controlled.

2.11.4 Contractors, subcontractors and all event staff need to appreciate the hazards and risk to others working on site and cooperate with each other to minimize identified hazards and risks. Effective cooperation can be achieved by encouraging participation in the preparation of the site safety rules and plans.

2.11.5 Effective communication ensures that everyone working on site understands the importance and significance of the health and safety objectives. Make sure contractors, subcontractors, vendors and event staff are kept informed of safety matters and procedures to be followed on site.

2.12 Monitoring Safety Performance

2.12.1 Monitoring is essential to maintain and improve health and safety performance. There are two ways of generating information on safety performance: "active" and "reactive" monitoring systems.

2.12.2 Active onsite monitoring systems of standards and practices can prevent accidents or incidents. Active monitoring can be achieved by carrying out reviews of the contractors on site during the load in and load out, as well as regular reviews of their project timeline and tasks. Examples would be reporting that a contractor is not operating a forklift in a safe and professional manner, or reporting that ground riggers are not maintaining control of the area underneath overhead work.

2.12.3 Reactive monitoring systems are triggered after an accident or incident has occurred. They include identifying and reporting injuries, losses such as property damage, and incidents with the potential to cause further injury, or weaknesses or omissions in safety standards.

2.12.4 Information obtained during inspections or as a result of incidents or property damage should be recorded in the event log book (which can be electronically maintained). This book can also be used to keep other records and provides a unified document storage location until the information is reviewed at a later date. The goal of keeping the information and reviewing it during the Post Production Phase is to help the event management team assess safety performance against the standards set in the event safety policy.

2.13 The Safety Coordinator

2.13.1 Event organizers will likely need competent help in creating and applying the provisions of health and safety policies. A competent person is someone who has sufficient training, expertise, experience or knowledge and other qualities that enable that person to devise and apply protective measures. The exception may be smaller events when the organizer is competent to devise and apply protective measures themselves, although on large events most organizers will not have the capacity performance-wise to simultaneously produce the event *and* actively execute the duties of the safety coordinator.

2.13.2 It is recommended the organizer appoint a suitably competent safety coordinator to help the organizer comply with health and safety legislation. The safety coordinator should report directly to the organizer to eliminate the "filtering" of information by third parties. A safety coordinator can assist with the:

- Preparation and monitoring of site safety rules;
- Selection, information sharing and monitoring of contractors;
- Liaison with contractors, event staff and the health and safety enforcement authority on site;
- Checking of safety method statements and risk assessments;
- Communication of safety information to contractors on site;
- Checking of appropriate training and certificates required of certain staff and equipment such as structures, electrical, and heavy equipment operators;
- Monitor and maintain observation over the site for evolving hazards, the mediation of those hazards and reporting procedures for inclusion into the event log book;
- Monitoring and coordinating safety performance; and
- Coordinating safety in response to a major incident.

2.13.3 The safety coordinator should have access to all safety documentation supplied by the contractors and organizer. The safety coordinator also needs to be easily available to workers on site from load in through load out. The safety coordinator should also be a member of the event safety management team.

2.13.4 The safety coordinator should not have other competing roles which would divert his or her attention during the event. For example, event organizers should generally not appoint themselves as the safety coordinator.

2.14 Auditing and Reviewing Safety Performance

2.14.1 Once the event is completed the organizer and the relevant team members should perform an audit reviewing the systems used and the documents in the log book to establish that appropriate safety management arrangements were in place, adequate risk control systems existed and that they were put into practice. It is good practice to schedule this audit after every event so any problems identified in the planning, organization or any matters that evolved during the event can be analyzed and corrected for any future events. In preparation for the audit, the evaluations of law enforcement, fire department, health authorities, first-aid providers and other local authorities, as well as those of the safety coordinator, contractors and event staff and security contractors should be presented by those entities or solicited and incorporated into the log book for review during the audit.

2.14.2 Before memories fade, event organizers, in conjunction with local authorities, should review the effectiveness of the safety management systems for that event.

2.14.3 Once the event is completed and the organizer has "closed the book" on the event, store all documents and log books in a secure environment.

2.14.4 If the event is to be repeated or a similar event is scheduled, it would be wise for the organizer to study the documents and log books from the preceding event and consider using them as a starting point for the next event.

2.15 Liaison with the Local Authorities and Emergency Services

2.15.1 The lead agency and other local authorities may require a preliminary meeting so the proposed event can be discussed. Members of the emergency services and health and safety inspectors should attend. It will be helpful for the organizer to ask the lead agency to provide a checklist of information required for that meeting for prior approval along with the timeline for submitting that information. The information you supply should be sufficient to enable the local authority to examine your safety management systems and check any necessary plans, calculations and drawings.

2.15.3 Local authorities will not usually require a copy of every safety-related document in advance of the event. They may, however, require evidence you have planned your event safely before the event takes place. Ensure that any safety documentation is easily available for review. Keep all information in a safe place such as the event log book and in one location onsite, e.g., the event office file cabinet, to ensure the information is not misplaced. If requested, it is the event organizer's responsibility to produce this information.

2.15.4 Make suitable arrangements for local authorities to contact the organizer quickly for matters that may need immediate attention or further clarification. Last minute changes by the organizer are not conducive to good safety planning and management. If last minute changes are necessary, they will need to be approved by all parties involved. Changing the organizer's event plan is significantly easier than trying to change the established systems of a multi-agency operations plan and any proposed last minute changes may be rejected on that basis.

2.15.5 To assist organizers during the planning process there is a series of excellent checklists in "Appendix A" of the FEMA IS-15 *Special Events Contingency Planning Job Aids Manual* (2010, p. A-1).

2.16 Public Entertainment Permits

2.16.1 It is usually necessary to obtain an event permit from the appropriate local authority for most events. Permanent venues usually have the necessary permits with specific conditions for different types of events. If you are organizing an event in a venue with an existing permit you will need to familiarize yourself with its specific requirements to ensure compliance.

2.16.2 The presence of an existing permit does not replace the need to consider and implement the practices recommended within this document.

2.16.3 Continue to coordinate with the local authorities and members of the emergency services once the necessary permits to stage the event have been granted. It is advisable to invite these organizations to your event safety team meetings to ensure that they are updated on aspects of the event safety management plan.

2.17 Life Safety Evaluation

2.17.1 A "life safety evaluation" is a written review dealing with the adequacy of life safety features related to fire, storm, collapse, crowd behavior, and other related safety considerations. A life safety evaluation may be required by the local authorities having jurisdiction and is recommended regardless. A life safety evaluation must comply with all of the following:
* The life safety evaluation must be performed by persons acceptable to the authorities having jurisdiction;
* The life safety evaluation must include a written assessment of the safety measure for conditions listed below in section 2.17.2;
* The life safety evaluation must be approved annually by the authorities having jurisdiction and updated for special and unusual conditions.

2.17.2 Life safety evaluations must include an assessment of all of the following conditions and related appropriate safety measures:
(1) Nature of the events and the participants and attendees;
(2) Access and egress movement, including crowd density problems;
(3) Medical emergencies;
(4) Fire hazards;
(5) Permanent and temporary structural systems;
(6) Severe weather conditions;
(7) Earthquakes;
(8) Civil or other disturbances;
(9) Hazardous materials incidents within and near the facility; and
(10) Relationships among facility management, event participants, emergency response agencies, and others having a role in the events accommodated in the facility.

2.17.3 Life safety evaluations must include assessments of both building systems and management features upon which reliance is placed for the safety of facility occupants, and such assessment must consider scenarios appropriate to the facility.

2.18 Beyond the Venue Perimeter

2.18.1 This section is intended to give a brief introduction to increase the organizer's awareness of the planning and systems in place beyond the venue's perimeter.

2.18.2 In 2003, the President of the United States issued Homeland Security Presidential Directive (HSPD)-5. It created a consistent nationwide template to enable federal, state, local and tribal governments and private sector and non-governmental organizations to work together effectively and efficiently. As part of that template, these systems were developed and put into place: the National Incident Management System (NIMS) and the Incident Command System (ICS). The government recommends all public service agencies use those systems to prepare for and respond to an incident during a special event. See Appendix A, *The National Incident Management System (NIMS) and Incident Command System (ICS)*, for more detailed information.

2.18.3 A risk assessment is a process and the resulting document that estimates the impact a hazard would have on people, services, facilities and structures, and matches these estimates with descriptions of specific mitigations for each of the identified risks. An event's risk assessment is an important tool to identify and mitigate risk at an event because many identifiable risks are preventable with reasonable planning and preparation. There are other tools and materials that may also be useful to an event organizer, especially during the planning stages of an event.

> **Many identifiable risks are preventable.**

2.18.4 A hazard vulnerability assessment (HVA) is an emergency management tool often used by communities, counties and states to identify, using current knowledge and past experience, the people, structures and areas that are vulnerable to hazards. A local HVA usually includes a hazards map and can help event organizers identify areas vulnerable to natural and other types of hazards. Vulnerability identification determines the facilities and people at risk and to what degree they might be affected, as well as how they might affect other surrounding areas. The HVA is usually available through the local emergency management and/or public service agencies. Although it is a public record, it may be considered a sensitive and/or protected document and should be handled with discretion and according to the requirements of the providing agency.

2.18.5 The HVA is part of a larger Emergency Operations Plan (EOP) that is developed by local, county and state jurisdictions. A jurisdiction's emergency operations plan is a document that:
- Assigns responsibility to organizations and individuals for carrying out specific actions at projected times and places in an emergency that exceeds the capability or routine responsibility of any one agency, e.g., the fire department;
- Sets forth lines of authority and organizational relationships, and shows how all actions will be coordinated;
- Describes how people and property will be protected in emergencies and disasters;

- Identifies personnel, equipment, facilities, supplies, and other resources available—within the jurisdiction or by agreement with other jurisdictions—for use during response and recovery operations; and
- Identifies steps to address mitigation concerns during response and recovery activities.

2.18.5.1 In the U.S. system of emergency management, local government must act first to attend to the emergency needs of the public. Depending on the nature and size of the emergency, state and federal assistance may be provided to the local jurisdiction. The local EOP focuses on the measures that are essential for protecting the public. These include warning, emergency public information, evacuation, and shelter.

2.18.5.2 States play three roles in this process: They assist local jurisdictions whose capabilities are overwhelmed by an emergency; they themselves respond first to certain emergencies; and they work with the U.S. Federal Government when federal assistance is necessary. The State EOP is the framework within which local EOPs are created and through which the federal government becomes involved. As such, the State EOP ensures that all levels of government are able to mobilize as a unified emergency organization to safeguard the well-being of state citizens.

2.18.6 To better understand the depth of planning that exists that can ultimately influence your event, it is recommended that organizers take the online version of the FEMA IS-15.b "Special Event's Contingency Planning" course (http://training.fema.gov/EMIWeb/IS/is15b.asp). This course communicates a working knowledge of the systems in place and in use by local public service agencies, and prepares the organizer and staff to better assist those agencies should an incident occur at their event.

3. Major Incident (Emergency) Planning

3.0.1 Planning any event can be complicated. Planning for the potential risks and hazards associated with an event is even is essential to the event's success. This chapter covers the issues that should be addressed in the early stages of planning, promoting or sponsoring an event.

3.0.2 By making decisions, agreeing on actions and gathering and documenting useful information before an incident occurs, emergency planning helps ensure that event organizers and emergency personnel are able to respond quickly and effectively.

3.0.3 Before scheduling the event, consider the risks to spectators and participants, community impact, and the emergency support required (personnel and logistics). Also identify the lead agency and members of the planning team.

3.1 Preparation

3.1.1 The first concern with emergency planning is to identify times when the event may most foreseeably strain existing public safety agencies. Even in the earliest stages of planning, the organizer should begin to make contingency plans. These plans should consider licensing and regulations, emergency response issues, identifying persons responsible for responding to particular types of hazards and risks, resources and expenses, and jurisdictions.

3.1.2 During the initial planning stages, each agency should review resources to ensure that all necessary equipment is available when the event will be held. If the agencies determine that additional equipment is needed, then they may acquire the equipment or supplies and be ready for the event. Given enough lead time, communities may be able to work together or pool equipment.

3.1.3 One way in which agencies work together is by adopting a "mutual aid" program. This program allows neighboring communities to pool resources and share liability for damages or loss of equipment. If one community needs a particular piece of equipment, it may borrow it from a neighboring community. The equipment will become an asset of the borrowing community and will be covered under their insurance until it is released and returns to its home organization. It is important that those involved in planning the event know the agreements established between neighboring communities and the assets that are available to assist in responding to any unforeseen incidents. These agreements may already be established and included as a part of the local emergency operations plan.

3.1.4 As recent events have shown, the consequences of a major incident at a music event can be catastrophic. A major incident will normally require a multi-agency approach in which the event organizer, law enforcement, ambulance service, fire authority, local authority, local emergency planning officer, stewards and first responders may play a part. It is therefore important to have a clear demarcation of duties that are agreed to and understood at the event planning stage. Once finalized, joint emergency procedures should be issued in writing to all relevant parties.

3.1.5 Procedures to deal with serious and imminent danger in the workplace including evacuation are an OSHA requirement.

3.1.6 The National Incident Management System (NIMS) and the Incident Command System (ICS)

3.1.6.1 On Feb. 28, 2003, the President issued Homeland Security Presidential Directive (HSPD)-5, Management of Domestic Incidents, which directs the Secretary of Homeland Security to develop and administer a National Incident Management System (NIMS). This system provides a consistent nationwide template to enable federal, state, local and tribal governments and private-sector and non-governmental organizations to work together effectively and efficiently to prepare for, prevent, respond to and recover from domestic incidents, regardless of cause, size, or complexity, including acts of catastrophic terrorism.

3.1.6.2 NIMS provides a set of standardized organizational structures—such as the Incident Command System (ICS), multi-agency coordination systems, and public information—as well as requirements for processes, procedures and systems designed to improve interoperability among jurisdictions and disciplines in various areas, including training, resource management, personnel qualification and certification, equipment certification, communications and information management, technology support, and continuous system improvement. It is recommended that NIMS and ICS should be followed to prepare for and respond to an incident during a special event. ICS can also be used to organize the functions related to planning an event.

3.1.6.3 For information on the Incident Command System, please see Appendix A, *The National Incident Management System (NIMS) and Incident Command System (ICS)*. For training on NIMS ICS, visit the FEMA Emergency Management Insitute Independent Study web site (http://training.fema.gov/is/nims.asp).

3.2 Major Incident Defined

3.2.1 OSHA defines an "incident" as an unplanned, undesired event that adversely affects completion of a task. However, the U.S. National Incident Management System (NIMS) adds the element of "emergency response" to its definition of an incident and defines it as follows:

> An "incident" is an occurrence or event, natural or human-caused, that requires an emergency response to protect life or property. Incidents can, for example, include major disasters, emergencies, terrorist attacks, terrorist threats, wildland and urban fires, floods, hazardous materials spills, nuclear accidents, aircraft accidents, earthquakes, hurricanes, tornadoes, tropical storms, war-related disasters, public health and medical emergencies, and other occurrences requiring an emergency response.

3.2.2 For use in the *Event Safety Guide*, the definition of incident is broken into two types. A "minor incident" refers to a simple, undesired event (a) that adversely affects a task or process,

(b) whose consequences can be managed through normal service delivery, and (c) which is not likely to escalate. A minor incident may or may not require the involvement of local authorities. Low-level crime, lost children and minor injuries may all fall into this category. A "major incident" refers to an incident that does or is likely to require the implementation of special or non-routine arrangements and resources from one or more emergency services. A major incident would typically involve the local authorities for:

- The initial treatment, rescue and transport of a large number of casualties;
- The involvement either directly or indirectly of large numbers of people;
- The handling of a large number of inquiries likely to be generated both from the public and the news media, usually to the police;
- The need for the large scale combined resources of two or more of the emergency services;
- The mobilization and organization of the emergency services and supporting organizations (e.g., local authority, to cater for the threat of death, serious injury or homelessness to a large number of people).

3.2.2.1 Event authorities should place emergency responders on a heightened level of alert while an incident is confirmed or when there is an increased risk of something happening.

3.2.3 Contingency plans for minor incidents that do not typically require the intervention of emergency services should be developed in collaboration with local authorities. A minor incident could develop into a major incident if not properly planned for and managed. Event organizers should therefore develop contingency plans to deal with minor incidents along with their major incident plans. Major incident plans should be developed in conjunction with the emergency services.

3.2.4 It is important to identify in the plans the situations in which it will be necessary to hand coordination of an incident over to the police or other agency. This could be before any actual major incident has taken place if it is thought that a handover might prevent an incident from developing. It is also important to jointly determine the procedures for declaring a major incident and who declares it.

3.2.5 Further information on major incident planning can be found in FEMA's online IS-15.b course titled "Special Events Contingency Planning for Public Safety Agencies," which can be found at http://training.fema.gov/EMIWeb/IS/is15b.asp. The manual for the course (FEMA's *Special Events Contingency Planning Job Aids Manual*, 2005, Updated 2010) is also an excellent resource and can be found online at http://emilms.fema.gov/is15b/assets/SpecialEventsPlanning-JAManual.pdf.

3.3 Planning

3.3.1 Hazard identification is the key to developing a reliable risk assessment. The ability to identify and mediate risk is essential for those persons charged with writing event risk assessments. The task may appear daunting to a person new to the concept of hazard identification and mediation. The more practiced one is at hazard identification and writing risk assessments, the more routine the process becomes. Although the writing of reports may become a regular thing to do, the writer of the report must recognize each hazard is unique and may not always fit a template. The event risk assessment is a good starting point for any major incident plan. This will help focus on areas to be considered, including:

- The type of event, nature of performers, time of day and duration;
- Audience profile including age, previous or expected behavior, special needs, etc.;
- Existence or absence of seating;
- Geography of the location and venue;
- Topography;
- Fire/explosion;
- Terrorism;
- Structural failure;
- Crowd surge/collapse;
- Disorder;
- Lighting or power failure;
- Weather, e.g., excessive wind/heat/cold/rain;
- Off-site hazards, e.g., industrial plant;
- Safety equipment failure such as CCTV and PA system; and
- Cancellation, delayed start, curtailment or abandonment of the event.

3.4 Preparation of Major Incident Plans

3.4.1 Consider the following matters when preparing a major incident plan:
- Identification of key decision-making workers;
- Command post or meeting location where key decision makers will convene;
- Conditions and procedure for stopping the event;
- Identification of emergency routes and access for the emergency services;
- People with special needs;
- Identification of holding areas for performers, workers and the audience;
- Details of the script or coded messages to alert and 'stand down' security and stewards;
- Alerting procedures;
- Public warning mechanisms;
- Evacuation and containment measures and procedures;
- Details of the script of PA announcements to the audience;
- Identification of rendezvous points for emergency services;
- Identification of ambulance loading points and triage areas including helipads for air-evacuations;
- Location of hospitals in the area prepared for major incidents and traffic routes secured to such hospitals;
- Details of a temporary mortuary facility;
- An outline of the roles of those involved including contact list and methods to alert them;
- Details of emergency equipment location and availability; and
- Emergency reporting forms, documentation and message pads.

3.4.2 The plan should provide a flexible response whatever the incident, environment or available resources at the time. It may be necessary to prepare variations of the plan to deal with specific issues. The plan should also build on routine arrangements and integrate them into the existing working procedures on site.

3.4.3 Experience has shown that a multi-agency approach to all planning will share the ownership of problems and lead to effective solutions. A planning team should be created from people and agencies who will be required to respond to any emergency or major incident.

3.4.4 To be effective, the major incident planning team should not be too large. It may be useful to have a number of specialist subgroups. Each organization, e.g., police, fire department, first-aid provider, etc., concerned with the event should give a clear undertaking as to their role and committed resources if a major incident happens. This will be in the form of a statement of intent.

3.4.5 The person leading the planning team must be competent to do so and have a broad appreciation of the issues. This person does not necessarily have to be the event organizer or one of their workers. However, they will be accountable for the plan's effectiveness and for the person chosen to lead the team. The event safety coordinator should be involved in the planning process. Keeping and retaining records of meetings and decisions is important.

3.4.6 The plan should be easily understood and without jargon. Instructions, particularly with respect of action to be taken, must be specific so that a named person/role/rank will carry out a specific function. A glossary of terms may assist. Much time can be saved if the layout of the plan allows for simple and quick updating. Revised copies should be easily identifiable from a date/numbering system.

3.4.7 Off-site implications will form an important part of the plan. Traffic issues will include emergency access and exits, as well as readiness for an off-site incident occurring with consequences for the event. This could include a coach crash or large numbers of visitors stranded. Where a venue is close to county or other administrative boundaries, liaison may be required by the emergency planning officers of the local authority and the ability to provide mutual aid determined. Consult the local (usually county) emergency management coordinator to learn about the existing local emergency plans and give a copy of your event major incident plan to the local emergency management coordinator.

3.4.8 Detailed, gridded site plans containing pertinent geographic and topographic features will be of great value during planning and in the event of a major incident. They will be particularly useful when calculating normal and emergency pedestrian flow.

3.4.9 Think about testing the plan to check its effectiveness and the competence of the individuals and teams who will operate it. Methods can include simulation exercises or tabletop exercises. Exercises need not be full scale and may be designed to test only one element of the plan at a time. Debriefing following an exercise is constructive and will dispel misunderstandings that may have arisen and strengthen future working relationships.

3.4.10 Once the plan has been agreed, each organization must ensure that the people responsible for putting the plan into practice are fully briefed. By doing so, problems can be prevented in the first instance, but if one occurs, properly briefed workers can stop a situation deteriorating. Communication exercises are recommended before the event. The training of event staff is also an important safety element. Guest services, security staff, and others likely to have an emergency role must be informed of and trained to implement their duties and major incident procedures. Brief relevant people connected with the event, including concessionaires and those supplying other services that could be in a position to provide important assistance.

3.4.10.1 When possible, having a plan approved at the highest level possible—not just at an operational level—is a good idea since these are the individuals who may have ultimate decision-making responsibility should a major incident occur.

3.4 11 A major barrier to effective briefing is the transient nature of guest services and security positions and shift working by the emergency services. This situation can be made more difficult when additional workers are hurriedly brought in. Methods of informing workers in these circumstances can include individual, team or group presentations, written instructions and training videos.

3.5 Emergency Service and Responsibilities of Local Authorities

3.5.1 Once a major incident has been declared the police will coordinate and facilitate the "onsite" and "offsite" response. However, in the case of a fire, the fire department should be responsible for dealing with an onsite response. The agency responsible for emergency medical services will initiate coordination of the overall medical response at the scene, nominating and alerting receiving hospitals, distributing casualties, providing emergency transportation, communications and liaison with the other agencies. Local authorities can provide a range of services in case of a major incident. Services may include reception centers, temporary emergency accommodation, feeding and access to a wide range of special equipment. Know what these capabilities include so duplication of efforts can be avoided.

3.6 Cordons

3.6.1 In the event of a major incident, cordons may be needed. Discuss with the police, fire department and the emergency medical services provider how this would be carried out on site. Place cordons according to the circumstances. They may need to be moved during the incident.

3.7 Organizational Structure of Major Incidents

3.7.1 The United States uses the National Incident Management System (NIMS) and the Incident Command System (ICS) for dealing with major incidents where there is a need to coordinate incident management at operational, tactical and strategic levels. Some event organizers already use this model and if one is managing an event in the U.S. it is recommended everyone be familiar with these systems and consider taking FEMA's online IS-15b, IS-100 and IS-200 courses to learn more about them (for a course list and course descriptions, visit http://training.fema.gov/is/crslist.asp?page=all). At this time, training on the NIMS ICS for event management staff is voluntary. However, all jurisdictions in the U.S. are required to use the system so it is advisable that event organizers, promoters, planners and managers do the same and become acquainted and proficient with the system.

3.7.2 The NIMS ICS offers a simple management structure that eases coordination between responding agencies. This structure includes four sections: Command, Operations, Planning, Logistics and Finance/Administration.

3.7.3 In ICS, Command comprises the Incident Commander (IC) and Command Staff. Command staff positions are established to assign responsibility for key activities not specifically identified in the General Staff functional elements. These positions may include the Public Information Officer, Safety Officer, and the Liaison Officer, in additional to various others, as required and assigned by the IC.

3.7.4 In ICS, the General Staff includes incident management personnel who represent the major functional elements of the ICS, including the Operations Section Chief, Planning Section Chief, Logistics Section Chief, and Finance/Administration Section Chief. Command Staff and General Staff must continually interact and share vital information and estimates of the current and future situation and develop recommended courses of action for consideration by the IC.

3.7.5 ICS is based on proven management tools that contribute to the strength and efficiency of the overall system. The following ICS management characteristics are taught in ICS training programs and should be part of event and major incident planning:
- Common Terminology;
- Modular Organization;
- Management by Objectives;
- Reliance on an Incident Action Plan;
- Manageable Span of Control;
- Pre-designated Incident Mobilization Center Locations & Facilities;
- Comprehensive Resource Management;
- Integrated Communications;
- Establishment and Transfer of Command;
- Chain of Command and Unity of Command;
- Unified Command;
- Accountability of Resources and Personnel;
- Proper Dispatch/Deployment; and
- Information and Intelligence Management.

3.7.6 For more information on the details of NIMS ICS, visit FEMA's online ICS Resource Center at: http://training.fema.gov/EMIWeb/IS/ICSResource/index.htm. This site includes an excellent ICS overview document, a list of ICS training courses, ICS job aids, ICS forms, ICS position descriptions, ICS glossary of terms, multiple reference documents, and useful ICS-related links.

3.7.7 Other than at small events, it is essential that an appropriate onsite facility be set aside as a designated incident command post (ICP). While the event is running, make sure this onsite facility is staffed continuously. Consider the location of this facility in the overall venue and site design (see Chapter 8, *Venue and Site Design*, and Chapter 6, *Communication*).

3.7.8 If there is a major incident, the emergency services may dispatch command vehicles to the scene. These vehicles must have access to and be able to park near the incident command post. Make sure to consider this factor in the overall venue or site design.

3.8 Communication

3.8.1 Advice on communications and emergency public announcements can be found in Chapter 6, *Communications*.

3.9 Media Management

3.9.1 Given the immediacy of social media, which turns every patron with a cell phone into a potential news reporter, the media make inquiries as soon as a major incident develops. Consider appointing a chief press officer and identify a media rendezvous point to help with media liaison. In the event of a major incident, the police media manager may be responsible for the coordination of the response to the media.

3.10 Scene and Evidence Preservation

3.10.1 Any major incident may lead to criminal and civil proceedings. The police, fire department, health and safety inspectors and local authorities carry out evidence gathering and investigations. In the first instance, the police will be responsible for preserving the scene and the evidence. Obviously, this action will not interfere with the saving of life. Make sure that you are clear as to which officers and inspectors will need access to information to carry out any necessary investigations.

3.11 Voluntary Agencies

3.11.1 Many voluntary agencies can provide high-quality aid at incidents and if they are available at your event consider involving them in your emergency planning.

3.12 Some Specific Scenarios

3.12.1 Cancellation of an Event

3.12.1.1 If an event needs to be cancelled after the audience has arrived, or a performance has begun, stopped and not re-started, there will be a wide range of issues to be managed. Even if there has not been an actual major incident, property may have been lost or abandoned and people stranded. There may also be an expectation for refunds or the re-issuing of tickets. Think about preparing statements which can be given to the audience together with a press release to the public.

3.12.2 Stopping and Starting an Event

3.12.2.1 Once the event has begun, unscheduled stopping of the event could present serious hazards. Any decision to do so must be taken after careful consideration and consultation with the major incident planning team. Likewise, deciding whether or not and when to evacuate the audience will require fine judgment. Both unscheduled stopping and evacuation are scenarios that must be pre-planned and as far as practical, tested and rehearsed. The major incident plan must state who makes the decision to stop or start the event.

3.12.3 Bomb Threats

3.12.3.1 If a telephone bomb threat is received, details of the call must be recorded as accurately as possible. (See *Bomb Threat Checklist*, Appendix B1.) It is essential that the information is immediately passed to the police for evaluation and response.

3.12.3.2 The police will advise on the validity of a threat. Generally, any decision to evacuate or move people will rest with the event organizer. The exception is where a device is found or where police have received specific information. In these circumstances the police may initiate action and the directions of the senior police officer present must be complied with. If a bomb is a real threat, care must be taken to be alert for secondary devices (e.g., devices specifically intended to injure or kill responders). These might be aimed at the emergency services or the moved/evacuated audience.

4. Fire Safety

4.0.1 The principle goal of this chapter is to identify the steps necessary to prevent loss or injury through fire. This includes looking at measures to avoid fire risks, effective response should an incident occur, planning escape routes and firefighting measures. Further details must be obtained from the local fire and building authorities having jurisdiction, but a good place to start is the local fire department or district with jurisdiction over the venue.

4.0.2 The threat of fire to humanity, equipment and structures in most venues is reduced significantly when appropriate, comprehensive fire safety measures are taken. However, never underestimate how easily and quickly fire can threaten not only the venue but also everyone in and around it. Do not ignore the lessons learned over the past 200 years in the United States where seemingly insignificant situations quickly turned into epic tragedies. Fire should always be considered one of the most significant threats to an event and, thus, fire safety should always be treated as a priority and managed accordingly.

4.1 Codes and Standards

4.1.1 A "fire code" is a set of standards established and enforced by government for fire prevention and fire and life safety. A similar and related document is based on requirements for the safety, health, and quality of life of building users and neighbors. This is referred to as a "building code."

4.1.2 Organizers or their designees must be familiar with, and comply with, the applicable fire and building codes or risk a wide variety of consequences ranging from a simple scolding from a local official, to a citation (e.g., local, state, or federal), to a serious injury and/or death of an employee, entertainer, or member of the audience. Although this may sound overly dramatic, it is a fact. To put it bluntly, many in the safety services rightfully argue that these codes were written with the blood of those who died in the many tragedies that occurred throughout the evolution of the codes. Compliance with these codes is one way to prevent having to relive these tragedies.

4.1.3 In states (and countries) where the power of regulating construction and fire safety is vested in local authorities (a.k.a., "home rule"), a system of model fire and building codes is used. Model fire and building codes have no legal status unless adopted or adapted by an authority having jurisdiction. The developers of model codes urge public authorities to reference model codes in their laws, ordinances, regulations, and administrative orders. When referenced in any of these legal instruments, a particular model code becomes law. This practice is known as adoption "by reference."

4.1.4 There are instances when local jurisdictions choose to develop their own fire and/or building codes. At some point in time all major cities in the United States had their own fire and building codes. However, due to ever increasing complexity and cost of developing fire and building regulations, virtually all jurisdictions in the country have chosen to adopt model codes instead. For example, in 2008 New York City abandoned its proprietary 1968 New York City

Building and Fire Codes in favor of a customized version of the International Building Code (IBC) and the International Fire Code (IFC). On the other hand, the City of Chicago remains the only municipality in America that continues to use a building code the city developed on its own as part of the Municipal Code of Chicago.

4.1.5 For years, there were four model fire codes that were basically regional in nature. Each of them was adopted in 8-12 states. That changed in the late 1990s when three of the model codes merged to form the International Code Council (ICC) and jointly developed the International Fire Code (IFC) and the International Building Code (IBC). Today the IFC and NFPA 1 (The National Fire Protection Association's [NFPA's] model fire code) serve as the two national model codes.

4.1.6 All codes include references to standards. "Standards" are defined as a required or agreed level of quality or attainment. When standards are referenced in model codes, the language of the referenced standards becomes part of the code in which they are referenced. For example, both the IFC and NFPA 1 model codes include a reference to a version of the American National Standards Institute (ANSI)/American Petroleum Institute (API) Standard RP 651, *Cathodic Protection of Aboveground Petroleum Storage Tanks*. Rather than duplicate this very technical 33-page document, the model codes simply reference RP 651, which incorporates the requirements of the standard into the code. This practice is also known as adoption "by reference."

Table 4-1

Non-Exhaustive List of Fire Protection-Related, Standard-Setting Organizations

AFSI	Architectural Fabric Structures Institute
API	American Petroleum Institute
ASME	American Society of Mechanical Engineers
ASTM	ASTM International, formerly known as the American Society for Testing and Materials (ASTM)
BHMA	Builders Hardware Manufacturers' Association
CGA	Compressed Gas Association
CGR	U.S. Coast Guard Regulations
CPSC	Consumer Product Safety Commission
DOC	U.S. Department of Commerce
DOL	U.S. Department of Labor
EN	European Committee for Standardization (EN)
ICC	International Code Council, Inc.
IEC	International Electrotechnical Commission
ISO	International Organization for Standardization (ISO)
NEMA	National Electrical Manufacturer's Association
NFPA	National Fire Protection Association
UL	Underwriters Laboratories, Inc.
USC	United States Code

4.1.7 A number of agencies, associations, and other types of organizations develop standards that can be referenced. Table 4-1 shows a non-exhaustive list of fire protection-related, standard-setting organizations, most of which are referenced in the model codes.

4.1.8 Since a complete presentation of both model codes is beyond the scope of this document, and since the codes that apply at any event and at any venue will vary depending on the location, this chapter will focus only on the most important aspects of fire safety at a music or entertainment event, which usually revolve around what is referred to in most fire codes as "means of egress" (e.g., a continuous and unobstructed path of vertical and horizontal egress travel from any occupied portion of a building or structure to a public way). Fortunately, the two U.S. model code sets are very similar in most details. Some of this commonality, especially as it relates to entertainment venues, will be described here.

4.1.9 Although building codes will not be specifically mentioned in this chapter, their application and use is implied because fire and building codes complement, supplement and often reiterate each other's content and intent. More details on elements of relevant building codes, which apply to both permanent and temporary structures, can be found in Chapter 19, *Structures*.

4.2 Selected Fire Safety Definitions

4.2.1 "Air-inflated structure" is a building where the shape of the structure is maintained by air pressurization of cells or tubes to form a barrel vault over the usable area. Occupants of such structures do not occupy the pressurized areas used to support the structure.

4.2.2 "Air-supported structure" is a structure wherein the shape of the structure is attained by air pressure, and occupants of the structure are within the elevated pressure area.

4.2.3 "Area of refuge" is an area where persons unable to use stairways (usually in wheelchairs) can remain temporarily to await instructions or assistance during emergency evacuation. At least one state's code uses the term "Areas of Rescue Assistance" as an equivalent term.

4.2.4 "Assembly occupancy" (Group A) is a specific classification of building occupancy. It includes, among others, the use of a building or structure, or a portion thereof, for the gathering together of persons for purposes such as civic, social or religious functions; recreation, food or drink consumption; or awaiting transportation (International Fire Code, 2009). More specifically, an "A-4" assembly occupancy includes arenas and skating rinks, and an "A-5" assembly occupancy includes amusement park structures, bleachers, grandstands, and stadiums, which are the most likely to serve as entertainment venues.

4.2.5 "Automatic sprinkler system," for fire protection purposes, is an integrated system of underground and overhead piping designed in accordance with fire protection engineering standards. The system includes a suitable water supply and a network of specially sized or hydraulically designed piping installed in a structure or area, generally overhead, to which automatic sprinklers are connected in a systematic pattern. The system is usually activated by heat from a fire and discharges water over the fire area.

4.2.6 "Exit" is that portion of a means of egress system that is separated from other interior spaces of a building or structure by fire-resistance-rated construction and opening protective as required to provide a protected path of egress travel between the exit access and the exit discharge. An example would be an appropriately constructed exit stair well in a three story building.

4.2.7 "Exit access" is that portion of a means of egress system that leads from any occupied portion of a building or structure to an exit.

4.2.8 "Exit discharge" is that portion of a means of egress system between the termination of an exit and a public way. In some parts of the world, this is equivalent to what is termed a "final exit."

4.2.9 "Fire extinguisher" (a.k.a., flame extinguisher or simply an extinguisher) is an active fire protection device used to extinguish or control small fires, often in emergency situations. Typically, a fire extinguisher consists of a hand-held cylindrical pressure vessel containing an agent which can be discharged to extinguish a fire. It is not intended for use on an out-of-control fire, such as one which has reached the ceiling, endangers the user (i.e., no escape route, smoke, explosion hazard, etc.), or otherwise requires the expertise of a fire department.

4.2.10 "Fire watch" is a temporary measure intended to ensure continuous and systematic surveillance of a building or portion thereof by one or more qualified individuals for the purposes of identifying and controlling fire hazards, detecting early signs of unwanted fire, raising an alarm of fire and notifying the fire department.

4.2.11 "Means of egress" (a.k.a., means of egress system) is a continuous and unobstructed path of vertical and horizontal egress travel from any occupied portion of a building or structure to a public way. A means of egress consists of three separate and distinct parts: the exit access, the exit, and the exit discharge (see these definitions above).

4.2.12 "Membrane structure" is an air-inflated, air-supported, cable or frame-covered structure (usually further defined in the building code).

4.2.13 "Occupant load" or "design occupant load" is the number of persons for which the means of egress of a building or portion thereof is designed. In some parts of the world, this is roughly equivalent to the term "occupant capacity," which is defined as the maximum number of people who can be safely accommodated at the venue. Occupant load is determined by the building and/or fire code officials having jurisdiction. These officials will compute the occupant load by establishing the maximum floor area allowances (in square feet) per occupant based on code requirements. For example, the IFC states that for assembly areas without fixed seating, chairs, or tables (standing only), a maximum floor area allowance per occupant is 5 square feet. Thus, the occupant load for an area that matched these characteristics and measured 100 feet wide by 100 feet long (10,000 square feet) would be 2000 occupants (10,000 / 5 = 2000). However, the building and/or fire official having jurisdiction must still authorize this number because they must also consider a number of other important factors such as means of egress issues, aisles, fences, exiting from multiple levels, stairs, and several others. Thus, it is important to note that occupant load is more than just a computation, it is a value reached by authorities having jurisdiction after considering a number of important factors.

4.2.14 "Panic hardware" is a door-latching assembly incorporating a device that releases the latch upon the application of a force in the direction of egress travel.

4.2.15 "Public way" is a street, alley or other parcel of land open to the outside air leading to a public street. The IFC also define "public way" as having a minimum clear width and height of not less than 10 feet (3.048 m). In some parts of the world, a public way is roughly equivalent to a "place of safety," which is defined as a place in which a person is no longer in danger from fire.

4.2.16 "Tent" is a structure, enclosure or shelter, with or without sidewalls or drops, constructed of fabric or pliable materials supported by any manner except by air or the contents that it protects.

4.3 Means of Egress

4.3.1 Whether the venue is in a building or outdoors, it is likely that some adaptation may be needed to accommodate a music event. This section describes some general means of egress concepts which may need to be addressed for buildings, sports and outdoor venues to safely accommodate a music event.

4.3.2 The proper design and construction of an adequate means of egress usually falls under the authority of the local building authority having jurisdiction. So, the building authority having jurisdiction must be consulted when designing and constructing a means of egress for any venue. The fire authority having jurisdiction also has some regulatory influence and so should be consulted during the design phase, as well.

4.3.3 All means of egress must comply with the requirements of NFPA 1, *Fire Code*, NFPA 101, *Life Safety Code*, and NFPA 101B, *Code for Means of Egress for Buildings and Structures*.

4.3.3 General Principles for Means of Egress

4.3.3.1 Every venue should be provided with exits that are sufficient for the number of people present in relation to their width, number and siting. Normally, no exit may be less than three feet (0.914 m) wide.

4.3.3.2 People should be able to move to safety along a clearly recognizable route by their own unaided efforts regardless of where a fire may break out at the venue. However, for some people with disabilities it will be difficult, if not impossible, to make their way to a place of safety without the assistance of others. Consider carefully the arrangements for these people. See Chapter 11, *Facilities for Persons with Special Needs*.

4.3.3.3 When evacuation is necessary, people often try to leave the way they entered. This is a well-documented phenomenon. If this is not possible (perhaps because of the position of the fire or smoke), they need to be able to turn away from the fire and find an alternative route to a place of safety. However, the audience may underestimate the risk or be reluctant to use exits with which they are unfamiliar. It is essential to train stewards (see Chapter 38, *Glossary of Useful Terms*, for definition) to recognize this fact and to ensure that the audience leaves promptly through the safest exit. This is yet another reason why ALL exits must be well marked/signed and lighted and why it may be important to notify the audience (through sound system or stewards) of the exits that should be used.

4.3.3.4 Information concerning fire safety for temporary structures used for entertainment purposes, which includes marquees, large tents and other membrane structures, can be found in local fire and building codes. However, before a permit to erect a tent or membrane structure will be issued (and permits will be required), expect that the fire authority having jurisdiction will require a certificate executed by an approved testing laboratory (approved by the fire authority having jurisdiction) certifying that the tents and membrane structures and their appurtenances, sidewalls, drops and tarpaulins, floor coverings, bunting and combustible decorative materials and effects, including saw dust when used on floors or passageways, are composed of material meeting the flame propagation performance criteria of NFPA 701, *Standard Methods of Fire Tests for Flame Propagation of Textiles and Films*.

4.3.3.5 Elevators, escalators and moving walks must not be used as a component of a required means of egress from any part of a building.

4.3.3.6 Any stairway, lobby, corridor or passageway (exit access, exit, or exit discharge), which forms part of the means of egress from the venue, must be unobstructed, of an approved minimum width, and constructed and arranged so as to provide a safe escape for the people using it. The aggregate capacity of stairways should be sufficient for the number of people likely to have to use them at the time of a fire. It is also necessary to consider the possibility of one stairway being inaccessible because of fire and the aggregate width should allow for this possible reduction. The width of any element of a means of egress must be computed, established and authorized by the local fire and building authorities having jurisdiction.

Fig. 4-1 – Marked fire aisle to door. When loading, moving and storing equipment and equipment containers, it is easy to forget that an aisle must remain open for egress. Marking the aisle with bright tape helps remind everyone that an aisle must be maintained to allow for access to the exit. Keeping the marked area completely clear is an important part of an effective safety culture. Photo courtesy of Steve Lemon.

4.3.3.7 Guy lines, guy ropes and other support members for a tent or membrane structure must not cross a means of egress at a height of less than 8 feet (2.438 m).

4.3.3.8 Stairways must have a width of not less than 36 inches (0.914 m) and a minimum headroom clearance of not less than 80 inches (2.032 m).

4.3.3.9 Stairways wider than about six feet (1.8 m) should normally be divided into sections, each separated from the adjacent section by a handrail, so that each section measured between the handrails is not normally less than three feet (0.914 m) wide. Consult local building and fire authorities having jurisdiction for specifics.

4.3.3.10 Ramps installed for wheelchair users must conform to local building and/or fire codes and possibly the Americans with Disabilities Act (ADA) of 1990 (see below for more details on ADA). Where ramps are used, keep in mind that:
- The slope should be constant and not broken by steps;
- Ramps must have landings at least as wide as the ramp and at least 60 inches long (1.525 m) located at the bottom and top of each ramp, points of turning, entrance, exits, and at doors;
- Where changes in direction of travel occur at landings provided between ramp runs, the landing must be at least 60 inches (1.525 m) by 60 inches (1.525 m) minimum;
- When the slope of a ramp is greater than one unit vertical in 20 units horizontal (5 percent slope), additional requirements may be applicable and should be discussed with the local authorities having jurisdiction;
- The maximum running slope for a ramp which is part of a means of egress must not exceed one unit vertical in 12 units horizontal (8 percent slope);
- The maximum running slope for a ramp which is NOT part of a means of egress must not exceed one unit vertical in 8 units horizontal (12.5 percent slope);
- The cross slope of a ramp measured perpendicular to the direction of travel must not be steeper than one unit vertical in 48 units horizontal (2 percent slope);
- The minimum clear width between handrails of a ramp which is part of a means of egress must be no less than 36 inches (0.914 m);
- The maximum vertical rise for any ramp must be no more than 30 inches (0.762 m);
- Ramps with a rise greater than 6 inches (0.152 m) must have handrails on both sides;
- Some elevated walking surfaces such as ramps are required to have guard rails (a.k.a., "guards") that minimize the possibility of a fall (consult with local authorities having jurisdiction);
- Ramps are required to have edge protection that minimize the possibility of going off the edge of the ramp with a wheelchair; and
- Each ramp surface must be of slip resistant materials that are securely attached.

4.3.3.11 As a general principle, if a building is used for public assembly, a door used for means of egress should open in the direction of travel. Also, the door should:
- Not open across the means of egress, thus reducing the width of the means of egress;
- Be hung to open through not less than 90 degrees and with a swing which is clear of any change of floor level;

- Be provided with a vision panel (window) if it is hung to swing both ways; and
- If protecting an exit, be a self-closing, rated, fire-resistant door.

4.3.3.12 Doors that protect exits (e.g., stairwell doors, fire-resistive doors, etc.) will be specially constructed fire-resistant doors and should not be propped or blocked open. To confirm that a door is a listed fire door ("listed" means equipment, materials, products, or services that are included in a list published by an organization acceptable to the fire and/or building code authority having jurisdiction; e.g., Underwriters Laboratories [UL], etc.), check for a small placard on the hinge edge of the door that indicates the rating of the door (e.g., 1 hour fire rating, 2 hour fire rating, etc.).

4.3.3.13 Where doors have to be kept fastened while people are present, they should be fastened only by pressure release panic hardware such as panic bolts, panic latches or pressure pads which ensure that the door can be readily opened by pressure applied by people from within traveling in the direction of egress.

4.3.3.14 It may be necessary for egress routes to be protected by fire-resistant construction and fire doors. All such doors, except those to cupboards and service ducts, should be fitted with effective self-closing devices to ensure the positive closure of the door. Rising butt hinges are not normally acceptable.

4.3.3.15 All fire doors should be regularly checked to ensure that they are undamaged, swing freely, and are closely fitted to frame and floor and that the self-closing device operates effectively.

4.3.3.16 Any door which for structural reasons cannot be hung to open outward may not count as a required exit door and must not be considered part of a means of egress. Consult the local building and fire authorities having jurisdiction. Some alternatives may be possible (e.g., locking it in the open position when the building is occupied) but must be approved by the local building and fire authorities having jurisdiction.

4.3.3.17 Doors, gates and turnstiles which are part of the means of egress, including all doors leading to exits, should be checked before and during the event to ensure that they are unlocked, or in circumstances where security devices are provided, can be immediately opened with panic hardware. Turnstiles or similar devices that restrict travel to one direction must not be placed so as to obstruct any required means of egress.

4.3.3.18 Security fastenings such as padlocks and chains should not, under any circumstances, be used when the venue is occupied; they should be placed on numbered hooks in a position which is not accessible to unauthorized people when the building is occupied. All fastenings should be numbered to match the numbered hooks.

4.3.3.19 Events featuring pyro, fog effects and/or other activities that may obscure one's vision during egress will slow and perhaps even obstruct egress. So, when any of these types of activities are planned, make sure all means of egress are fully clear and available or add additional egress routes.

4.3.4 Exit and Directional Signs

4.3.4.1 In an emergency, it is essential that all available means of egress are used. Clearly indicate all available exit routes so that members of the audience and workers are aware of all the routes to leave the venue in an emergency. In addition, the provision of easily visible exit signs in full view of everyone present will give a feeling of security in an emergency.

4.3.4.2 All fire safety signs, notices and graphic symbols must conform to the applicable fire and building codes. Exit and exit access doors must be marked by an approved exit sign well illuminated and readily visible from any direction of egress travel. The path of egress travel to exits and within exits must be marked by readily visible exit signs to clearly indicate the direction of egress travel.

4.3.4.3 Where an exit cannot be seen or where people escaping might be in doubt as to the location of an exit, provide directional exit signs at suitable points along the egress route. Such signs should be sufficiently large, fixed in conspicuous positions, and wherever possible be positioned approximately six feet (1.829 m) above the ground level.

4.3.4.4 Although the model codes require that exit sign lettering be no less than six inches (0.1524 m) tall, exit signs that must be seen from distances greater than 100 feet (30.48 m) must be larger to be seen. Therefore, it is recommended that exit signs that must be seen from greater than 100 feet (30.48 m) be at least eight inches (0.2032 m) in height.

4.3.4.5 Exit signs and signs incorporating supplementary directional arrows should be illuminated whenever people are present. Signs at outdoor events should be weatherproof and clearly visible above people as well as illuminated at night, if necessary.

4.3.4.6 The means of egress, including the exit discharge, must be illuminated to at least one foot-candle (11 lux) at all times the building space is occupied. In addition, if used outside the hours of daylight, or in the absence of natural daylight, all parts of the venue to which the audience has access should be provided with normal lighting and emergency lighting (see Chapter 17, *Electrical Installations and Lighting*).

4.3.5 Indoor Venues

4.3.5.1 Buildings designed for public assembly must have suitable and sufficient means of egress for their designed purpose. However adaptations, such as the provision of a stage, temporary stands, or a significant increase in the number of people to be accommodated, need to be considered and may require extra measures.

4.3.5.2 Where additions to the existing means of egress are needed, make sure that the local building and fire departments are consulted early and that:
- Exits are suitable and sufficient in size and number;
- Exits are distributed so that people can turn their back on any fire which may occur;
- Exits and exit routes are clearly indicated; and
- Escape routes are adequately illuminated (see also Chapter 17, *Electrical Installations and Lighting*).

4.3.5.3 Regarding buildings that were not designed for public assembly, it is unlikely that such places were designed to accommodate large numbers of people. Thus, it is almost certain that additional means of egress will be required to accommodate a music event. In this situation, consult the local building and fire departments at an early stage.

4.3.5.4 In deciding whether the means of egress are reasonable, authorities will consider:
- The occupant load of the building;
- The type and nature of the use of the building;
- Means of egress factors (e.g., width and number of exits required, etc.);
- Whether temporary stands and/or stages will be constructed within the building;
- The number of levels from which occupants will exit;
- Exit and directional signs; and
- The normal and emergency lighting provided in the venue.

4.3.6 Sports Stadiums

4.3.6.1 A sports stadium which has been issued an occupancy permit from the building and/or fire authorities having jurisdiction should already have adequate means of egress from the normal spectator areas. However, do not let this impede a full confirmation that they all meet the necessary requirements. Additional means of egress may also be needed if the playing field area is to be occupied by the audience and/or by temporary structures, such as a stage or stands. Some configurations may require additional permits and/or authorization from local authorities having jurisdiction. Where such a permit or authorization is required, or when in doubt, check with the relevant local authority as early as possible.

4.3.6.2 The U.S. Department of Justice provides a four-page pamphlet for ADA accessible stadiums that organizers may find informative. It can be found online at http://www.ada.gov/stadium.pdf.

4.3.7 Outdoor Venues

4.3.7.1 Generally, the same occupancy load requirements that apply to a building also apply to outdoor venues. That is, the local building and fire authorities having jurisdiction will establish (and require the organizer to post) maximum occupancy loads for specific configurations of various venues. Remember also that occupancy load is not simply a computed value (see definition above). The local authorities having jurisdiction have the flexibility to consider exactly how the space will be used, access and distance to egress, obstacles such as chairs and tables, and a number of other safety-related issues. Make sure to fully inform the authorities regarding the use of the space and nature of the audience so that the number at which they arrive is the most realistic possible.

4.3.7.2 Outdoor venues such as parks, fields and yards of stately homes will normally have boundary fences at their perimeters. To provide means of egress which will allow for an orderly evacuation to take place, ensure that:
- The number and size of exits in the fences, etc., are sufficient for the number of people present and are distributed around the perimeter;
- Exits and gateways are unlocked and staffed by stewards throughout the event; and

- All exits and gateways are clearly indicated by suitable signs which are illuminated if necessary.

4.3.7.3 At the planning stage, consult the fire and building authorities having jurisdiction about any proposals for means of egress.

4.3.7.4 Investigate whether there are any wild fires or controlled burns scheduled in the area and whether there is an elevated fire risk within 50 miles of the venue. Notify all members of the planning team of these findings and revise the fire safety plan accordingly. Discuss these issues with the local fire authority including how the fire safety plan should be modified in response.

4.4 Classification of Fires

4.4.1 In firefighting, fires are identified according to one or more fire classes. Each class designates the fuel involved in the fire, and thus the most appropriate extinguishing agent. The classifications allow selection of extinguishing agents along lines of effectiveness at putting the type of fire out, as well as avoiding unwanted side-effects. For example, nonconductive extinguishing agents are rated for electrical fires, so to avoid electrocuting the firefighter. NFPA 10, *Standard for Portable Fire Extinguishers*, classifies fires as follows:

- Class A fires (designation symbol is a green triangle) involve ordinary combustible materials like paper, wood and fabrics, rubber. Most of the time, this type of fire is effectively quenched by water or insulating by other suitable chemical agent.

- Class B fires (designation symbol is a red square) mostly involve flammable liquids (like gasoline, oils, greases, tars, paints, etc.) and flammable gases. Dry chemicals and carbon dioxide are typically used to extinguish these fires.

- Class C fires (designation symbol is a blue circle) involve live electrical equipment like motors, generators and other appliances. For safety reasons, non-conducting extinguishing agents such as dry chemicals or carbon dioxide are usually used to put out these fires.

- Class D fires (designation symbol is a yellow decagon [star]) involve combustible metals such as magnesium, sodium, lithium, or potassium. Sodium carbonate, graphite, bicarbonate, sodium chloride, and salt-based chemicals are used to extinguish these fires.

- Class K fires (designation symbol is a black K) are fires in cooking appliances that involve combustible cooking media (vegetable, animal oils or fats).

4.4.2 Class A Fires

4.4.2.1 Class A fires are the most likely type of fire to occur in the majority of venues. Water, foam and dry chemical are the effective agents for extinguishing these fires. Water and foam are usually considered to be the most suitable extinguishing agents and the appropriate equipment are therefore hose reels, water-type extinguishers or extinguishers containing fluoroprotein foam (FP), aqueous film-forming foam (AFFF), or film-forming fluoroprotein foam (FFFP).

4.4.3 Class B Fires

4.4.3.1 Where there is a risk of fire involving flammable liquid, dry chemical or carbon dioxide fire extinguishers may offer the best result. But, foam (including FP, AFFF and FFFP) can be quite effective in dealing with a fire involving exposed surfaces of contained flammable liquid.

4.4.3.2 Care should be taken when using carbon dioxide extinguishers as the fumes and products of combustion may displace oxygen and become hazardous in confined spaces.

4.4.3.3 Dry chemical (and dry powder) extinguishers can produce a vision obscuring cloud of "smoke" and can effect visibility and breathing if used in a crowd of people or in a confined space. Obscuring the vision of members of the audience can induce panic so these extinguishers should be used with care, especially when used in proximity to large crowds.

4.4.3.4 A solid stream of water should never be used to extinguish a class B fire because it can cause the fuel to scatter and spread the fire.

4.4.4 Class C Fires

4.4.4.1 Electrical fires are fires involving energized electrical equipment. The U.S. system designates these "Class C;" the Australian system designates them "Class E." This sort of fire may be caused by short-circuiting machinery or overloaded electrical cables. These fires can be a severe hazard to firefighters using water or other conductive agents: Electricity may be conducted from the fire, through water, the firefighter's body, and then earth. Electrical shocks have caused many firefighter deaths.

4.4.4.2 Electrical fires may be fought in the same way as an ordinary combustible fire, but water, foam, and other conductive agents are not to be used. While the fire is or possibly could be electrically energized, it can be fought with any extinguishing agent rated for electrical fire. Carbon dioxide (CO_2), FM-200 and dry chemical extinguishers such as PKP and even baking soda are well suited to extinguishing this sort of fire. Once electricity is shut off to the equipment involved, it will generally become an ordinary combustible (class A) fire.

4.4.5 Class D Fires

4.4.5.1 Certain metals can be flammable or combustible. Fires involving such materials are designated "Class D." Examples of such metals include sodium, titanium, magnesium, potassium, uranium, lithium, plutonium, and calcium. Magnesium and titanium fires are common. When one of these combustible metals ignites, it can easily and rapidly spread to surrounding ordinary combustible materials and pose a significant hazard.

4.4.5.2 With the exception of the metals that burn in contact with air or water (e.g., sodium), masses of combustible metals do not represent unusual fire risks because they are very difficult to ignite and have the ability to conduct heat away from hot spots so efficiently that the heat of combustion cannot be maintained. This means that it will require a lot of heat to ignite a mass of combustible metal. Metal fire risks exist when sawdust, machine shavings and other fine particles of metal are present. Generally, these fires can be ignited by the same types of ignition sources that would start other types of fires.

4.4.5.3 Water and other common firefighting materials used on metal fires can explode and make these fires worse because the high heat involved raises the temperature of the applied water to boiling so fast it does not have enough time to cool the fire. Metal fires should be fought with "dry powder" extinguishing agents that extinguishes by separating the four parts of the fire tetrahedron. It prevents the chemical reactions involving heat, fuel, and oxygen and halts the production of fire sustaining "free-radicals," thus extinguishing the fire.

4.4.5.4 Today a wide range of powder agents may be effective on class D fires including sodium chloride (Super-D, Met-L-X), copper based powder (Copper Powder Navy125S), graphite-based powder (G-Plus, G-1, Lith-X, Pyromet), and sodium carbonate based powder (Na-X).

4.4.5.5 Metal fires represent a unique hazard because people are often not aware of the characteristics of these fires and are not properly prepared to fight them. Therefore, even a small metal fire can spread and become a larger fire in the surrounding ordinary combustible materials.

4.4.6 Class K Fires

4.4.6.1 Fires that involve cooking oils or fats are designated "Class K" under the American system, and "Class F" under the European/Australasian systems. Class K fires involve combustible cooking media such as oils and grease commonly found in commercial kitchens. Though such fires are technically a subclass of the flammable liquid/gas class (B), the special characteristics of these types of fires are considered important enough to recognize separately. The new cooking media formulations used from commercial food preparation require a special wet chemical extinguishing agent that is especially suited for extinguishing and suppressing these extremely hot fires that can re-flash. Saponification (a process that produces soap, usually from fats and lye) can be used to extinguish such fires. Appropriate fire extinguishers may also have hoods over them that help extinguish the fire.

4.4.6.2 Wet chemical fire extinguishers are tested and approved for Class K fires. They contain a potassium acetate based, low pH agent that was originally developed for use in pre-engineered cooking equipment fire extinguishing systems. The agent discharges as a fine mist which helps prevent grease splash and fire re-flash while cooling the appliance. The Class K extinguisher (a.k.a., Purple K or K-Guard) is a good choice for use on all cooking appliances including solid fuel char-broilers.

4.5 The Fire Triangle/Tetrahedron

4.5.1 The concept of a fire triangle is a simple model for understanding the necessary ingredients for most fires. The triangle represents the three elements required for a fire to ignite: fuel, heat (the ignition source) and an oxidizing agent (usually oxygen in the air). Fire prevention is about ensuring that these three things do not combine to start a fire.

4.5.2 Heat is required to ignite a fire, and for the fire to continue to burn. Heat can be removed by a number of substances that cool the fire below a temperature at which the fuel can continue burning. This is usually water which absorbs a great deal of heat as it turns to steam.

4.5.3 Fuel is required for a fire to ignite and continue burning. Fuel can be consumed by fire and thus reduce enough for the fire to extinguish. The fuel can be manually removed so that the fire cannot access it and continue, much like is done in wildland firefighting where controlled burns remove the fuel and oxygen. Fuel can also be separated from the fire with chemical elements such as aqueous film forming foam (AFFF) which can be applied to the surface of ignited flammable liquid fuel.

4.5.4 A fire cannot begin without sufficient oxygen, which is found in adequate amounts in the earth's atmosphere to support most combustion. If available oxygen reduces, the combustion process slows but enough heat and fuel may remain to quickly reignite should air (with oxygen) be reintroduced. It should be noted that special precautions are required when enriched oxygen is present, such as when using gas cutting gear, or when a material contains its own source of oxidizer, such as pyrotechnics and some combustible metals.

4.5.5 The fire tetrahedron represents the addition of a fourth component—the chemical chain reaction—to the three already present in the fire triangle. Once a fire has started, the resulting exothermic chain reaction sustains the fire and allows it to continue until or unless at least one of the elements of the fire is blocked. Foam can be used to deny the fire the oxygen it needs. Water can be used to lower the temperature of the fuel below the ignition point or to remove or disperse the fuel. Halon (any of several halocarbons used as fire-extinguishing agents) can be used to remove free radicals (an uncharged molecule, typically highly reactive and short-lived, having an unpaired valence electron) and create a barrier of inert gas in a direct attack on the chemical reaction responsible for the fire.

4.5.6 "Combustion" is the chemical reaction that feeds a fire more heat and allows it to continue. When the fire involves the high heat of burning metals like lithium, magnesium, titanium, etc. (known as a class-D fire [see above]), the metals can react faster with water than with oxygen and thereby more energy is released. Putting water on such a fire can result in the fire getting hotter or even exploding. Carbon dioxide extinguishers are ineffective against certain metals such as titanium. Therefore, inert agents such as dry sand must be used to break the chain reaction of metallic combustion. In the same way, when any one of the four elements of the tetrahedron is removed, combustion stops.

4.6 Portable Fire Extinguishers

4.6.1 The following paragraphs give advice on fire extinguishers for use in the early (incipient) stages of a fire before the arrival of trained fire fighters. Some venues designed for public assembly may have a fire suppression systems, fire protection equipment, and fire alarm systems in place (e.g., a sprinkler system, standpipe system with hose reels, fire alarm system, etc.). But, fire extinguishers are usually also required, and will be especially important where such systems are not installed.

4.6.2 All venues must be provided with appropriately located portable fire extinguishers. This provision should be determined at the planning stage in consultation with the local fire authority having jurisdiction.

4.6.3 A fire extinguisher typically consists of a hand-held cylindrical pressure vessel that contains an agent which can be discharged to extinguish a fire (Fig. 4-2). The model fire codes state that portable fire extinguishers must be installed, maintained and used in accordance with NFPA 10, *Standard for Portable Fire Extinguishers*. OSHA 1910.157 also offers guidelines for the placement, use, maintenance, and testing of portable fire extinguishers provided for the use of employees.

4.6.4 The types of fire extinguishers are matched to the types of fires they are intended to extinguish. For example, an ABC extinguisher is capable of extinguishing class A, class B and class C fires. See the section above (*Classification of Fires*) for details regarding which type of extinguisher is best for each class of fire.

Fig. 4-2 - A good multi-purpose fire extinguisher for A, B, and C fires (dry chemical based).

4.6.5 Fire extinguishing capacity is rated in accordance with ANSI/UL 711, *Rating and Fire Testing of Fire Extinguishers*. The ratings are described using numbers preceding the class letter. As an example, consider an extinguisher rating of 1-A:10-B:C. The inclusion of A, B and C means that this extinguisher can be effective on class A, B and C fires. The number preceding the A multiplied by 1.25 gives the equivalent extinguishing capability in gallons of water. In this example, the extinguisher with this rating (1-A) is equivalent to 1.25 gallons of water on a class A fire. The number preceding the B indicates the size of a class B fire in square feet that an ordinary user should be able to extinguish. In this example, the "10-B" indicates that a 10 square foot class B fire can be extinguished by an ordinary user. There is no additional rating for class C, as it only indicates that the extinguishing agent will not conduct electricity. An extinguisher will never have a rating of just C.

4.6.6 In the United States, fire extinguishers, in all buildings other than houses, are generally required to be serviced and inspected by qualified personnel (e.g., a fire protection service company) at least annually. Some jurisdictions require more frequent service for fire extinguishers. The servicer places a tag on the extinguisher to indicate the type of service performed (e.g., annual inspection, recharge, new fire extinguisher, etc.) and the date of service. Event organizers can easily check these tags to confirm that the extinguisher to which each is attached has been serviced in the past year. If not, they should be replaced or properly serviced.

4.6.7 The typical steps for operating a fire extinguisher (described by the acronym "PASS") are as follows:

    P—Pull the safety pin
    A—Aim the nozzle at the base of the fire, from a safe distance (about 6 feet [2 m] away)
    S—Squeeze the handle
    S—Sweep the extinguisher from side to side while aiming at the base of the fire

4.6.8 To be useful, portable extinguishers must be located in a conspicuous location where they will be readily accessible and immediately available for use. These locations must be along normal paths of travel, unless the fire authority having jurisdiction determines otherwise. In addition, portable fire extinguishers must not be obstructed or obscured from view and must be installed on manufacturer-provided or approved brackets securely anchored to the mounting surface. Generally, the maximum travel distance to an extinguisher should be 50 feet (15.24 m).

4.6.9 Appropriate types of portable fire extinguishers must be properly installed in (at least) the following areas:
- In residential areas where people live and sleep;
- Within 30 feet (9.144 m) of commercial cooking equipment;
- In areas where flammable or combustible liquids are stored, used or dispersed;
- Fuel tank vehicles (for flammable or combustible liquids): minimum rating of 2-A:20-B:C;
- In areas where quantities of any type of combustible materials are stored;
- Where flammable solids such as magnesium are stored, machined, grinded or otherwise processed, an appropriate Class D extinguisher within 75 feet (22.860 m) or other extinguishing materials (scoop, shovel and bucket or extinguishing material) within 30 feet (9.144 m);
- Where liquefied petroleum (LP) gases are stored or used;
- In structures under construction, at each stairway on all floor levels where combustible materials have accumulated, in every storage and construction shed, and where special hazards exist;
- In all tents and membrane structures;
- In vehicle repair and maintenance facilities;
- Special hazard areas, including areas such as laboratories, computer rooms and generator rooms; and
- Where required by the applicable building and/or fire codes.

4.6.10 Most class D extinguishers will have a special low velocity nozzle or discharge wand to gently apply the agent in large volumes to avoid disrupting any finely divided burning materials. Class D agents are also available in bulk and can be applied with a scoop or shovel.

4.6.11 Some extinguishers containing dry chemical extinguishing agents may be confused with extinguishers containing dry powder extinguishing agents. The two are not the same. Mistakenly using a dry chemical extinguisher in place of a dry powder extinguisher can be ineffective and may increase the intensity of a metal fire. Always match the type of extinguisher with the class of fire anticipated.

4.6.12 Carbon dioxide (CO_2) is a clean gaseous extinguishing agent that displaces oxygen. The highest rating for a 20 pound (7.7 kg) portable CO_2 extinguisher is 10B:C. CO_2 extinguishers are not intended for class A fires as the high-pressure cloud of gas caused by the rapidly expanding frozen liquid stored in the pressurized vessel can scatter burning materials. CO_2 is also not suitable for use on fires containing their own oxygen source, such as metals. Although CO_2 may work well on a person's clothing on fire, such use should be avoided where possible as it can cause frostbite and, in an enclosed space, is dangerous because it displaces the oxygen needed for breathing. CO_2 extinguishers will have the characteristic "horn" on the discharge end of the device, which makes a CO_2 extinguisher easy to identify.

4.6.13 IMPORTANT: Fire extinguishers are not a substitute for trained firefighting professionals with state-of-the-art equipment. Fire extinguishers are, rather, a supplement to be used as a prevention measure that give people in the hazard area time to escape. If the fire is past its early stages or if the use of more than one extinguisher is required, the safest recourse is to evacuate the hazard area, direct others to do the same, and call the fire department.

4.6.13.1 To prevent the indiscriminant use—and perhaps misuse—of a fire extinguisher, it is best if all members of staff are trained in the proper selection and use of fire extinguishers.

4.6.14 Other Types of Fire Extinguishing Equipment

4.6.14.1 In North America, a "standpipe" is a type of rigid water piping which is built into multistory buildings or structures in a vertical position, to which fire hoses can be connected, allowing manual application of water to the fire. Standpipe systems are often integrated with automatic sprinkler systems and both are often required in newly constructed assembly occupancies. Standpipes may also be equipped with hose cabinets that allow building occupants and/or fire fighters to use the hose and nozzle provided to manually apply water to a fire. Organizers should consult with local fire and building authorities having jurisdiction for more information regarding where these are required. NFPA 14, *Standard for the Installation of Standpipes and Hose Systems*, describes the requirements for these types of systems and equipment.

4.6.14.2 There are three classes of standpipe systems:
- Class I – a system providing 2-1/2 inch (64 mm) hose connections to supply water for use by those trained in handling heavy fire streams.
- Class II – A system providing 1-1/2 inch (38 mm) hose stations to supply water for use primarily by the building occupants or by the fire department during initial response.
- Class III – A system providing 1-1/2 inch (38 mm) hose stations to supply water for use by building occupants and 2-1/2 inch (64 mm) hose connects to supply a larger volume of water for use by those trained in handling heavy fire streams.

4.6.14.3 In addition to class, standpipes are also be characterized by type:
- Automatic dry – normally filled with pressurized air that uses a dry pipe valve to admit water into the system piping automatically upon the opening of a hose valve.
- Automatic wet – has a water supply that is capable of supplying the system demand automatically.
- Manual dry – does not have a permanent water supply attached to the system and requires the fire department to provide water through the fire department connection.
- Manual wet – connected to a water supply for the purposes of maintaining water in the system only but which does not have a water supply capable of delivering pressurized water to the system. This type of system also requires the fire department to provide water through the fire department connection.

- Semiautomatic dry – a dry system that uses a device to admit water into the system piping upon activation of a remote control device located at the hose connection. A remote control device must be provided at each hose connection.

4.6.14.4 Automatic sprinkler systems may also be dry or wet systems and are required to be installed in most types of new buildings and structures designed for assembly. NFPA 13, *Standard for the Installation of Sprinkler Systems*, describes the requirements for these types of systems. OSHA 1910.158 also offers guidelines that apply to all small hose, Class II, and Class III standpipe systems installed to meet the requirements of a particular OSHA standard.

4.6.15 Stages

4.6.15.1 Stages greater than 1000 square feet (93 square meters) must be equipped with a Class III wet standpipe system with 1-1/2 inch and 2-1/2 inch (38 mm and 64 mm) hose connections on each side of the stage. The 1-1/2 inch (38 mm) connection must be equipped with sufficient lengths of 1-1/2 inch (38 mm) hose to provide fire protection to the stage area. This hose must be equipped with an approved adjustable fog nozzle and be mounted in a cabinet or on a rack. In addition, where stages are larger than 1,000 square feet (93 square meters) in area, are greater than 50 feet (15.240 m) in height, and include combustible hangings or curtains, automatic sprinklers may be required.

4.6.15.2 Stages have numerous fire safety requirements due to a long, rich history of fire-related tragedies associated with theatrical stages. Organizers should consult with local fire and building authorities having jurisdiction to determine all the requirements that must be incorporated into stages.

4.7 Fire-Fighting Equipment Provision

4.7.1 With indoor venues specifically designed for public assembly, the scale of provision required in connection with the normal use of the building should be adequate. However, if additional facilities are to be provided, e.g., a stage, concessions on the field, changing rooms, etc., there may be a need for additional equipment.

4.7.2 Indoor venues not designed for public assembly should be cause for the greatest concern as existing provisions may be minimal. However, there may be some provision (e.g., hose cabinets in a warehouse) and provided that the maintenance is satisfactory, this should be taken into account. In deciding what firefighting equipment is appropriate, consider both the structure and the contents of the building including the scale of both. The general principle is that no one should have to travel more than 50 feet (15.24 m) from the site of a fire to reach an extinguisher. Position extinguishers on exit routes near exits.

4.7.3 The provision of firefighting equipment for outdoor venues will vary according to the local conditions and what is brought onto the site. There will need to be equipment for tackling fires in vegetation, vehicles, structures, and tents.

4.7.4 Arrangements may need to be made to protect fire protection equipment located outdoors from the effects of frost, vandalism and theft. Where necessary, provision must be made to clear accumulations of snow, ice and other forms of weather-induced obstructions from fire protection equipment located outdoors. The location of such equipment should be well marked with prominent signs. Further advice should be sought from the fire authority or local authority.

4.8 Special Risks

4.8.1 Besides the recommendations and requirements provided above regarding the provision of fire protection equipment and procedures, address special fire and life safety risks according to the following guidelines:

- Where welding or similar hot work is undertaken, a minimum of one 2-A:20-B:C portable fire extinguisher must be readily accessible within 30 feet (9.144 m). In addition, a fire watch must be established during hot work activities and for 30 minutes after the work ends.
- Portable fire extinguishers with a minimum rating of 2-A:20-B:C must be provided where motor vehicle fuel is dispensed or stored. They must be located such that an extinguisher is not more than 75 feet (22.860 m) from pumps, dispensers, or storage tank fill-pipe openings.
- When open to the public, smoking, open flames, devices emanating flame, fire, flammable or combustible liquids, gas, charcoal or other cooking device are not permitted inside or within 20 feet (6.096 m) of a tent or membrane structure unless specifically authorized by the fire authority having jurisdiction.
- If pyrotechnics or flame effects will be used, the requirements of NFPA 160, *Standard for the Use of Flame Effects Before an Audience*, should be met and may be required;
- If tents, membrane structures, grandstands, or bleachers are involved, comply with the requirements of NFPA 102, *Standard for Grandstands, Folding and Telescopic Seating, Tents, and Membrane Structures*.
- Portable fire extinguishers installed at heights above occupied areas (e.g., in occupied trusses, on catwalks, on followspot platforms, etc.) should be secured with a reasonable length of safety cable.
- Portable fire extinguishers with a minimum rating of 2-A:20-B:C should be installed, and may be required, in dressing rooms, where scenery is stored, electrical intake rooms (where breaker boxes are located), boilers rooms, mobile concessions, near portable generators, in occupied trusses, at intersections of catwalks and on followspot platforms.

4.9 Means of Giving Warning in Case of Fire

4.9.1 The following paragraphs give general recommendations on the means for giving warning if there is fire. More detailed information may be obtained from NFPA 72, *National Fire Alarm and Signaling Code*.

4.9.2 The purpose of a fire-warning system is to provide information to everyone present so that all can be safely evacuated before escape routes become impassable through fire, heat or smoke. The means for giving warning should be suitable for the particular venue, taking into account its size, layout, planned means of egress and the number of people likely to be present.

4.9.3 Fire-alarm and warning systems should comply with NFPA 72, *National Fire Alarm and Signaling Code*. This standard requires that a sign or signal that needs a power supply to operate should also have a back-up power supply. Existing systems designed or installed to an earlier standard may be acceptable subject to satisfactory testing, electrical certification and approval by the local authority having jurisdiction (see Chapter 6, *Communication*, for further advice on emergency public announcements).

4.9.4 An indoor venue designed for public assembly which has previously been approved for music events will have an approved means for giving warning in case of fire. However, it will be necessary for the fire authority having jurisdiction to be consulted at an early stage to ensure that the system is appropriate.

4.9.5 Buildings not designed for public assembly such as warehouses, aircraft hangars, agricultural buildings, etc., may have a warning system which is unsuitable for a music event or no fire-warning system at all. It will therefore be necessary to either modify the existing system to use the building for the event or provide a temporary warning system.

4.9.6 If a temporary warning system is installed (and this may be the more appropriate action to take), the provision of a radio-transmission system has a number of advantages as it will not require the laying of electrical wiring or modifications to a building. Static call-points can also be replaced by mobile call-points carried by stewards so that the alarm can be raised instantly at the point of discovery of any fire. It is, however, still necessary for any system to comply with NFPA 72, *National Fire Alarm and Signaling Code*. The fire and building authorities having jurisdiction should be consulted as to the suitability of the system for the venue.

4.9.6.1 In an exposition or trade show environment, fire suppression equipment has been required under the floor of the second story of a two story booth.

4.9.7 For some buildings not designed for public entertainment, an alarm system incorporating automatic fire detection may be required, particularly in circumstances where a fire could reach serious proportions before discovery. The fire and building authorities having jurisdiction must be consulted regarding these requirements.

4.9.8 Although there is less likelihood of people becoming trapped by fire when the event is staged outdoors it will still be necessary to provide a fire-warning system for temporary and moveable structures such as roofed structures and tents. Campsites should have fire watches and campers should be provided with fire safety advice.

4.10 Curtains, Drapes and Other Decorative Materials

4.10.1 The use of curtains, drapes, and temporary decorations could affect the safe use of a means of egress, and drapes across an exit are explicitly prohibited by the model codes. However, because of ambient light that may interfere with the production (particularly in arenas), black-out curtains may be necessary to control unwanted light entering the performance space. Any proposal to use combustible decorative materials should be requested in writing of the fire authority having jurisdiction and should be accompanied by full details. Providing samples of the materials proposed to be used may also be required. Where a building is already being used for public assembly the use of these materials will probably have been approved.

4.10.2 The use of decorative vegetation (e.g., fresh cut trees, flowers, etc.) is generally prohibited in assembly occupancies. However, if the use of such material is approved by the fire authority having jurisdiction, there would likely be many limitations to the approval. For example, such materials would not be permitted to obstruct or reduce the width of any required means of egress, be located near heat vents, or be located near any open flame.

4.10.3 Although there are some exceptions where automatic sprinklers are installed, in assembly occupancies (and several other types of occupancies), curtains, draperies, hangings and other decorative materials suspended from walls or ceilings must be approved and meet the flame propagation performance criteria of NFPA 701, *Standard Methods of Fire Tests for Flame Propagation of Textiles and Films*. Combustible decorative materials are also limited to no more than 10 percent of the specific wall or ceiling area to which it is attached. However, artwork and/or teaching materials on the walls of corridors is permitted to be up to 20 percent of the wall area.

4.10.4 Where the weight of the material is more than one pound (0.45 kg) in assembly occupancies, exposed foam plastic materials and unprotected materials containing foam plastic used for decorative purposes or stage scenery or exhibit booths are required to have a maximum heat release rate of 100 kW when tested according to UL 1975, *Fire Tests for Foamed Plastics Used for Decorative Purposes*.

4.10.5 Where motion picture screens are used in assembly occupancies, the screen must meet the flame propagation performance criteria of NFPA 701, *Standard Methods of Fire Tests for Flame Propagation of Textiles and Films*.

4.10.6 Combustible scenery of cloth, film, vegetation (dry), and similar materials must comply with one of the following:
- They must meet the flame propagation performance criteria of NFPA 701, *Standard Methods of Fire Tests for Flame Propagation of Textiles and Films*; or
- They must exhibit a heat release rate not exceeding 100 kW when tested in accordance with NFPA 289, *Standard Method of Fire Test for Individual Fuel Packages*, using the 20 kW ignition source.

4.10.7 Where required to be flame resistant, decorative materials must be tested by an approved agency (approved by the fire authority having jurisdiction) and meet the flame propagation performance criteria of NFPA 701, *Standard Methods of Fire Tests for Flame Propagation of Textiles and Films,* or such materials must be noncombustible.

4.10.8 In ancient Rome, the stage area in front of the scaenae frons (the elaborately decorated background of a Roman theatre stage) was known as the "proscenium," meaning "in front of the scenery." Today, the area of a theatre surrounding the stage opening is referred to as the proscenium. A proscenium arch is the arch over this area. Where required by the fire authority having jurisdiction, the proscenium opening must be protected by a listed, minimum 20-minute opening protective assembly, a fire curtain complying with NFPA 80, *Standard for Fire Doors and Other Opening Protectives*, or an approved water curtain complying with NFPA 13, *Standard for the Installation of Sprinkler Systems*. In addition, proscenium opening protection provided by other than a fire curtain must activate upon automatic detection of a fire and upon manual activation.

4.10.9 Scenery and stage properties not separated from the audience by proscenium opening protection must be of noncombustible materials, limited-combustible materials, or fire-retardant treated wood.

4.11 Fire Risk Assessment

4.11.1 Underlying plans to keep people and property safe from fire hazards is the idea of a fire risk assessment. This is nothing more than a systematic analysis of the precautions required to prevent fires and, should one occur, how event staff and management will respond to protect life safety and minimize the harm or loss caused by the fire.

4.11.2 Like any other type of risk assessment, a fire risk assessment can be recorded in any format. However, it is useful to follow the basic theme set out below (Table 4-2), where the separate elements of fire prevention and incident response are dealt with as specific topics and the organizer does not have to describe the entire plan in one narrative.

4.11.3 The guidance given here is as a recommendation only. The event organizer must determine an appropriate level of provision that is acceptable to the authorities having jurisdiction.

4.11.4 For smaller events, a single fire risk assessment may be sufficient to cover all activities. For larger events, with multiple stages, production, camping, catering and trade areas, it may be necessary to draw up a number of fire risk assessments to ensure all locations and activities are properly considered. Regardless, the goal of a fire risk assessment is to prevent the combination of the elements in the fire triangle/tetrahedron usually by separating fuels from ignition sources.

Table 4-2

Key Considerations in a Fire Risk Assessment

Fuels and/or Combustible Materials	What is there in the venue or event site that can burn?
Ignition Sources	How might these fuels ignite?
Detection and Alarm	If a fire does start, how will people become aware and how can they raise the alarm or summon help?
Escape and Evacuation	What arrangements are required to get people away from danger?
Fire Fighting	What resources are appropriate for immediate firefighting (extinguishers etc.) and how can professional assistance quickly be brought to the scene?
Training	What level of skill is required by various staff, and how are people effectively briefed on procedures?

4.11.5 Fuels

4.11.5.1 Many materials should be considered potential fuels. This includes obvious substances like generator and vehicle fuel or fuel gases such as propane used for cooking (flammable products), and less obvious materials such as paper, trash, and tents (combustible products). The organizer must ensure that all flammable and combustible materials are properly controlled and separated from potential ignition sources.

4.11.5.2 This means using tent fabrics, set, dressings and fabrics which meet the fire retardant standard(s) required by the authorities having jurisdiction. Organizers have a duty to limit and control such materials through proper sourcing of supplies and the preferential use of lower risk substances. Using diesel rather than gasoline generators is a good example (diesel is combustible and gasoline is flammable [more volatile]), as is ensuring arrangements for regular collection of trash—especially from trade and catering stalls where paper and food waste can quickly accumulate. Particular attention should be paid to the location. For example, flammable and/or combustible material beneath a seating bleacher is far more significant than overflowing trash in an open field.

4.11.5.3 The "fuels" section of a fire risk assessment will outline what will be done to eliminate or minimize the presence of flammable and combustible materials at the event site, and how those which cannot be avoided will be properly stored and controlled.

4.11.6 Ignition Sources

4.11.6.1 Ignition sources are the means whereby sufficient heat is added to ignite the fuel and cause a self-sustaining fire. Potential sources include:
- Smoking;
- Cooking or campfires;
- Electrical faults;

- Poorly located appliances, such as lighting fixtures;
- Blocking of ventilation ports on equipment;
- Vehicle exhaust;
- Naturally occurring (e.g., lightning);
- Pyrotechnics and special effects; and
- Intentionally set fires.

4.11.6.2 Each potential ignition source will require its own control measures, and the organizer must establish that appropriate and proportionate steps are taken. In some instances this may mean discussion with stage vendors about the use of pyrotechnic effects, in others it may mean the use of fencing and security to prevent access to sensitive areas.

4.11.7 Detection and Alarm

4.11.7.1 In fixed venues it is normal to encounter automated smoke /heat/carbon monoxide detection systems which will trigger an integrated alarm—both within in the building and/or to emergency services. At a temporary event venue, the organizer will have to replace all of these elements and come up with an acceptable way (to the authorities having jurisdiction) to identify an incident and rapidly raise the alarm with staff, the public and the fire department.

4.11.7.2 Particular attention must be paid to high risk activities or to locations where a fire may quickly take hold, or develop without being spotted, such as in sleeping accommodations. In such instances, battery operated alarm fittings or a permanent fire watch by designated crew may be required.

4.11.7.3 Arrangements should include continual security and staff supervision, communication systems for reporting an alarm, code words, manual alarms, klaxons (alarm horns), PA announcements, fire watch towers and so on. The organizer must determine what is appropriate to meet the needs of the event and the level of threat from fire, then get the procedures approved by the authorities having jurisdiction.

4.11.7.4 Experience from previous catastrophic fires shows that fatal conditions can develop very rapidly, and the organizer needs to have in place a proper and effective plan for stopping a show and initiating an evacuation.

4.11.8 Escape and Evacuation

4.11.8.1 Considerable detail has been given above on the requirements of means of egress (e.g., escape and evacuation routes). It is imperative that adequate and properly positioned exits are provided to enable all occupants of the premises to leave in safety.

4.11.8.2 Complications arise for events held in premises with more than one story, or where the exit route is long or complex, or leads past additional risk areas. The guidance of the authorities having jurisdiction is of critical importance.

4.11.8.3 Determining the occupancy load of a premises is not simply limited to managing the density of people within the space, but also includes consideration of the exit capacity (i.e., how many people can safely get out of that space in an emergency).

4.11.8.4 Any exit route must be properly signed and illuminated and provide clear, unobstructed passage along its entire length. This means avoiding positioning stage or scenic elements, or storing flight cases and equipment—even temporarily—in any designated means of egress.

4.11.9 Fire-Fighting Measures

4.11.9.1 The organizer needs to determine what extinguishers or other facilities are appropriate to the type of fire which may occur, and what skills and training are required for staff members to effectively handle an incident.

4.11.9.2 It should be emphasized that while portable extinguishers can be vital tools to tackle a developing fire, they do not in themselves make an event safe. Extinguishers must be of the correct type and positioned to allow quick access. They should be located to allow occupants to defend exit and escape routes (i.e., at exit doors as well as close to equipment or materials which present the initial fire risk).

4.11.9.3 Portable extinguishers can help tackle small fires and prevent them growing to the point they pose a life safety risk or require the evacuation of a venue. However, to use them safely staff must understand the basics of how to use them and the limitations of both the equipment and the user. Trying to handle a situation already out of control may quickly turn staff into casualties and further delay the reporting of the fire.

4.11.10 Training and Briefing

4.11.10.1 An integral part of effective fire safety planning is ensuring all relevant people understand the basics of fire prevention and are aware of what to do should an incident occur. This does not mean that all event personnel need to become trained firefighters. They simply need to recognize fire risks and be familiar with their own role in an emergency.

4.11.10.2 Alongside long-term training programs for staff, the organizer should consider short site briefings to familiarize people with the layout of the show and any particular issues they should be mindful of. The location of exits, extinguishers, alarm points, etc., can be explained very quickly and a few minutes spent on discussing fire safety and emergency actions can transform the speed and effectiveness of response. Those few moments may save lives.

4.12 Americans with Disabilities Act of 1990

4.12.1 The extent to which the Americans with Disabilities Act (ADA) Standards for Accessible Design (http://www.ada.gov/ 2010ADAstandards_index.htm) and accessible stadiums (http://www.ada.gov/stadium.pdf) may apply to buildings and structures that incorporate fire and life safety features, equipment and devices is beyond the scope of this document and must be established by competent legal counsel. However, organizers and venue managers would be prudent to familiarize themselves with these requirements and be constantly mindful of their intent as they relate to event venues: to improve accessibility (to all aspects of the venue, including safety features) and enjoyment of those with disabilities and certain physical limitations who wish to attend.

4.12.2 See Chapter 11, *Facilities for People with Special Needs*, for more information about the Americans with Disabilities Act and its applicability.

4.13 Emergency Plans

4.13.1 A <u>fire evacuation plan</u> is recommended and may be required. It must include at least the following:

> 4.13.1.1 Emergency egress or escape routes and whether evacuation of the building is to be complete or, where approved, by selected floors or areas only.

> 4.13.1.2 Procedures for employees who must remain to operate critical equipment before evacuating.

> 4.13.1.3 Procedures for assisted rescue for persons unable to use the general means of egress unassisted.

> 4.13.1.4 Procedures for accounting for employees and occupants after evacuation has been completed.

> 4.13.1.5 Identification and assignment of personnel responsible for rescue or emergency medical aid.

> 4.13.1.6 The preferred and alternative means of notifying occupants of a fire or emergency.

> 4.13.1.7 The preferred and any alternative means of reporting fires and other emergencies to the fire department or designated emergency response organization.

> 4.13.1.8 Identification and assignment of personnel who can be contacted for further information or explanation of duties under the plan.

> 4.13.1.9 A description of the emergency voice/alarm communication system alert tone and preprogrammed voice messages, where provided.

4.13.2 A <u>fire safety plan</u> is recommended and may be required by the authorities having jurisdiction. It must include at least the following:

> 4.13.2.1 The procedures for reporting a fire or other emergency.

> 4.13.2.2 The life safety strategy and procedures for notifying, relocating or evacuating occupants, including occupants who need assistance.

4.13.2.3 Site plans indicating the following:
- The occupancy assembly point;
- The locations of fire hydrants; and
- The normal routes of fire department vehicles access.

4.13.2.4 Floor plans identifying the locations of the following:
- Exits;
- Primary evacuation routes;
- Secondary evacuation routes;
- Accessible egress routes;
- Areas of refuge;
- Exterior areas for assisted rescue;
- Manual fire alarm boxes;
- Portable fire extinguishers;
- Occupant-use hose stations; and
- Fire alarm annunciators and controls.

4.13.2.5 A list of major fire hazards associated with the normal use and occupancy of the premises, including maintenance and housekeeping procedures.

4.13.2.6 Identification and assignment of personnel responsible for maintenance of systems and equipment installed to prevent or control fires.

4.13.2.7 Identification and assignment of personnel responsible for maintenance, housekeeping and controlling fuel hazard sources.

5. Medical, Ambulance and First Aid Management

5.0.1 This chapter lists the responsibilities of the event organizer to ensure that medical, ambulance and first-aid assistance are available to all those involved in an event. The event organizer must minimize the effects of an event on the healthcare provision for the local population and, wherever possible, reduce its effect on the local hospital facilities and emergency medical services (EMS).

5.0.2 The number of people requiring medical treatment at any music event will vary considerably as will the most reasonably foreseeable conditions requiring medical treatment during the event.

5.1 Planning

5.1.1 Plan the provision of medical, ambulance and first-aid services along with the statutory services and appoint a competent organization to provide medical management. This organization need not be the sole provider of resources at the event, but must demonstrate competence in operating the medical arrangements. The appointed organization should be experienced in the medical management of similar events, and must accept responsibility for providing an appropriate management and operational control infrastructure and coordinate the activity of other medical providers. Ensure that the appointed medical provider coordinates and has direct communication access with other statutory services and first-aid providers on site. Respective roles and responsibilities should be set out in a medical, ambulance and first-aid plan.

5.1.2 For events that foreseeably experience serious medical incidents or a large number of people requiring treatment, it is prudent to consult with the local hospital and EMS services so that they can advise both the event organizer and the local authorities on the likely effect of the event on pre-hospital accident and emergency services and so the local authorities can prepare themselves accordingly.

5.1.3 A manager from the medical provider should be appointed to take overall control and coordination of first-aid provision. This person should also be readily available during the event. The event organizer and the appointed medical provider should liaise with all interested parties which may include the local authorities, health board, ambulance service or competent first-aid providers, as appropriate.

5.1.4 It is recommended that the final details of the event are confirmed in writing to the appointed medical provider as soon as possible.

5.1.5 Consider the availability of medical, ambulance, and first-aid provision during the load-in and load-out of the event (see section on *First-Aid for Employees and Event Workers*, below).

5.1.6 Consider the need for medical, ambulance and first-aid arrangements for any audience members lining up before the gates or doors open and when they leave at the end of the event.

5.1.7 The location of first-aid facilities must be available to all those attending. Provide adequate signage and consider printing the location of first-aid facilities on tickets for the event. In addition, stewards should be aware of the nearest facility.

5.1.8 At events with overnight campsites, the medical, ambulance and first-aid services should be available while the campsite is open. Because of the likely range of conditions requiring medical care, also consider general medical services through the appointed medical provider during the times the campsites are in operation.

5.1.9 Where practical, consider the provision of suitable sterile, unoccupied or unused routes for the exclusive use of emergency vehicles.

5.1.10 The location of responders is important when assessing the response times for the arrival of emergency care to individual casualties at any location within the event.

5.1.11 Only in exceptional circumstances should ambulance vehicles be allowed to enter audience areas. Ambulances should not move from their designated position except on the instruction of their control unless compromised on grounds of safety. At events with high audience densities consider the use of foot patrols or golf carts to remove casualties.

5.1.12 The appointed medical provider should have arrangements in place to ensure that cover is maintained at the correct level throughout the event. If a casualty needs to be removed from the site by ambulance, arrangements must be in place to replace that vehicle or to transport the casualty using an ambulance dedicated to offsite patient movement (if there is the need for ambulances on site).

5.1.13 At certain events, an area for medical evacuation by helicopter may be required and a suitable landing site, either at the site or nearby, should be identified, prepared and maintained. Advice from the local airspace manager, qualified pilot or airport should be solicited. In the U.S., call 1-800-WX-BRIEF to speak with someone at a nearby Flight Service Station. The person who answers the phone should be able to answer your questions or refer you to someone who can.

5.2 Communications

5.2.1 At large events, there may be a need for a separate medical radio channel connecting the ambulance service with ambulance workers, key medical workers, mobile response teams, and key first-aiders. A protocol for the use of radio equipment, including consistent call signs, must be agreed before the event. A communications plan detailing medical communications links should be produced and held at both the medical control point or incident control room and central ambulance control.

5.2.2 If there is more than one medical facility, there should be a designated main medical facility with an external telephone line (which does not go through a switchboard) and a list of appropriate numbers. All other medical facilities should have an internal telephone or radio link to the main position.

5.3 Documentation

5.3.1 An event log should be maintained, which should include any actions or decisions taken by the manager of the medical provisions and the reasons for those actions.
Event logs, report forms and records contemporaneously completed at an event may be required later to assist in the reporting of accidents and injury to workers and audience members. They will also be essential pieces of evidence if an incident leads to litigation, so event organizers are advised to train their staff how to provide complete, accurate, presentable incident reports and to support them with photographs and video where appropriate.

5.3.2 Ensure that the appointed medical provider maintains a record of all people seeking treatment. In some locations, for consistency and ease of documentation, suitable patient report forms may be supplied by the ambulance service. This record should include details such as: name, address, age, gender, presenting complaint, diagnosis, treatment given, the onward destination of casualties (e.g., home, hospital, own GP), and the signature of person responsible for treatment. The only people who may be shown patients' records are those that are involved in the treatment or those that have legal authority.

5.3.3 Regular reports of the total number of casualties and the type of medical complaint should be provided to the event organizer during the event. This report should identify the person making the report and event conditions that may be contributing so that action can be taken.

5.3.4 Consideration should be given to being able to confirm to friends/relatives at the event whether a missing individual has received/is receiving treatment or has been taken to the hospital.

5.4 Definitions and Competencies for Medical Workers

5.4.1 First-aiders, ambulance and medical workers must all:
- Be at least 16 years old (first-aiders under 18 years old must not work unsupervised);
- Have no other duties or responsibilities and be dedicated to their first-aid duties;
- Have identification;
- Have all necessary personal protective equipment (PPE) and appropriate clothing;
- Have relevant experience or knowledge of requirements for first-aid at major public events;
- Be physically and psychologically equipped to carry out the assigned roles.

5.4.2 A "physician" is a professional (doctor) who has been educated, trained, and licensed to practice the art and science of medicine. In the United States and Canada, the term physician also describes all medical practitioners holding a professional medical degree. Each state in the United States, and each province in Canada, has its own requirements for licensing physicians— a requirement for legally practicing medicine in North America.

5.4.3 A physician working at a live event should be familiar with—or better, be well experienced at—specific subject matter in which all physicians are not necessarily trained. A physician working at an event should:

- Have a working knowledge of the National Incident Management System (NIMS) Incident Command System (ICS);
- Be familiar with, and have access to, local/county/regional emergency plans such as mass casualty, emergency operations, and disaster plans;
- Have experience with handling multiple, simultaneous emergencies in the pre-hospital (out of hospital) setting;
- Be familiar with the operation of the local emergency medical services (EMS) and casualty transport (ambulance) service, which may not both be the same entity; and
- Be familiar with the training and capabilities of the local EMS responders.

5.4.4 A "registered nurse" (RN) is a nurse who has graduated from a nursing program at a college or university and has passed a state licensing exam. To be effective in a live event, a qualified RN should have knowledge and recent experience in dealing with emergencies in the pre-hospital or emergency environment.

5.4.5 A "paramedic" is a person who is trained to give emergency medical treatment or to assist physicians in providing medical care. In the United States, a paramedic is a state-certified healthcare professional who may exercise the full authority of his or her certification only while working under the license of a medical director physician. Paramedics provide advanced levels of care for medical emergencies and trauma. A paramedic's required competencies and capabilities vary from state to state but usually include the administration of limited medications via intramuscular, subcutaneous, sublingual and intravenous routes; cardiac monitoring and defibrillation; insertion of advanced airways (e.g., endotracheal intubation, etc.); treating medical emergencies such as hypoglycemia, imminent child birth, trauma, apnea/dyspnea, shock, allergic reactions, etc.; and, selected emergency invasive techniques such as needle reduction of a tension pneumothorax (collapsed lung) and cricothyrotomy (emergency surgical airway).

5.4.6 Paramedics operate away from a hospital on written standard operating procedures approved by a specific medical director physician (a.k.a. "protocol). This set of standard procedures describes what a paramedic is permitted to do medically in certain situations. A paramedic away from a hospital may also establish direct communication with the medical director, or his/her designee, to receive specific instructions (orders) via radio or, more commonly, telephone.

5.4.7 Many states also refer to a paramedic as an "Emergency Medical Technician – Paramedic" (EMT-P), which should not be confused with lesser trained and qualified levels of EMT described below.

5.4.7.1 "ALS" (Advanced Life Support) is a term often used to describe emergency, high level medical interventions performed by someone trained as a paramedic or higher (e.g., nurse, physician, etc.). "BLS" (Basic Life Support) is the term used to describe emergency medical interventions performed by someone trained as a paramedic or lower (e.g., Basic EMT, first aider, etc.) and include basic first aid, CPR and/or similar basic first aid skills.

5.4.8 An "emergency medical technician" (EMT) responds to emergency calls, performs certain medical procedures and transport patients to hospital in accordance with protocols and guidelines established by physician medical directors. They may work in an ambulance service (paid or volunteer), as a member of technical rescue teams/squads, or as part of an allied service such as a fire or police department. EMTs are trained to assess a patient's condition, and to perform such emergency medical procedures as are needed to maintain a patent airway with adequate breathing and cardiovascular circulation until the patient can be transferred to an appropriate destination for advanced medical care. Capabilities include cardiopulmonary resuscitation (CPR), defibrillation, controlling severe external bleeding, preventing shock, body immobilization to prevent spinal damage, and splinting of bone fractures.

5.4.9 "Certified first responders" (CFRs)(a.k.a. "Emergency Medical Responders") in the U.S. can either provide emergency care first on the scene (police/fire department/park rangers) or support emergency medical technicians and paramedics, provide basic first aid, CPR, automated external defibrillator use, spinal immobilization, oxygen administration, and assist in emergency childbirth (in some areas they are trained in the use of suction and airway adjuncts). CFRs can also assist with administering glucose, aspirin, and epi-pens and are trained in packaging, moving and transporting patients. CFR is considered a higher level of medical training than basic first-aid (first-aider) and a lesser level of training than an EMT. The American Red Cross now offers the "Emergency Medical Response" course, which fits this definition. In the U.S. in 2012, the term "Emergency Medical Responder" began replacing the term "Certified First Responder."

5.4.10 A "first-aider" is a person who holds a current certificate of first-aid, usually at the advanced first-aid level, such as provided by the American Red Cross. To be effective at a live event, the first aider should have prior training or experience in providing first aid at crowd events. In the United States, there is no universal schedule of first aid levels that are applicable to all agencies that provide first aid training. Training is provided typically through the American Red Cross, but may also be conducted by local fire departments and, for CPR and automated external defibrillator (AED) use, the American Heart Association (AHA).

5.4.11 An "appointed medical provider" is a competent organization chosen by the event organizer, to provide overall management of medical, ambulance and first-aid services at an event.

5.5 Medical, Ambulance and First-Aid Provision

5.5.1 Following the risk assessment and agreement on levels of medical, ambulance and first-aid coverage, event organizers would do well to memorialize those conclusions to which the relevant parties indicate their agreement. Ensure that a suitable skills mix exists and that medical, ambulance and first-aid providers are located effectively throughout the site.

5.5.2 The decision on the level of medical provision and whether the ambulance service will be directly involved, or not, at any particular event will depend on a near infinite number of factors. Some of the more foreseeably important factors to consider include:
- Size of audience;
- Nature and type of event and entertainment;

- Nature and type of audience - including age range;
- Location and type of venue - outdoor or indoor, standing or seated, overnight camping and the size of the site;
- Duration of event - hours or days;
- Seasonal/weather factors;
- Additional activities and attractions;
- Proximity/capability/capacity of local medical facilities;
- Intelligence from other agencies regarding previous experience of similar events;
- Availability and potential misuse of alcohol or drugs (illicit, recreational, or controlled);
- External factors including the complexity of travel arrangements;
- Time spent in queues;
- Availability of facilities on site including hospitality and other social services;
- Range of possible major incident hazards at or associated with the event (structure collapse, civil disorder, crushing, explosion, fire, chemical release, food poisoning); and
- Availability of experienced first-aiders.

5.5.3 Tables are provided at the end of this chapter, which show a way of calculating the quantities of medical, first aid and ambulance provision suggested for various event types. These tables are borrowed from the U.K.'s Event Safety Guide (1999, Second Edition, pp. 121-124) and are included only as a means of estimating the number of personnel and equipment that might be needed. Use these reference tables with caution and at your own risk.

5.5.4 First-Aiders

5.5.4.1 The recommended minimum number of first-aiders at small events where no special risks are considered likely is 2:1000 for the first 3,000 attending.

5.5.4.2 At indoor venues or stadiums, first-aid facilities are likely to have been agreed. However, the historical number of first-aiders provided at an existing venue does not replace the need to carry out an assessment for each event. Some venues will be in multiple use. In such cases, the overall provision of medical, ambulance and first-aid resources should take account of all activities taking place within that venue.

5.5.4.3 No matter how well the event organizer knows the event demographics and likely risk of medical incidents, they are not medical professionals. For this reason, local medical care providers including the EMS staff for the event should be consulted for their recommendations regarding staffing levels and availability of medical equipment both on- and off-site.

5.5.5 Physicians

5.5.5.1 The risk assessment may indicate the need for the provision of physicians on site. Medical professionals should be available to treat not only patrons, but also workers and performers.

5.5.5.2 Depending on the size of the event and reasonably foreseeable medical incidents, one suitably experienced physician is often able to fulfill the role of Medical Group Supervisor (or

Medical Branch Director, as the size of the event requires) with overall responsibility for the management of medical resources at the scene of a major event.

5.5.6 Psychiatric Care

5.5.6.1 At lengthy or large events, consideration should be given to any requirement for a psychiatric care team including psychiatrists, psychiatric nurses and drug advisers. This team may need to liaise with the local authority, social services department, hospital authorities and the police.

5.5.7 Nurses

6.5.7.1 Qualified nurses may be required to care for patients requiring longer-term management on site. Unless trained as part of a mobile-response team, nurses should undertake the specific duty of staffing the main medical facility, working as a team with the physicians, paramedics and first-aiders in the triage and treatment of casualties.

5.5.8 Paramedics and EMTs

5.5.8.1 Paramedics and EMTs may need to be positioned in the pit area, medical facilities (first-aid stations) or areas of perceived risk, or deployed in immediate response to emergencies arising throughout the event area.

5.5.8.2 Non-medically trained staff may assist paramedics, EMTs and first-aiders in the transport of those with non-urgent medical conditions or with minor injury.

5.5.9 Medical Cover in Pit Areas

5.5.9.1 The risk assessment may indicate that medical cover may be required within the pit area. Medical workers in this area should be suitably experienced and trained to provide advice on casualty handling to stewards, appropriate triage to casualties and, where required, can facilitate the rapid evacuation of any casualties to a medical facility. As a minimum, the area in front of the stage should have the following equipment quickly available:
- Rescue board and neck collars;
- Oxygen therapy and resuscitation equipment; and
- Assorted splints.

5.5.10 Onsite Medical Facilities (First-Aid Stations)

5.5.10.1 The number, location and suitability of medical facilities should be planned. If there is more than one medical facility, one should be designated as the main medical facility. Primary medical facilities, including those in the pit area, will refer those requiring further treatment to the second-line main medical facility. The main medical facility may be equipped as a medical center or field hospital. If there is a major incident, according to local major incident procedures, a medical facility will be established or designated as the casualty clearing station.

5.5.11 Maps and Plans

5.5.11.1 Detailed gridded maps or site plans with position of medical facilities clearly marked must be available before the event. This should include the surrounding roads and access routes.

5.5.12 Structures

5.5.12.1 At outdoor events, if a suitable permanent structure is not available, provide suitably equipped mobile first-aid units or tents with appropriate flooring. At indoor events, position the medical facility in or next to the main arena.

5.5.13 Staffing Plan

5.5.13.1 An appropriate number of competent first-aiders should staff each medical facility and, as appropriate, EMTs, nurses, and physicians, some of whom should be available to offer assistance within audience areas. At large outdoor events ensure that a proportion of mobile first-aiders are strategically positioned or asked to patrol a defined area, in consultation with the EMS provider, if present. All workers must be clearly identified. Mobile first-aiders should be in constant radio contact with their supervisor.

5.5.14 Mobile Response Teams

5.5.14.1 At high-risk events, consider the use of a suitably equipped mobile response team with an appropriate skills mix and means of transport to attend medical emergencies where their specific skills are required.

5.5.15 Position

5.5.15.1 At larger events, provide a medical facility near to the stage area with unrestricted access to this position from the pit area. Other medical facilities are situated on the perimeter of the audience area enabling unrestricted access and exit for ambulances without entering audience areas.

5.5.16 General Considerations for the Main Medical Facility

5.5.16.1 As a minimum requirement, the main medical facility should be:
- Designated as a "no smoking area;"
- Of an adequate size for the anticipated number of casualties and readily accessible for the admission of casualties and ambulance crews;
- Large enough to contain at least two examination couches or ambulance stretchers or cot, with adequate space to walk around, and an area for the treatment of sitting casualties;
- Accessible at ground level and have a doorway large enough to allow access for an ambulance cot or wheelchair;
- Maintained in a clean and hygienic condition, free from dust and with adequate heating, lighting and ventilation;
- Provided with adequate first-aid and medical equipment and screens, etc., including resuscitation equipment, patient-care consumables and where appropriate, a defibrillator, all of which should be separate from those contained in ambulances. An agreement should be reached during the planning stage about who will provide such items;
- Within proximity of an easily accessible wheelchair-user's toilet and workers' facility;
- Provided with a supply of running hot and cold water. If this is not possible, provide adequate fresh clean water in containers;
- Provided with a supply of drinking water over a sink or hand-wash basin or suitable receptacle;

- Provided with a worktop or other suitable surface for equipment and documentation, e.g., folding tables;
- Provided with suitable secure storage facilities for drugs and equipment used by the medical providers;
- Next to appropriate paved areas or parking facilities for ambulances or associated emergency vehicles.

5.5.16.2 The workers at medical facilities should be made aware of the arrangements for social/well-being provision so that people can be suitably redirected to those facilities.

5.6 Clinical Waste

5.6.1 Specific arrangements for the disposal of clinical waste must be planned. Special bio-hazard containers (Fig. 5-1) for the disposal of needles ("sharps") or appropriately marked "bio bags" for the disposal of dressings or other contaminated materials will be required. Suitable arrangements must also exist for the disposal of non-clinical waste at medical facilities.

Fig. 5-1 – The international symbol for biological hazard.

5.7 First-Aid for Employees and Event Workers

5.7.1 Employers are responsible for ensuring that first-aid facilities, equipment and personnel are provided for all event personnel (employed/paid or not) if they are injured or become ill at work. In an ICS organization, this is referred to as the Medical Unit, which is defined as the functional unit within the Logistics Section responsible for the development of the Medical Emergency Plan, and for providing emergency medical treatment of incident personnel. It is recommended that the event organizer establish a Medical Unit at each event to attend to the medical needs of event personnel.

5.7.2 To decide on the level of first-aid provision necessary, an employer should make an assessment of the first-aid needs appropriate to the circumstances of the workplace. Employees who are appointed as first-aiders must have successfully completed the necessary training with an approved training organization. It is also good practice to have an "incident book" available in which to record incidents which require first-aid treatment. It is strongly recommended to have a written agreement between the various employers, e.g., contractors, subcontractors and others working at the event, to ensure that the first aid provided meets all their needs and to avoid misunderstandings.

5.7.3 Plan the welfare of the medical, ambulance, nursing and first-aid workers. At any event which lasts more than four hours, provide rest areas, sanitary and dining facilities. Where possible, separate these areas from the audience facilities.

5.7.4 Further guidance on Health and Safety (First Aid) Regulations is contained in the U.S. Department of Labor's Workplace Safety & Health laws, which are described online at: http://www.dol.gov/dol/topic/safety-health/index.htm#.UOCoho7FXR0.

5.8 Medical, Ambulance and First-Aid Provision

5.8.1 It is recognized that medical coverage at events can be organized in different ways and that the most appropriate model will vary according to the medical provider and the nature of the event. The following tables set out one method of estimating a reasonable level of resource.

5.8.2 It is emphasized that these figures may require modification as some providers may choose to substitute medical staff or paramedics for first-aiders. In any case, the suggested levels of resource are intended only as general guidance and should not be regarded as prescriptive. The tables are not a substitute for a full risk assessment of the event. Figures do not take account of dedicated medical personnel for performers or VIPs and do not incorporate the availability of alcohol at the event as a relevant factor.

- Use Table 5-1 to allocate a score based on the nature of the event.
- Use Table 5-2 to allocate a score based on available history and pre-event intelligence.
- Use Table 5-3 to consider additional elements, which may affect the likelihood of risk.
- Use Table 5-4 to indicate a suggested resource requirement.

Table 5-1

Event Nature - From the U.K.'s Event Safety Guide (1999, Second Edition, pp. 121-124)

Item	Details	Score
(A) Nature of event	Classical performance	2
	Public exhibition	3
	Pop/rock concert	5
	Dance event	8
	Agricultural/country show	2
	Marine	3
	Motorcycle display	3
	Aviation	3
	Motor sport	4
	State occasions	2
	VIP visits/summit	3
	Music festival	3
	Bonfire/pyrotechnic display	4
	New Year's celebrations	7
	Demonstrations/marches/political events	
	Low risk of disorder	2
	Medium risk of disorder	5
	High risk of disorder	7
	Opposing factions involved	9
(B) Venue	Indoor	1
	Stadium	2
	Outdoor in confined location, e.g., park.	2
	Other outdoor, e.g., festival	3
	Widespread public location in streets	4
	Temporary outdoor structures	4
	Includes overnight camping	5
(C) Standing/seated	Seated	1
	Mixed	2
	Standing	3
(D) Audience profile	Full mix, in family groups	2
	Full mix, not in family groups	3
	Predominately young adults	3
	Predominately children and teenagers	4
	Predominately elderly	4
	Full mix, rival factions	5
Add A+B+C+D	Total score for Table 5-1	

Table 5-2

Event Intelligence - From the U.K.'s Event Safety Guide (1999, Second Edition, pp. 121-124)

Item	Details	Score
(E) Past History	Good data, low casualty rate previously (less than 1%)	-1
	Good data, medium casualty rate previously (1% - 2%)	1
	Good data, high casualty rate previously (more than 2%)	2
	First event, no data	3
(F) Expected numbers	< 1,000	1
	< 3,000	2
	< 5,000	8
	< 10,000	12
	< 20,000	16
	< 30,000	20
	< 40,000	24
	< 60,000	28
	< 80,000	34
	< 100,000	42
	< 200,000	50
	< 300,000	58
Add E+F	Total score for Table 5-2	

Note: As attendance numbers may vary throughout the longer events, resource requirements may need to be adjusted accordingly.

Table 5-3

Sample of Additional Considerations - From the U.K.'s Event Safety Guide (1999, Second Edition, pp. 121-124)

Item	Details	Score
(G) Expected queuing	Less than 4 hours More than 4 hours More than 12 hours	1 2 3
(H) Time of year	Summer Autumn Winter Spring	2 1 2 1
(I) Proximity to definitive care (nearest suitable Emergency Medical facility)	Less than 30 min by road More than 30 min by road	0 2
(J) Profile of definitive care	Choice of Emergency departments Large Emergency department Small Emergency department	1 2 3
(K) Additional hazards	Carnival Helicopters Motor sport Parachute display Street theatre	1 1 1 1 1
(L) Additional on-site facilities	Suturing X-ray Minor surgery Bandaging Psychiatric / GP facilities	-2 -2 -2 -2 -2
Add G+H+I+J+K Subtract L	Total score for Table 5-3	

5.8.3 Calculation

5.8.3.1 To calculate the overall score for the event, add the total scores for Tables 5-1 + 5-2 + 5.3 above to give an overall score for the event.

5.8.3.2 Use the score from the above calculation to gauge the levels of resource indicated for the event.

5.8.3.3 Note: The following shows the resources that may be required to manage an event based on assessment of factors set out in the previous tables. This table, in conjunction with the medical chapter, is intended for guidance only. It cannot encompass all situations and is not intended to be prescriptive.

5.8.3.4 The score refers to the suggested resources that should be available on duty at any time during the event and not the cumulative number of personnel deployed throughout the duration of the event.

Table 5-4

Suggested Resource Requirement

Score	Ambulance	First Aider	Ambulance Crew[1]	Physician	Nurse	Group Supervisor[2]	Support Crew[3]
<20	0	4	0	0	0	0	0
21-25	1	6	2	0	0	Visit	0
26-30	1	8	2	0	0	Visit	0
31-35	2	12	8	1	2	1	0
36-40	3	20	10	2	4	1	0
41-50	4	40	12	3	6	2	1
51-60	4	60	12	4	8	2	1
61-65	5	80	14	5	10	3	1
66-70	6	100	16	6	12	4	2
71-75	10	150	24	9	18	6	3
>75	15+	200+	35+	12+	24+	8+	3

Note 1: An ambulance crew, as a minimum, consists of two EMTs; however, it most often includes at least one paramedic, which is preferred.

Note 2: A Group Supervisor is an Incident Command System (ICS) position that is responsible for a group—the medical group, in this case. Groups are established to divide the incident into functional areas of operation. The maximum ratio of personnel to supervisor is 7:1 (span of control), although 5:1 is a preferred ratio.

Note 3: "Support crew" is a collection of personnel and equipment necessary to support the personnel operating in the medical group. Members of a support crew (or support task force, as the medical group grows) might include administrative support, facilities management personnel (e.g., lighting, restrooms, etc.), and equipment management personnel.

6. Communication

6.0.1 Effective communication is of prime importance if an event is to run smoothly and safely. Communication requirements of all the organizations involved in the event, assessed individually or jointly, need to be examined thoroughly. This includes examining the general and operational management of the event, handling routine health, safety and welfare information and communicating effectively in the event of a major incident.

6.0.2 This chapter explores key communication issues from two main perspectives: internal communication, and public information and communication.

6.1 Internal Communication

6.1.1 Communication During the Event Planning Phase

6.1.1.1 The communication network during this phase is wide and involves a range of communication activities and information requirements:
- Intelligence gathering about the event characteristics, etc.
- Seeking appropriate licenses
- Preparation of detailed plans for arrangements on and off site
- Commercial arrangements - ticketing policy, publicity, contracts, etc.

6.1.1.2 Everyone involved in the planning of an event will need to keep proper records of decisions and ensure that relevant information is communicated to others. It is particularly important that "statement of intent" documents are clear in their definition of roles and the responsibilities of different agencies and individuals.

6.1.2 Preparation of Key Support Documentation

6.1.2.1 Clear language is crucial in providing reliable communication. Avoid jargon and acronyms wherever possible. Where they are necessary, it is worth including a glossary of terms within the main planning documents.

6.1.2.2 Agree on special terminology to be used by people preparing plans, documents and communication procedures in relation to:
- Naming different control points and control workers;
- Labeling different types of rendezvous and collection points;
- Providing unique reference labels for key locations within and around the venue;
- Naming conventions for categories of people involved on site;
- Compatible terminology for assessing risks and grading levels of urgency;
- Contact protocols for establishing communication.

6.1.2.3 Wherever possible, plans should say who does what, not just what is to be done. For example, "the incident control room must be informed," is not as helpful as, "the duty officer must inform the incident control room."

6.1.2.4 The Federal Government provides helpful guidance, tips and examples about many aspects of written communication on their Plain Language web site (http://www.plainlanguage.gov).

6.1.2.5 Relevant maps and site plans are crucial. Visual data should show key routes for vehicles and people, and restrictions on access. A gridded site plan for the venue and its immediate surroundings is recommended. Discrepancies can result in delayed responses, misdirected resources and communication channels being unnecessarily blocked with requests for clarification and attempts to sort out the confusion.

6.1.2.6 Pay attention to labeling features and functions consistently in different documents. If a feature occurs more than once (e.g., if there are several first-aid points) each should have a unique reference. Consult before altering plans so that the consequences of changes can be considered.

6.1.2.7 Consider establishing a single point-of-contact who would receive, collate, cross-check and spread information and documents relevant to the event.

6.1.2.8 Ensure that major incident plans are compatible with emergency plans drawn up by local authority and emergency services. Make sure relevant information is easily available to people in control rooms at remote locations.

6.1.3 Communication During the Event

6.1.3.1 A physical command center should be established during the event as a centralized hub of communication suitable for the event.

6.1.3.2 Consider the following matters in relation to your event:
- Power supplies for emergency communication equipment;
- All key personnel must be connected by radio or other communication device;
- Provide key items of documentation and stationery in all control rooms such as site plans, key contact details, alerting cascades, message pads, log sheets;
- Display frequently-used information clearly (site plans, key contacts, etc.) and make sure facilities such as white boards or flip charts are available for writing up incident-specific information as it arises;
- The need to maintain and operate emergency communications from an alternative site;
- Production intercom systems should also be powered from a stand-alone emergency power system, as communications between personnel during a power outage can be critical to synchronizing stage activities.

6.1.5 Off-Site Links

6.1.5.1 Provide details of the event to local emergency services.

6.1.5.2 Consider arrangements for communicating with outside organizations that are affected by the event such as local businesses.

6.1.6 Radio Communication

6.1.6.1 Identify and coordinate all radio frequencies for the event.

6.1.6.2 Each organization requiring radio communication will need to consider what operational channels are necessary for identified functions or areas. In addition, emergency services will have to consider the need for command channels at large events.

6.1.6.3 Radio is an important medium for general operational requirements and a prime medium for responding to emergencies. Pre-event checks are therefore essential. Carry out full perimeter tests to ensure coverage is adequate. At an outdoor site, appropriate positioning of masts, antennae and repeaters may require research and testing.

6.1.6.4 The issuing of full ear-defending headsets should be considered for key workers in high-noise areas.

6.1.6.5 Fully charge all batteries at the start of the event. Adequate numbers of spare batteries and charging facilities are essential.

6.1.6.6 All employees equipped with radios should be formally trained in the use of the assigned device(s), all proper operating protocols and procedures, and a procedure for dealing with an "open mic," a situation that locks everyone out of a channel or talk group. An open mic procedure could be as simple as, "Should there be an open mic for more than 30 seconds, switch to backup channel (or talk group) X until further notice"

6.1.7 Telephone Equipment

6.1.7.1 Provide and clearly mark external lines for emergency telephone contact between the venue control points and off site emergency services. The location of this equipment must be known to key venue management personnel. Cell phones can quickly become unusable when a major incident occurs, and if an emergency call must be made it is important to know where the hard line is. Do not use external telephone lines designated for emergency use for other communication.

6.1.7.2 Field telephone networks (or internal telephone networks in a venue such as a sports stadium or arena) provide vital links between on-site communication controls and other key points around the venue. Cell phones are widely used and provide extra communication options. However, they should not be relied upon for important links and especially not used for emergency communication.

6.1.8 Closed Circuit Television (CCTV)

6.1.8.1 A high-resolution CCTV system can provide real time situational awareness at any event, but may be most useful at large, well-attended and wide area events. When designed and deployed properly, a CCTV system allows event management to view and assess crowd dynamics at key areas such as stages, ingress and egress points, campsites and parking lots. A CCTV system that includes recording capability can also be highly beneficial in support of security services and potential litigation that may arise.

6.1.8.2 Certain fundamental questions regarding CCTV are worth asking and addressing early in the event planning stage:
- Will the use of CCTV make the event safer?
- Where should cameras be located?
- Will there be sufficient light?
- Who should have control over the devices and technology?
- Who should have viewing access?

6.1.8.3 An effective location for the CCTV control center and operator(s) is adjacent to the event command post. All security, health and safety elements of the event should have direct communication with the CCTV operator.

6.1.8.4 A qualified CCTV operator should have relevant experience in live event surveillance, any specific technology in use, and in assisting security and safety personnel in discovering and reporting adverse changes in crowd dynamics and other potentially undesirable and/or unsafe situations.

6.1.8.5 Ideally, the CCTV system should have the capability to be able to be viewed remotely.

6.1.8.6 The CCTV system should have the capability of searching and reviewing recorded footage to assist in the location of captured activities that could assist in identifying suspects or assisting in potential litigation.

6.1.8.7 In terms of the design and deployment of CCTV technologies, service providers should initiate and maintain communication with all operations and security elements that might benefit from CCTV monitoring.

6.1.8.7.1 Determine which areas of the event should be covered and what type of camera to install that will best achieve the goals of that coverage.

6.1.8.7.2 Determine any areas that may need observational and archival support.

6.1.8.8 Areas to consider deployment of CCTV coverage include, but are not limited to, the following:
- Stages - Monitor front of stage crowd dynamics, monitor density of crowd in stage arena. Assist security dispatch needs.
- Arena - Monitor crowd flow and density; assist security dispatch needs.
- Gates and Entrances - Monitor ingress and egress; assist security with identifying ticket scalping and other front of venue issues
- Camp Grounds - Monitor compliance of campers with safety regulations; assist security and campsite operations with camper flow.
- Parking - Monitor ingress and egress; assist security with surveillance; assist parking management with the determination of vehicle flow and capacity.
- Production Gates - Create a record of all vehicles entering and exiting the site.

6.1.8.9 CCTV images can greatly enhance the potential to identify problems in a crowd resulting from surges, sways, excessive densities or public disorder.

6.1.9 Communication Procedures

6.1.9.1 There must be a clear framework of information flow procedures - people need to know who should inform whom of what, when, and by what means.

6.1.9.2 Prime concerns are:
- Tight radio discipline with proper use of call signs and contact protocols;
- Making the purpose/function of a message clear (is it a question, warning, request for action, command, prohibition?);
- Concise and precise information;
- Cross-checking that messages have been received and interpreted correctly;
- Relaying message content clearly and unambiguously;
- Keeping accurate records of communication activity;
- Keeping accurate logs of decisions and actions.

6.1.10 Message Delivery and Acknowledgement

6.1.10.1 Workers must be aware of the possible consequences if messages are not properly communicated and understood. There will be marked differences in levels of local knowledge among workers at and around the event and so procedures for acknowledging or reading back messages should be introduced.

6.1.11 Situation Reports

6.1.11.1 Develop procedures for providing information from the scene of an incident or emergency. Note that a practiced format helps the person providing information to include necessary details for an appropriate response, a familiar communication pattern helps people receiving information to anticipate and recognize items; this assists the receiver to note the information ready for subsequent use or relay.

6.1.11.2 A situation report format must work equally well for any type of incident. It is particularly important to include the following items of information in such a report:
- Identification: call signs, names of calling and called parties;
- Location: exact details of where the incident is taking place;
- Incident: precise details of what is involved;
- Requirements: details of services, equipment and agencies required.

6.1.11.3 One example format: CHALET
- Casualty – number and types of injuries;
- Hazard – what hazards are present (e.g., fire, toxic gas);
- Access – best route to approach the incident;
- Location – specific location;
- Emergency services – what services are present, what services are required;
- Type of incident – description of the incident.

6.1.12 Record Keeping

6.1.12.1 Keeping records and logging information throughout the event is a key activity. Logs must show key events and actions in sequence and are a valuable tool for keeping workers informed of the progress of any incident.

6.1.13 Training, Briefing and Preparation

6.1.13.1 All organizations have a responsibility for training their workers appropriately, covering everything from using appropriate radio discipline to keeping a decision log. There must be proper briefings for all workers about their duties for the event. This includes briefing workers offsite who need to be aware of special arrangements for an event, e.g., those in incident control rooms.

6.1.14 Emergency Communications to Public Vendors

6.1.14.1 In the unlikely event of an evacuation, ensure that ALL parties (vendors, guests, staff) are notified.

6.2 Public Information and Communication

6.2.1 Types of Information

6.2.1.1 The information requirements of the audience range from performance details, ticketing arrangements, travel options, recommended routes, location of facilities, venue layout and welfare information right through to urgent contact messages or emergency instructions. Anticipating public information needs has an important bearing upon welfare and safety. Consider what information the audience will require if the event is cancelled or curtailed and how to provide that information. Well-informed people are less likely to be frustrated, aggressive or obstructive. Advance information on how to get to the venue, where to go on arrival or what will or not be allowed, all reduce frustration and irritation. If there is a need to communicate rules and restrictions, people are more likely to comply if they are aware of the reasons behind them.

6.2.2 Communication Channels

6.2.2.1 Communication methods include:
- Publicity material and tickets
- Media (press, radio, TV)
- Route-marking
- Signs
- Notices, information displays
- Screens, scoreboards
- Face-to-face contact
- Emergency public announcements
- PA systems
- SMS text alerts
- Interactive event web-site

- Twitter Feeds - Public, with special hash tags (#) for outbound communications that are for public information (weather, lost person, etc.), and separate hash tags (typically more obscure codes) for event staff communications.
- Facebook Feeds - Public, for outbound communications that are for public information (weather, lost person, etc.)

6.2.3 Alarms

6.2.3.1 Audible alarms are useful alerting devices but convey little information. The activation of an audible alarm will most often need to be followed by an explanation about what to do, or simply information that it has been a false alarm.

6.2.4 Public Address (PA) Systems

6.2.4.1 PA systems are a vital method of communication with the audience. Output should be clear and intelligible for everyone of normal hearing in all parts of the venue, including people in the immediate surrounds. Ensure that the PA announcer has a good view over as much of the venue as possible and good communication links with control points. In the event of a major incident, override facilities must allow announcements to be made over the PA system without interference from other sound sources. Agree to the circumstances in which this will happen in your major incident plans. The PA system should be fully tested before the event. In the event of an emergency where power is cut off a contingency plan to make general announcements must be considered (is there an emergency power source available to power the essential portions of the PA system for announcements, is there another means to make announcements to the general audience).

6.2.4.2 The PA System must remain operational until an "All Clear" has been provided.

6.2.5 Screens, Scoreboards

6.2.5.1 Video screens and scoreboards are a useful communication method for putting out public messaging. They can provide information without interruption to a performance. For urgent public announcements, however, they can reinforce the message and give information to those who have hearing difficulties.

6.2.6 Battery Operated Megaphones

6.2.6.1 Provide megaphones at strategic points in the venue for use by stewards and police for urgent communication and as a back-up in case the PA system fails. Train workers how to use them and where they are located. Keep batteries fully charged.

6.2.7 Staff (Face-to-Face Contact)

6.2.7.1 Direct contact between personnel and the public is obviously a vital communication channel, particularly in the safety chain. Approachable and helpful staff have an important role in creating a positive relationship with the audience. Their role in giving people clear and concise directions and assistance in an emergency can be a vital one.

6.2.7.2 Staff with any safety role should be easily identifiable by jackets/vests or other high-visibility items of clothing. These allow the public to seek them out as a source of assistance and to recognize their authority when appropriate. If people are being directed along a route of safety, staff in high-visibility clothing can help indicate the way much more clearly.

6.2.7.3 When problems are being dealt with, high-visibility clothing also helps colleagues, supervisors or CCTV controllers to pick them out and spot when they may be in difficulty or need support. In some cases, for workers who do not normally need to be visually conspicuous but may need to be identifiable for certain contingencies, reversible jackets that are high visibility on one side are worth considering.

7. Weather Preparedness

7.0.1 Severe weather can strike any geographical location at any time. Severe weather threats can take many forms, including tornadoes, hurricanes, lightning, hail, strong straight-line winds, flooding, blizzards, and many others. Event organizers, venue managers, stage rigging crews, and performing artists should be as prepared as possible. Severe weather preparedness is a fundamental responsibility of all persons involved with an event in protecting the safety of guests, event personnel, employees, performing artists as well as the assets of the venue.

7.0.2 Alongside such severe weather threats, which may require an immediate emergency response, consideration needs to be given to less severe activity which nonetheless may challenge the viability of an event or the welfare of attendees. Such instances may include extensive rain which compromises parking areas, or unexpected hot weather leading to medical and welfare concerns for an exposed audience.

7.0.3 Determining an appropriate reaction to severe weather threats can requires input from entities knowledgeable about the structure of the venue, the size and distribution of the spectators and participants, the weather conditions involved and the amount of advance warning time.

7.0.4 The starting place for severe weather planning and preparedness is to first assess the potential threats and then recognize and understand the inherent strengths and weaknesses of the venue's infrastructure and available resources.

7.0.5 A professional assessment of any potential weather threats to the venue should be performed by individuals trained or certified in the appropriate field of study. This typically includes meteorologists, engineers and venue management working together on a weather plan.

7.1 Chain of Command

7.1.1 The event command structure should clearly establish the responsibilities related to severe weather including planning, and incident management. For example, the individual responsible to make decisions to react to a severe weather threat must be designated and clearly understood by all relevant parties, and the authority of this individual should be final.

7.1.2 Having good communication channels is key and must be arranged before any venue event. In the event of severe or adverse weather, one primary point of contact must be available to contact a designated weather organization or professional meteorologist. Life safety must at all times be the first priority.

7.2 Weather Planning

7.2.1 Have a plan.

7.2.2 A proper plan will clearly define trigger point criteria and the action required at those points for any severe weather threat. Threats may include but not be limited to: thunderstorms, lightning, wind, heat, hail, tornadoes, etc. The establishment of these trigger points requires input from technical providers, meteorologists, engineers, structure suppliers and others deemed necessary to complete the process.

7.2.3 An example of a weather decision matrix related to an event production is shown in Fig. 7-1. Organizers are reminded that each event needs to establish their own matrix relevant to their event. The organizer will also need to consider a similar set of triggers and actions to protect event attendees and a similar decision matrix to protect the lives and safety of the audience should be established. Having a list of well-defined weather triggers, associated with a coloring scheme provides quick access to venue staff and decision makers.

Threat	Alert Method	Concourse	FOH	Pyro	Back Line	Video	Audio	Lighting	Stage	Catering
Thunderstorms	Radio alert from production	Double check security of anchors	Secure and cover gear	Safe all pyro	Secure and cover gear	Secure and cover gear	Secure and cover gear	Secure and cover gear	Monitor proximity	Monitor
Lightning Inside 6 miles	Radio alert from production	Take shelter	Take shelter	Safe all pyro	Take shelter	Take shelter	Take shelter	Take shelter	Take shelter	Evacuate to fixed structure
Non-Severe Hail < 3/4 inch	Radio alert from production	Take shelter	Shelter in place	Safe all pyro	Secure and cover gear	Secure and cover gear	Secure and cover gear	Secure and cover gear	Monitor	Monitor
Severe Hail >3/4 inch	Radio alert from production FTF voice communication	Shelter in fixed structure	Shelter in fixed structure	Safe all pyro	Shelter in fixed structure	Shelter in fixed structure	Shelter in fixed structure	Shelter in fixed structure	Shelter in fixed structure	Evacuate to fixed structure
Surface Winds 15-25 MPH	Radio alert from production	Carps proceed to concourse, collapse and stow pop up tents	Secure gear and monitor conditions	Safe all pyro	Secure and cover gear	Clear wall swing obstructions	Monitor PA swing	Monitor rig swing	Monitor	Monitor
Surface Winds 25-40 or Tornado Watch Issued	Radio alert from production	Evacuate to fixed structure	Secure gear and monitor conditions	25 MPH threshold for pyro – no pyro	Secure and cover gear	Video wall to the deck	Land PA	All truss to the deck	Secure and cover gear	Secure gear and monitor
Surface Winds > 40 MPH or Tornado Warning	1 long air horn blast + radio communication	Evacuate to fixed structure	Evacuate to fixed structure	Evacuate to fixed structure	Evacuate to fixed structure	Video wall to the deck	Land PA	All truss to the deck	Evacuate stage to fixed structure	Evacuate to fixed structure
Surface Winds > 60 MPH or Tornado Activity	1 long air horn blast + radio communication	Immediate retreat to shelter	Immediate retreat to shelter	Immediate retreat to shelter	Immediate retreat to shelter	Immediate retreat to shelter	Immediate retreat to shelter	Immediate retreat to shelter	Everything to the deck; immediate retreat to shelter	Immediate retreat to shelter
Common sense is key – Protect yourself and your family first – We can rebuild the show.										

Fig. 7-1. Weather Decision Matrix related to event production.

7.2.4 Predicting the reaction of people to severe weather threats is difficult. Many things must be taken into consideration when trying to foresee the behavior of the crowd in a severe weather situation, particularly if the threat becomes reality.

7.2.5 The organizer must generate an atmosphere of trust through clear and accurate communication with everyone entering the event site. Communication should include frequent updates as deemed necessary by venue management. If visitors feel as if their best interests and safety are a priority, they will be more likely to respond in a cooperative and predictable manner.

7.2.6 Proper planning will identify ways to prepare for and mitigate problems associated with moving people to safe shelter.

7.2.7 Sheltering will vary based on the threat, venue, and shelter type. For example, a large tent may be appropriate to shelter for heavy rain, however this same structure may require evacuation in the event of extreme winds. Depending on the severe weather threat, potential locations for shelter could include: cars, shelter in place at the venue, tents, permanent structures, under bleachers or grandstands, or even simple evacuation from the event site. Sheltering locations and capacity must be clearly identified in the plan.

7.2.8 The path to safe shelter needs to be clearly identified and free of any equipment, material or debris that might slow down or hinder the movement of people. It is suggested that safe shelter location also be marked with appropriate signage.

7.2.9 Delaying or cancelling an event must be considered an option based upon established trigger criteria and adverse weather conditions. A proper plan will clearly provide the appropriate actions to venue staff based on the decision to delay or cancel an event.

7.3 Weather Monitoring

7.3.1 Given the complexity of weather forecasting, use of professional weather consulting services is strongly encouraged. Weather consulting services typically provide 24/7 access to professional meteorologists and can be used as a direct point of contact during adverse weather conditions. Such sources can focus specifically on your event's physical address and threats specific to your event with expert meteorologists making precise forecasts. Such service providers are able to offer customized data such as lightning strike detection and proximity, surface level winds, accurate radar interpretation and other weather related phenomena.

7.3.2 The Storm Prediction Center (SPC) (http://www.spc.noaa.gov) is part of the National Weather Service (NWS) and the National Centers for Environmental Prediction (NCEP). The mission of the SPC is to provide timely and accurate forecasts and watches for severe thunderstorms and tornadoes over the contiguous United States. The SPC also monitors heavy rain, heavy snow, and fire weather events across the U.S. and issues specific products for those hazards.

7.3.3 While the SPC issues products for the entire country, the local NWS (http://www.nws.noaa.gov) weather forecast offices (WFO) issue weather information for their County Warning Area (CWA). They are considered the experts in their region and publish area specific forecasts, warnings and advisories.

7.3.4 The National Hurricane Center (NHC) is a component of the National Centers for Environmental Prediction (NCEP; http://www.ncep.noaa.gov/). The NHC mission is to save lives, mitigate property loss, and improve economic efficiency by issuing the best watches, warnings, forecasts, and analyses of hazardous tropical weather and by increasing understanding of these hazards.

7.3.5 Given the complexity of weather forecasting, use of professional weather consulting services is strongly encouraged. Such sources can focus specifically on your event's physical address and threats specific to your event with expert meteorologists making precise forecasts. Such service providers are able to offer customized data such as lightning strike detection and proximity, surface level winds, accurate radar interpretation, etc.

7.3.6 Additionally, there are subscription services available online that provide real-time lightning and weather data as well as phone applications that can aid in situational awareness. Many of these services provide alerts as well, based on user-defined criteria.

7.3.7 An event should utilize suitably located anemometers (wind speed measuring devices) in order to monitor the weather at that specific site. Device quantity and location can be determined by consulting a meteorologist. These devices can also keep a record of weather conditions, which is useful information. Weather data collected on site can be shared with hired meteorologists and is proven to be helpful in the forecast process.

7.3.8 On-site weather monitoring should be used in conjunction with other monitoring and forecasting services. Management, event staff, and other production personnel should always maintain situational awareness but they should never attempt to perform the roles of professional meteorologists. Providing a meteorologist on site may also be beneficial, depending upon the venue.

7.4 Communication

7.4.1 During severe weather, communication amongst venue staff needs to be efficient, accurate and targeted. Communication is to be associated to trigger criteria and actions defined in the plan. As time is of the essence during a severe weather situation, all communication should, as much as possible, be rehearsed during training and exercises.

7.4.2 The sequence of communication needs to be considered, allowing technical departments, security and event staff and food and beverage/merchandising to make preparations before the audience is advised.

7.4.3 Performers can be of great value in calmly asking guests to take action, assuming that they are willing and able to relay this important information. They should not be solely relied upon to communicate this important information, but should be asked to do so if their special connection with audience members could make the process smoother.

7.4.4 Any weather information or action announcement should be accurate and unambiguous. Communication should continue throughout the implementation of an evacuation or sheltering process which may require utilizing additional means such as bullhorns, radio, social and media. Other technologies that can be used to relay information include mobile apps, SMS alerts and visual displays. All weather information must be consistent through all communication channels.

7.5 All-Clear and Determination of Event Continuation/Cancellation

7.5.1 Once a weather situation has moved past the event, an "All-Clear" should be declared. The organizer needs to consult with local authorities, technical staff, security and other relevant parties to determine if the event can proceed safely. Particular attention must be given to temporary structures which may have been compromised by the weather.

7.5.2 Organizers should have professional relationships with first responders. An effective severe weather response is likely to include coordinated efforts with first responders.

7.5.3 In the event that there is serious damage to the interior or exterior of the venue, the best decision is to terminate or cancel the event. If there is serious damage, evacuation of people in shelters or remaining at the event site should be conducted in a safe manner. This may require delaying movement until it is safe to do so or using safe routes.

7.6 Post-Incident Analysis

7.6.1 It is valuable to have a post-incident review as soon as possible. Information gained may be helpful for future events and incident audits by local authorities or insurance providers. Typically engineers, meteorologists and other technical professionals work together provide these services.

7.7 Training and Resources

7.7.1 Training is an essential component of an event or venue's preparedness for severe weather. Preparedness planning must include all components of the event's operations and production; and, training must be provided to managers and supervisors, security, event and operations staff, artists, production staff, and anyone else who may be involved in event operations.

7.7.2 The International Association of Venue Managers (IAVM) offers a helpful guide titled *Severe/Hazardous Weather Preparedness Plan and Guideline.* The guide gives venue and event managers important tools and was written by venue management professionals with the assistance of top experts from the National Weather Service, the insurance industry, and legal counsel specializing in public assembly venue issues. Visit https://www.iavm.org/ for more information.

7.7.3 The IAVM also offers a 2-day course titled *Severe Weather Preparedness & Planning for Public Assembly Venues and Events*, which held annually at the National Weather Center in Norman, Oklahoma. The course teaches how to develop a severe weather preparedness plan for all types of public assembly venues, fairs, expositions, and other events where large crowds assemble. Visit https://www.iavm.org/ for more information.

7.7.4 The NOAA National Weather Service offers an excellent, 8-page toolkit titled *Lightning Safety: Large Venues* that includes advice on what to do before, during, and after a severe weather event. It also includes multiple venue preparation checklists. Visit http://www.lightningsafety.noaa.gov/ to access the free document.

7.7.5 The NOAA National Weather Service offers a program called "StormReady®" that better prepares communities through advanced planning, education and awareness to save lives from the onslaught of severe weather. Businesses, schools, and other non-governmental entities often establish severe weather safety plans and actively take part and promote severe weather safety awareness activities. An entity that meets the principles and guidelines and completes the large/public assembly venue preparedness toolkit for the StormReady program may be recognized as a StormReady "Supporter." Visit http://www.stormready.noaa.gov for more information.

7.7.6 Private weather consulting firms usually offer onsite services and training programs, in addition to, and in association with, weather planning and software. Typically this involves weather planning and technical training from professional meteorologists and engineers. Private training can provide a tailored approach to venue weather planning.

8. Venue and Site Design

8.0.1 This chapter summarizes some of the factors to consider when laying out your venue or site. In sports arenas the dimensions of ice hockey playing surfaces differ even though there is a prescribed specification to their dimensions. Outdoor stadiums vary greatly with their field dimensions. Theaters, amphitheaters, auditoriums all differ in size, shape and layout. Open field venues may have the available space to do almost any event; however, there are always terrain, foliage, road and other considerations that make every site unique.

8.0.2 Organizers should always consider the viability of the proposed event in a venue or proposed site. Part of that analysis should include the ability of the site or venue to accommodate the demands of the proposed event, including venue capacity, sightlines, terrain, running water, drainage, sewage, parking, restrooms, power and other utilities, fencing, foliage, insect issues, odors from other properties, and Internet access.

8.0.3 The information below can be assessed by walking the site, studying the appropriate mapping and seeking advice and information from the landowner, venue management or local authorities. Such information is essential before beginning detailed site design.

8.0.4 In an entertainment scenario, the general principle behind venue design is to provide an arena in which the audience can enjoy the entertainment in a safe and comfortable atmosphere. The requirement for certain safety provisions, the type, number and specification of facilities and services will depend on the type of event and the outcome of the risk assessment.

8.0.5 The final design of a site will be dependent on the nature of the entertainment, location, size and duration of the event. It will also need to take account of the existing geographical, topographical and environmental infrastructure.

8.1 Site Suitability Assessment

8.1.1 It is important for the event managers who conduct the preliminary assessment to be familiar with the venue or site to determine its suitability for the planned event. The main areas for consideration are: available space for the proposed audience and its required infrastructure, temporary structures, utilities, backstage facilities, parking, camping and rendezvous points. You may already have a proposed capacity in mind, together with some ideas of the concept of the entertainment. Rough calculations of the available space are useful at this stage.

8.1.2 Factors to consider include the following:
- Ground conditions: Are they suitable? Even and well-drained open sites are preferable. Avoid steep slopes and boggy areas.
- Potential weather conditions should be considered. With open field sites especially, there may be a "100-year" condition regarding temperatures, storms or floods to be incorporated into the plans.

- Traffic and pedestrian routes and emergency access and exits: What routes already exist? Are they suitable to handle the proposed capacity? Is a separate emergency access possible? If not, can other routes be provided? Are roads, bridges, etc., structurally sound? For further information see Chapter 12, *Transportation Management*.

- Position and proximity of noise-sensitive buildings: Are there any nearby? Is it possible to satisfy both the requirements of the audience and the neighbors? A noise propagation test may be advisable.

- Geographical location: Where is the site located? How far away is the hospital, fire station, public transport, parking, major roads, local services and facilities? Such information can be valuable when assessing the suitability of the site and determining the extra facilities that need to be accommodated within the site.

- Topography: How does the land lie on its surroundings? Does it form a natural amphitheater? Where does the sun rise and set? Could any natural features assist in noise reduction? Are there any natural hazards/features such as cliffs, lakes and rivers?

- Location and availability of services: Water, sewage, gas, electric, telephone or towers supporting overhead power lines or cables. Are there any restrictions or hazards? Can they be used? Is the event site within the "consultative distance" of a hazardous installation or pipeline?

- Insects and wildlife should also be considered in this suitability analysis. If the analysis is taking place in December and the proposed event is in July, there are potential infestations that may not be immediately apparent.

8.2 Predesign Data Collection and Appraisal

8.2.1 The next step in site design is to collect all the available data together and appraise it. The site design should be based on the site suitability and risk assessments.

8.2.2 Ensure that you have considered the following factors:
- Proposed occupant capacity;
- Artist profile;
- Audience profile;
- Duration and timing of event;
- Venue evaluation;
- Whether alcohol is on sale;
- Whether the audience is standing, seated or a mixture of both;
- The movement of the audience between the entertainment and/or facilities; and
- Artistic nature of the event, single stage, multiple-stage complex, etc.

8.2.3 The above information can then be used to determine the provisions and facilities needed within the site; for example, stages, tents, barriers, toilets, first aid, concessions, exits, entrances, hospitality area, sight lines, power, water, sewerage, gas, delay towers, perimeter fencing, backstage requirements, viewing platforms and waste disposal requirements. Once all the information is collated, detailed site design can begin.

8.3 Site Plans

8.3.1 Once the basic outline has been determined, detailed scaled site plans should be produced. Often, many versions will be produced as amendments are made and as further information is obtained. Ensure your site plans are kept up to date and current revisions are given to relevant departments, especially members of your event safety team. Make sure that alterations are not made to the site plans after capacity levels have been determined and tickets placed on sale as the alterations may affect sight lines and therefore available viewing areas. Plans may already exist for permanent existing venues.

8.4 Site-Design Considerations

8.4.1 Venue Capacity/Occupant Capacity

8.4.1.1 *Critical Crowd Densities* (FEMA, 2010)
8.4.1.1.1 The objective should be to prevent large accumulations of patrons, particularly within short time periods, in confined spaces—especially if they cannot see what is happening.

8.4.1.1.2 A study by noted crowd behavior expert John Fruin (1981) identified critical crowd densities as a common characteristic of crowd disasters. Critical crowd densities are approached when the floor space per standing person is reduced to about 5.38 square feet (0.5 square meters).

8.4.1.1.3 Considering the various movements or the positions that spectators will occupy, Mr. Fruin listed the following crowd movement characteristics:
- Pedestrians require average areas of 24.73 square feet (2.297 square meters) per person to attain normal walking speed, and to pass and avoid others.
- At 10 square feet (0.929 square meters) per person, walking becomes significantly restricted, and speeds noticeably reduced.
- At 4.95 square feet (0.46 square meters) per person, the maximum capacity of a corridor or walkway is attained with movement at a shuffling gait and movement possible only as a group. This would be characteristic of a group exiting a stadium or theater.
- At less than 4.95 square feet (0.46 square meters) per person average, individual pedestrian mobility becomes increasingly restricted.
- At approximately 3 square feet (0.2787 square meters) per person, involuntary contact and brushing against others occurs. This is a behavioral threshold generally avoided by the public, except in crowded elevators and buses.
- Below 2 square feet (0.1858 square meters) per person, potentially dangerous crowd forces and psychological pressures begin to develop.

8.4.1.1.4 Mr. Fruin (1981) wrote that "the combined pressure of massed pedestrians and shock-wave effects that run through crowds at critical density levels produce forces which are impossible for individuals, even small groups of individuals, to resist."

8.4.1.1.5 The above information shows that you may need to provide a monitoring system, such as observers in strategic locations or closed circuit television monitoring of crowd movements that will warn event personnel that they must take action to prevent a major incident.

8.4.1.2 Crowd Throughput Capacities (FEMA, 2010)

8.4.1.2.1 In his writings on crowd disasters, Mr. Fruin (1981) identifies several areas regarding spectator throughput in entry to a performance. For planning purposes, he suggests:

(a) Ticket Collectors/Scanners - Ticket collectors/scanners should be in a staff uniform or otherwise identifiable. Ticket collectors/scanners faced with a constant line can throughput a maximum of:
   - One patron per second per portal in a simple pass-through situation.
   - Two seconds per patron if the ticket must be torn and stub handed to the patron.
   - More complicated ticketing procedures (and/or answering the occasional question) will protract time per patron.
   - There is currently no published time estimate for the scanning of barcodes or other electronic ticketing scan system, so it is recommended organizers budget at least two seconds per patron. Keep in mind the type of scanner and networking system used along with the bandwidth available for the system can have a bearing on the scan speed of some electronic scanning devices.

(b) Doorways - A free-swinging door, open portal, or gate can accommodate up to one person per second with a constant queue. Revolving doors and turnstiles would allow half this rate of throughput, or less.

(c) Corridors, Walkways, Ramps - Have a maximum pedestrian traffic capacity of approximately 25 persons per minute per 1 foot (0.3048 m) of clear width, in dense crowds.

(d) Stairs - Have a maximum practical traffic capacity of approximately 16 persons per minute in the upward direction. Narrow stairs (less than 5 feet or 1.524 m) will lower the maximum flow.

(e) Escalators and Moving Walkways - A standard 3.94 feet (0.366 square meters) wide escalator or moving walkway, operating at 118 feet (35.9664 m) per minute can carry 100 persons per minute under a constant queue.

8.4.1.3 The venue capacity depends upon the available space for people and the number of emergency exits. The latter is the subject of a calculation involving the appropriate evacuation rate, i.e., width of available exit space and appropriate evacuation route.

8.4.1.4 Some of the site will be taken up by unoccupied structures. The rest of the site will need to be considered in calculating occupant capacity even though a direct view of the entertainment may not be possible for all locations. Any space where the audience does not have a reasonable view of the performance should be deducted from the available area or a lesser density used in calculations. Areas with partial or total cover to the audience if there is severe weather should be identified and the effects of audience migration to these areas considered.

8.4.1.5 In venues where seating is provided, the major part of the occupant capacity will be the lesser of the two figures determined by the number of seats and exit provision. In other cases a calculation based on the acceptable occupant density should be carried out. Generally, 7 square feet (0.65 square meters) of available floor space per person is used for the prime viewing areas of outdoor music events. There may be zones within the viewing areas where the quantity of square feet per person will be increased due to queue lines, pedestrian traffic aisles, etc.

8.4.1.6 Double-check the preliminary occupant capacity calculation and exit requirements once all initial infrastructure requirements and facilities are in place on the site design. Further detailed information on occupant capacities can be found in Chapter 4, *Fire Safety*.

8.4.1.7 Once you have a proposed capacity figure, meet with the local authority with jurisdiction (typically the fire department) over the determination of the venue capacity and review your numbers. During this meeting you will also need to present your exiting plans along with dimensions of those exits.

8.4.1.8 It is essential for effective crowd management during emergencies and evacuations that a well-planned and efficient means of escape exists for all occupants of the venue. As a guide, consider the following:
- For outdoor open field sites, a maximum 15 minute evacuation time is recommended;
- For stadiums, the maximum exit times should be between 2 minutes 30 seconds and 8 minutes;
- For indoor venues, the maximum evacuation time should be between 2 and 3 minutes depending on conditions and depending on the venue layout and arrangement;
- Calculate exit door flow rates at 40 persons per minute for a single, fully open door (36 inches or 0.9 m wide) and 60 persons per minute for wider passages;
- Use 24 inches (0.610 m) when calculating the width of a person over the width of wider passage ways. For example, if a passage way is 24 feet wide, 24 inches goes into 24 feet 12 times. Therefore, 12 is the maximum estimated number of people that can fit down that passage way at a time. Multiply the 12 persons wide by the 60 persons per minute mentioned above, and the result is a maximum exit capacity of 720 persons per minute. Note: This formula is intended only to acquaint one with the basics of egress space estimation and is not an official means by which egress capacity is measured. Chapter 4, *Fire Safety*, includes details more about means of egress. But, as always, work with the local fire and building authorities having jurisdiction to establish exactly what is required in terms of means of egress.

8.4.2 Exit Requirements

8.4.2.1 The exit numbers for a venue depend directly on the occupant capacity, the width of the means of egress and the appropriate evacuation time for the type of structure. More details on this can be found in Chapter 4, *Fire Safety*.

8.4.2.2 Place exits around the perimeter and ensure that they are clearly visible, directly and indirectly by signage. Ensure they are free from obstruction on either side. The final exit destination should be assessed and be safe, i.e. into open spaces, assembly areas, etc., rather than into a main road or into traffic flows. It is important to examine these areas when carrying out your overall event risk assessment. Exit gates should operate efficiently and effectively. Where practical, provide separate exits for pedestrians and service and concession vehicles. Wheelchair access and exit will also need to be considered.

8.4.2.3 All hinged exit gates should swing outward, with the flow of exiting traffic.

8.4.3 Venue Access

8.4.3.1 Venue access is a function of the design and location of transportation and parking facilities and the design of access roads. Such facilities must be able to handle the peak demand as determined from the arrival profile (see Chapter 12, *Transportation Management*).

8.4.3.2 The layout of the access routes depends upon the location of facilities. Distribute routes around the site to minimize the load and ensure that the routes do not converge. Routes should be simple, easy to follow, direct and avoid routes from crossing one another.

8.4.4 Entrances

8.4.4.1 The entrances provide the means for supervising, marshaling and directing the audience to the event. They may be used as an exit or they may be separate. It may be necessary to provide separate entrances for performers, workers, guests, etc.

8.4.4.2 The design and location of entrances depends on the numbers of entrances required, where they are placed and the capacity to be handled at each entrance. There should be sufficient numbers of entrances to cope with the peak demand and achieve a smooth and orderly flow of people through them. The direction from which people are likely to come, the maximum number of people from each direction and the flow rate through the entrance are important issues which determine the number of entrances required. For purpose-built venues, these will already have been considered and approved.

8.4.4.3 Flow rates depend on the type, design and width of the entrances and whether pat-downs, magnetometers or wand searching takes place. The desired entry time is the time taken to allow everyone access to the venue. This will depend on the type and duration of the event and the audience profile. The possibility of severe weather may affect the desired time. Any queuing system to manage people at the entrance needs to be planned and carefully designed.

8.4.4.4 Once it is known how many entrance lanes there will be and the estimated time-per-person to enter through a lane, an organizer can estimate with reasonable confidence how long it will take to load an audience into a venue. There are other factors outside the venue entrance lanes which can impact the people arriving such as how fast the parking lots can load with vehicles, traffic outside the venue perimeter (and the organizer's control), etc.

8.4.4.5 The organizer should know what factors can impact the time required to load the venue and closely monitor those factors. In a live performance situation, the artist's management may need to be advised and the performance delayed for a reasonable period.

8.4.5 Sight Lines

8.4.5.1 It is important the audience has a clear line of vision to the stage to avoid movement toward the center of the venue. The widest possible sight lines help to reduce audience density in front of the stage and help to minimize surging and the possibility of crushing injuries. The stage width, height and position of PA wings, suspended show elements, sound delay towers, spotlight towers, the location and dimensions of the "mix" or control riser or tower, etc., all affect sight lines and have an impact on the audience capacity. Design sight lines to create areas of clear space on the immediate stage left and right. This allows movement and emergency access.

8.4.6 Video Screens

8.4.6.1 Large distances between the stage and the back of the viewing area lead not only to poor visibility and reduced entertainment value, they can also contribute to crushing and overcrowding. Strategically placed video or projection screens can be effective. Delay screens located at some distance from the stage encourage a proportion of the audience to use a less crowded part of the site. Screens near the stage can help to stop people pushing toward it. Screens may require substantial foundations and support so sufficient space should be allowed in any site design. Not all types of screens operate in daylight and if the intention is to use a screen in these conditions, make sure that an appropriate type is used.

8.4.7 Seating Arrangements

8.4.7.1 Where there is a risk of over-excitement among audience members, consider holding an all-seated event as this may help to prevent crowd surges and crushing at the front of the stage area. Spacing requirements and aisle widths, etc., should be obtained from the venue or the authority with jurisdiction over the event, e.g., the local venue fire marshal.

8.4.7.2 If temporary seating is provided, seating needs to be adequately secured. Temporary seating and the means of securing seats must be approved by the local authority. An example of securing seats is using two tie wraps on the front legs of two adjoining chairs and two tie wraps on the rear legs of those two adjoining chairs.

8.4.8 Slopes

8.4.8.1 Ensure that you have fully considered the effects of any slopes at your venue in your risk assessment. It may be necessary to consider providing exit steps or ramps with non-slip surfaces. The area in front of the stage should be flat to prevent tripping and crushing.

8.4.9 Observation Points

8.4.9.1 At some outdoor music events, crowd observation points may be considered necessary. These should be strategically placed to maximize the view of the audience. Establish safe entrances and exits to these observation points.

8.4.10 Production Infrastructure and Backstage Requirements

8.4.10.1 The production infrastructure will depend on the type, size and duration of the event. Typically, production offices, refreshment facilities, accommodation (for workers and artists), dressing rooms, large vehicle parking and access, storage space, equipment, etc., need to be accommodated, usually backstage. Carefully consider the number of units required, fire hazards, access routes and circulation space, generators, first-aid posts, ambulance, fire and police requirements. Try to keep performers' areas separate from production and working areas.

8.4.11 Fire and Ambulance Requirements

8.4.11.1 Fire and ambulance requirements such as parking areas, first-aid posts, rendezvous points, triage areas, etc., need to be carefully assessed and positioned in the appropriate places. Design the site so they are readily accessible and can be easily identified. Fire apparatus should be able to access all parts of the site and be able to get within 50 feet (15 m) of at least one exterior door of all structures and 150 feet (46 m) of any wall of all structures (*NFPA 1, 2012,* 18.2.3.2). Establish emergency access routes that are always kept clear so that emergency vehicles can access even the most remote parts of the site. Temporary trackways (reinforced road surfaces) may be necessary for wet, difficult ground. Consider separate gated entrances and exits, of sufficient height and width, for fire and ambulance vehicles.

8.4.12 Police and Security Positions

8.4.12.1 The presence of law enforcement and their number and positioning of security staff will depend upon the nature and type of entertainment provided (see Chapter 9, *Crowd Management*).

8.4.13 Site Workers

8.4.13.1 For large events a significant number of workers will be on site and will need their own facilities such as catering, toilets, showers, offices, sleeping accommodation, etc. Such facilities may form a separate compound or be distributed between backstage and/or main area. Carefully plan such requirements to incorporate them safely into the site design.

8.4.14 Hospitality Area

8.4.14.1 The level of hospitality will vary with the event size. Accommodation and facilities may need to be provided for only a few people requiring no more than a small meeting area through to very large sophisticated complexes catering for several thousand people. Tents and viewing platforms may be required. The exact requirements need to be planned and incorporated into the overall site and venue design. Often such large numbers are forgotten in the capacity calculations but need to be included.

8.4.15 Noise Considerations

8.4.15.1 The overall site design and layout should maximize the audience's enjoyment and protect the neighbors from noise (see Chapter 22, *Sound: Noise and Vibration*). Consider the stage location and other sound sources, in relation to nearby noise-sensitive properties and the topography of the site. Use slopes and natural barriers to their maximum effect. It may be advantageous to use a distributed sound system suspended from delay towers. Carefully consider the location and construction of such towers to control sight lines, avoid crushing points and prevent unauthorized "viewing" platforms.

8.4.16 Catering and Merchandising

8.4.16.1 Food and beverage and merchandise managers will always want points of sale to be in the highest traffic areas available. Satisfying the needs of these two moneymakers is very important, but should not be done at the cost of public safety. Position merchandising and catering concessions away from critical access routes and in less densely occupied areas of the venue. Some units will have highly flammable products such as propane and require careful positioning (see Chapter 14, *Food, Drink and Water*, and Chapter 24, *Merchandising and Special Licensing*). Consider circulation space and potential queuing arrangements, which should not obstruct pathways.

8.4.16.2 Plan on the merchandiser asking to be located near the entry and exit points as they will want the exposure those sites allow. Place them so as to allow free and adequate flow of the audience in and out but close to the entrances. Plan accordingly as this will often cause greater crowds to collect in the area.

8.4.17 Perimeter Fencing

8.4.17.1 Whether a perimeter fence is required depends on the type and nature of the event. Fences may be necessary to prevent trespassers from entering the site and for safe audience management.

8.4.17.2 Some events may not require a fence, just a manned stake and visual tape barrier, whereas others may need a sophisticated, substantial fence or multiple arrangements. Assess the crowd loading on such structures and the climbing potential.

8.4.17.3 A typical arrangement for large music events is an opaque inner fence with an outer fence, providing a moat in which security can patrol. To minimize the climbing of the inner fence for those who have breached the outer, a 16 foot gap is usual to prevent the run-up approach to jumping the fence. Three fences may be used which can easily form an emergency vehicle route. Carefully consider the ground conditions, obstructions, support legs, bracing and entrance and exit requirements.

8.4.18 Front-of-Stage Barrier Requirements and Arrangements

8.4.18.1 A front-of-stage barrier may be required if significant audience pressure is expected. The risk assessment for the event, relating to the performer's popularity and the audience capacity and profile, should help determine if one is required and if so, what type and design. For most large music events, some form of front-of-stage barrier will be necessary. The barricade needs to be of a design such that the force of crowd pressure cannot cause patrons to be crushed against it without a method of extrication (see Chapter 23, *Barriers*, for further information).

8.4.19 Signage

8.4.19.1 The location and size of all signage is critical when designing a site. For indoor/permanent venues such signage is normally in place for emergency exits, extinguisher points, entrances, parking areas, emergency vehicle points, etc. For supplementary facilities and all outdoor sites, this will not be the case and must be designed into the signage plan.

8.4.19.2 Event signage is one of the most important elements of an event. It is often the first impression the attendee has of the event and the first opportunity the event has to inform and make an impression on the attendee. When planning for signage, coordinators must consider the following: artwork, materials used to create the signage, the required tie down methods, required ballast, vehicular and foot traffic flow, language requirements, distance the signage will be from the person seeing it and potential weather issues. Fig. 8-1 shows a windy day during a daylight public ingress period.

8.4.19.3 The effective use of signs provides a rapid way of conveying orientation, directions and emergency information. It assists in audience flow. Signage should be clearly visible, easily understood, and lit in the dark.

8.4.19.4 From a site-design perspective, signage size and position is very important. Large outdoor venues will require signage larger than usual so that it can be seen from a distance. Fixture points may have to be constructed, such as scaffold towers, etc. Safety signs must comply with the standards of the local authority with jurisdiction over the event, e.g., the fire department or department of building and safety.

Fig. 8-1 - This is an example of a reasonably well laid out entrance. Positives include: visibility from the parking lot; signage clearly identifies the entrance, prohibited items, and points of interest; solid surface to stand on while awaiting entry. Negatives include: the stage and video IMAG are visible from the entrance; and a member of the event staff placed near the two "Entrance" flags would be helpful to serve the arriving public. Photo courtesy of Steve Immer.

8.4.20 Welfare Facilities

8.4.20.1 The number and type of assistance and information facilities, sanitary accommodation, and water supply, depend upon the event. But, once numbers have been agreed, they must be considered in the venue or site design.

8.4.20.2 Distribute sanitary accommodation around the site in a manner that does not block sight lines and serves the greatest need, e.g., near bars and catering concessions. It is prudent to plan access for the pump truck to service the units even if you do not intend to do so during the event. Ensure they are clearly visible and well signed and that queuing areas do not obstruct any gate, emergency route, etc. Water supply is normally situated next to sanitary accommodation. If water or sewage tankers are used, consider the space requirement and ground drainage.

8.4.20.3 Information points vary from a notice board to a billboard. Size and location must be considered. The best positions are near the main entrance into the site, but not too close to any gate or emergency access route, as people using or waiting near the facility could cause an obstruction. Try to locate assistance and information points in quieter areas.

8.4.21 Excess Visitors

8.4.21.1 An open public event will occasionally exceed the anticipated occupancy. If this is a possibility, contingency arrangements should be made to manage excess visitors. The design of a holding and/or queuing area and related facilities may need to be accommodated within the design.

8.4.22 Cable and Hose Routing and Access

8.4.22.1 Cabling and hose routing is often omitted from many site plans.

8.4.22.2 Hoses and cables should not be deployed across the ground, driving, or walking surfaces without guards to protect the hoses or cables from physical damage and to reduce the likelihood of individuals tripping over exposed hose or cable protectors.

8.4.22.3 Hose and cable management systems should also be deployed across all working personnel and vehicle routes. Hose or cable protection can be subterranean, surface, or suspended overhead in a manner that meets vehicle clearance requirements, electrical code (NFPA 70) requirements, and 2010 ADA ramp angle requirements.

8.4.22.4 Where hose and cable protectors rise above the local surface elevation by more than 1/2 inch (12mm), elevation changes should be clearly marked with contrasting colors and illuminated to not less than 10 foot candles (fc) (105 lux) in off-stage show support areas and public areas outside of the event seating or viewing area.

8.4.22.5 For onstage performance areas and within the event seating / viewing areas, they must be illuminated to not less than 1/5 foot candle (fc) (2 Lux). Hose / cable management troughs and subterranean pathways must be structurally sufficient to prevent collapse under anticipated vehicular and personnel traffic.

8.4.22.6 The organizer should coordinate the anticipated concentrated wheel loading of cranes, dollies, tractors, trucks, cars, forklifts, carts, wheelchairs, and other event support equipment with the elected means of hose or cable protection deployed. For more details refer to Chapter 17, *Electrical Installations and Lighting*.

8.4.23 Incident Command Post (ICP)

8.4.23.1 According to the NIMS Incident Command System (ICS), the Incident Command Post, or ICP, is the location from which the Incident Commander oversees all incident operations. Should an incident occur at an event, a pre-established ICP should be identified and equipped to provide a location from which a minor or major incident could be managed by event authorities. It should incorporate communications capabilities with internal and external resources; adequate power supply during a power outage; work space, sanitation facilities, and supplies for everyone who may need to work there; and, adequate security to control access. Personnel who can make high level decisions at the event must have immediate access to it and should be located there during an incident. Consideration should be given to making the ICP large enough to incorporate and support emergency response decision makers should the incident require it.

8.4.23.2 According to NIMS ICS, there is only one ICP for each incident, and every incident or event must have some form of an Incident Command Post.

8.5 Helicopter Use at the Event Site

8.5.1 When designing an open field site, public service agencies will require access for emergency vehicles, and this may include one or more helicopters. It is incorrect to assume that a helicopter can land anywhere and it is a dangerous myth that a helicopter can land on any open area on or near an event site. The safe landing of an aircraft takes careful planning and consideration. If planning to use helicopters, it would be prudent to refer to FAA Document AC150/5390-2C, *Heliport Design*.

8.5.2 The Incident Command System (ICS) defines a "helispot" as a designated location where a helicopter can safely take off and land. In contrast, a "helibase" is the main location for parking, fueling, maintenance, and loading of a helicopter operating in support of an incident or event. A helibase is usually located at an airport or airfield, whereas a helispot is usually a temporary, off-airport location established for takeoff and landing. Although terms like "helipad" and "landing zone" are sometimes used, a helispot is what is usually built at an event site and is the term used in this discussion. Please note that the FAA uses the term "heliport" to describe a place where helicopters take off and land. In this discussion, a heliport is the same as a helispot.

8.5.3 There are many variables a flight crew will consider before landing including security of the space, terrain, soil, lighting, loose debris, trees/foliage, electrical towers, wires and more.

8.5.4 When including a helispot into a site plan, organizers and site planners need to consult the authority having jurisdiction over the event, agencies and/or vendors who may land helicopters on the site and the flight crews expected to use the facility. It may also be prudent to research Federal Aviation Administration (FAA) requirements, although flight personnel will be a good source of this type of information. (see FAA Document AC150/5390-2C, *Heliport Design*.)

8.5.5 Consider whether the evacuation of ill and/or injured patients may be required. Transporting patients to the helispot will require careful consideration and may include access by ambulance.

8.5.6 If transporting talent by helicopter, the success of the overall event will depend entirely on suitable facilities and safe air operations. If the event plan calls for transporting talent by air, organizers must also make contingency transportation plans for the artists. Meteorological conditions and flight visibility can be unpredictable and limit the use of aircraft.

8.5.7 If the event will incorporate air operations, it is strongly recommended that a competent air operations team be assembled to manage the system. This team should be included in the drafting of the event's health and safety policy and their operations will have a significant impact on the risk assessment.

8.5.8 A well-planned helispot will have certain construction requirements including landing area dimensions appropriate for aircraft used, construction/clearing of hard surface for landing, spacing between multiple helispots, lighting, fencing, security, refueling capability (not recommended at the event site), fire suppression, accommodations for flight and air operations crews and more.

8.5.9 It is ultimately the pilot's responsibility to make a safe landing and the decision to commit to a landing is exclusively that of the pilot in command. Thus, the aircraft flight crew should be involved in site planning.

8.6 Final Site Design

8.6.1 Once all the necessary details and requirements have been completed, each should be drawn to scale on its own layer on a site plan with spacing requirements, etc. The final plan should be reassessed to check the occupant capacity (with sight lines and circulation space) and emergency services, worker and audience entry and exit. Power generation and distribution positions can now be completed. Other useful layers for a master site plan include entrances and exits, cash-handling locations, first aid and ambulance locations, security and law enforcement locations, fire extinguisher locations, food and beverage locations, merchandise locations, plumbing and the water distribution system, facility and site infrastructure, vehicular routes inside the perimeter, terrain and modifications to the terrain as well as any site restoration requirements etc. The list of items to place on a drawing can be extensive and not always necessary. Care should be taken to manage site drawings so a particular drawing is not overpopulated with too much information.

8.6.2 In order to keep all departments on the same page, many organizers place a grid over the drawings. This allows users to view and discuss locations by referring to their grid coordinates on the plans. This can save time in critical moments or during an emergency.

9. Crowd Management

9.0.1 In "The Focal Guide to Safety in Live Performance" (1993, George Thompson, Ed.), John Shaughnessy describes "Crowd Management" as, "the business of ensuring that the demands of a large body of people in one place are analyzed and met by a combination of forward planning, engineering response, adequate information systems and alert general management." The author recommends avoiding the term "Crowd Control" and suggests it be replaced with "crowd management."

9.0.2 Shaughnessy (in Thompson, 1993) also cites the following list as being of particular concern while planning for effective crowd management:
- Barriers and fencing
- Means of access and escape
- Public address capability
- Arrangement of staging, structures and other event infrastructure
- Emergency and general event lighting
- Sightlines
- Production detail
- Enforcement of event policies
- Law enforcement, security, stewarding
- Medical and first aid
- Provisions of emergency services
- Evacuation plans

9.0.3 The overall safety and enjoyment of patrons attending any type of music event or public assembly attraction will depend largely on effective crowd management. This is not simply achieved by attempting to control the audience, but by trying to anticipate their behavior and the various factors which can affect it. It is vital to implement a complete system rather than attempt to control only certain elements of obvious concern.

9.0.4 Many factors in event or venue planning and design discussed throughout this publication will have a bearing on crowd management. Examples include the venue design itself—whether a fixed venue or temporary structure—which must allow maximum entry and exit flow and to support generous crowd movements within the venue, restrooms and concession stand access, emergency response access, etc. Always consider accessibility to all areas for persons with special needs in the planning process.

9.1 Audience Profile and Crowd Demographic

9.1.1 Two important aspects to be considered in crowd management are the audience profile (the type of audience a particular artist attracts) and crowd demographic (the social statistics of the expected crowd, such as age group, and predominant gender).

9.1.2 Many factors contribute to the potential for crowd movements and therefore need to be considered at the venue and site-design stage, including:

- The parking layout and relationship to venue entrance(s);
- The box office line management and direction;
- The multiple entrance line management and control;
- Multiple-stage entertainment;
- Provision of satellite stages, platforms and stage thrusts;
- Sound and video towers;
- Sight-line obstructions or restricted views;
- Multiple-barrier systems and pens;
- Location of facilities;
- The psychological state of the audience; and
- Special effects.

9.1.3 The way in which crowds behave and respond is a combination of many factors. Crowd dynamics will depend largely on the activities of the crowd and this will be influenced by the demographics of the crowd and the artists performing.

9.1.4 Matters to be addressed include:

- The performance of the artists or groups (e.g., diving into audience, throwing items into the audience and performing in audience areas)
- The audience profile (e.g., male/female split, age of audience, alcohol or drug consumption, physical behavior such as moshing, body surfing, slam dancing, aerialists and stage diving).

9.1.5 It is important for security and event personnel to be able to recognize and understand in advance what are "normal" activities for the anticipated audience. They can then prepare accordingly by increasing, if necessary, additional perimeter fencing, restrooms, security personnel, medical support and equipment, etc.

9.2 Entry and Exit of the Audience

9.2.1 Before the audience enters the venue, checks must be made of all fire and emergency doors, gates, and equipment. In addition, the following should be confirmed:

- All exits are clearly marked, unlocked and staffed;
- Escape routes are clear with appropriate clearly marked signage;
- Fire-fighting equipment and personnel and alarms are in full working order;
- A PA system for use in emergencies can be heard clearly in all parts of the venue; and
- If these checks are to be carried out by security, clear instructions must be given and supervised.

9.3 Planning Considerations

9.3.1 There is a tendency to link sporting events with concert events simply on the basis of similar crowd capacities. However, the two types of events differ in venue configuration, crowd behavior and the general event management. Some sports teams will draw problematic fans as will some types of artists or concerts. Organizers should do their homework on the target audience of the event they are producing and apply during the planning process the knowledge gained. This information should also be freely shared with the venue and all health, safety and security managers.

9.4 Entrances and Exits

9.4.1 Ensure that entrances and exits have clearly posted signage, which reflect venue policies and procedures and operate efficiently. Consider the needs of children and people with disabilities and separate entrances and exits for pedestrian access from entry routes used by emergency services and concession vehicles. Provide information to the audience about any restricted exits that are not in use while the event is in progress (see Chapter 8, *Venue and Site Design*, for more information on entrances and exits).

9.4.2 Pre-Opening Considerations

9.4.2.1 Events that are general admission (GA) should expect attendees to arrive earlier than an event with reserved seating. In some stadium or open field events, the GA ticket holders may arrive even the day prior to the event. Organizers should consider how many persons may arrive early and plan for what additional infrastructure requirements will be needed by the presences of those persons, e.g., sanitation facilities, waste management and security.

9.4.3 Entrance Preparations

9.4.3.1 One of the more exciting moments on event day is the opening of "doors" and preparing for that moment is no small task. Streamlining the entry process for the crowd's smooth entry is important and should include:

- Informational signage alerting arriving fans to the event's "prohibited items" policy that may cause a delay at a the venue's entry point;
- A soft ticket check as patrons enter the chutes leading to the venue entry point not only verifies the fan has a ticket- it makes sure they know where the ticket is and can shave seconds per person at the entry;
- Event staff with megaphones walking the lines offering information and reminders on event policies can save time and also passively alerts ticket holders there is a strong staff presence at the event;
- If the event is selling alcohol, ID checks and wrist-banding can be done in the line just before the venue entry point;
- Methods of streamlining the ticket holder search process at the entry point can include additional lanes, additional tables for bag checks and additional positions for pat-down searches; however without trained and experienced staffing there to support the additional load capacity the effort is wasted;

9.4.3.2 All streamlining processes generally require additional staffing which, of course, is not free. Organizers must balance the increased efficiency of adding staff with the additional cost as they consider their options.

9.4.4 Opening Time

9.4.4.1 Problems may occur at entry points if large numbers of people seek to enter at the same time. This could result in potential injuries from crowd surges. It is therefore recommended that entrances be opened 1-2 hours before the event is due to start and the audience is informed of this by tickets, websites, radio, social media and/or other means. If significant crowding is likely to occur before that time, consider opening gates before the published time, providing that on-site services are ready. Admission can be staggered by providing early supporting acts or other activities. Another suggestion is to provide light entertainment (i.e., radio station promotions in the parking lot to distract and keep patrons entertained).

9.4.4.2 It is important to appreciate that when entrances are opened early, the audience demands will increase on facilities such as waste clearing, sanitary accommodation and concessions.

9.4.4.3 In his chapter on Crowd Management in *The Focal Guide to Safety in Live Performance* (1993, G. Thompson, Ed.), John Shaughnessy recommends that when using turnstiles for entry, that the maximum rate of flow used in calculating throughput should be 660 persons per hour, per turnstile (5.45 seconds per person). This rate is accepted by many around the world as the standard maximum rate of flow through turnstiles.

9.4.4.4 Crowd pressure at the entrances can be reduced by:
- Keeping all other activities, including mobile concessions, well clear of entry points;
- Arranging for adequate queuing areas away from entrances;
- Creating holding areas away from entrances to relieve the pressures on these points;
- Ensuring that barriers, fences, gates and turnstiles are suitable and sufficient for the numbers using them;
- Locating ticket sales and pick-up points away from the entrance;
- Providing a sufficient number of trained and competent event staff to maintain line control and provide accurate information to patrons;
- Arranging for a short-range PA system and megaphones to be made available at entrances to notify people of any delay.

9.4.5 Opening the Entrances and Arrangements for the Front-of-Stage Area

9.4.5.1 When entrances are opened at non-seated events or general admission, the audience tends to rush toward the front, which can cause tripping accidents and injuries. Carefully consider how the area in front of the stage will be managed and secured. If a standing area is provided in front of the stage, make sure entrances do not lead directly to this area from stage right or left.

9.4.5.2 One recommended method of easing the initial rush toward the stage and preventing slipping or tripping accidents is to provide a line or lines of security and/or event personnel across the arena through which the audience can move toward the stage in an orderly manner. This may be supplemented by PA announcements to keep the audience informed about what is happening and encouraging them to slow down and be safe

9.4.5.3 When allowing the public onto the floor of the venue, especially in a general admission situation, a good rule is to always feed from the opposite end of the venue from the stage, even if you have to temporarily close off the venue's right and left floor access into the front of stage area until the area is occupied, then open those access points. Once people arrive at the front of stage area, they should be allowed to stand—or preferably encouraged to sit—as close as possible to the stage barricade. If the audience is held back from approaching the barricade and allowed to move forward later, the audience will assume the concert is about to start and can cause a crush toward the barrier resulting in possible injuries.

9.5 Ticketing

9.5.1 Ticketing policies can have a direct effect on audience safety. Consider the following:

Fig. 9-1 - Chutes formed with bike rack barriers lead to bag check tables at the entrance to a venue. Photo courtesy of Steve Lemon.

- Where a capacity or near-capacity attendance is expected for an event, admission should be by advance ticket sales only;
- Tickets for seats which offer restricted views, or are uncovered, are marked accordingly, and the buyer forewarned;
- Tickets for seats with severely restricted views are not sold;
- The ticket stub retained by the audience member after passing through a ticket control point should clearly identify the location of the accommodation for which it has been issued;
- A simplified, understandable ground plan is shown on the reverse side;
- If there is more than one entrance, introduce color coding of tickets corresponding to different entrances and ensure audience members are proportionally divided between entrances;
- All sections of the venue, all aisles, rows and individual seats, are clearly marked or numbered, as per the ticketing information.

9.6 Admission Policies

9.6.1 As stated above, the admission policies can have a direct effect on the rates of admission and the management of entrance areas and audience accommodation in general. Specific points to be considered include:

9.6.2 Cash Sales

9.6.2.1 To ensure a steady flow of audience into the venue when entry is by cash, set the admission price at a round figure. This avoids the need for handling large amounts of small change.

9.6.3 Ticket-Only Sales

9.6.3.1 The advantage of confining entry to ticket-only is that the rate of admission should be higher than for cash sale. If tickets are sold at the event, provide separate sales outlets wherever possible. Ensure that these outlets are clearly signposted and positioned so that queues do not conflict with queues for other entry points.

9.6.4 Reserved (or Numbered) Seat Ticket Sales

9.6.4.1 Selling tickets for specific numbered seats has its advantages: seats are more likely to be sold in blocks and the system allows different categories of audience members (e.g., parent and child) to purchase adjacent seats and enter the venue together. This policy helps to avoid random gaps and ensures that in the key period before the event there will be less need for ushers to direct late-comers to the remaining seats, or move members of the audience who have already settled.

9.6.5 Unreserved Seat Sales

9.6.5.1 Selling unreserved seats has the advantage of being easier to administer. However, people are prone to occupy seats in a random pattern, and it can be hard to fill unoccupied seats before the start of the event. For this reason, when seats are sold unreserved, a reduced number of seats made available for sale may be necessary (in the region of 5-10% of total capacity, according to local circumstances).

9.6.6 No Ticket Sales on Site

9.6.6.1 If all tickets have sold out in advance, or if tickets are not sold on site, every effort should be made to publicize this fact in the media. In addition, place signs advising people of the situation along all approaches to the event, to avoid an unnecessary build-up of crowds outside. This is a preferred method for likely sell-out concerts.

9.6.7 Ticket Design

9.6.7.1 Ticket design can have a direct effect on the rate of admission. Clear, easy-to-read information will speed the ability of the entry-point steward/usher to process the ticket. Similarly, if anti-counterfeiting features are incorporated (as is recommended), ensure that there are simple procedures in place for the event staff to check each ticket's validity. If digital tickets are displayable on smart phones, ensure that the staff can interpret all valid ticket formats to speed the seating times and reduce confusion.

9.6.8 Admission of Young Children

9.6.8.1 It may not be appropriate to allow young children, particularly those under the age of five years, to attend certain events because they may be trampled or crushed. If they are not to be allowed in, clearly advertise this fact in advance. Where young children are allowed, consider arrangements for baby carriers and strollers, and at large events, dedicated children's areas may be useful. Consider contingency planning for dealing with this element of the audience, such as relocation to a specific area. Ensure a procedure is in place for stewards to assist with such relocation.

9.6.9 Re-Entry

9.6.9.1 The practice of "re-entry" generally enables audience members to leave the event for a short time and return later that day. An example of this is when a visitor to a theme park visits the park in the morning during the cooler hours, then takes a break and departs the venue during the heat of day, then returns again for the evening hours when it is cool again. It is suggested that most venues not allow re-entry and restrict venue policy to "exit, no return" status unless granted a hand stamp, etc., which can be selectively used. Unrestricted access to re-entry can allow patrons to consume alcohol and or drugs in parking lots and vehicles and return to the venue in an intoxicated state, which can increase the potential for problems.

9.6.10 Guest/VIP/Restricted Areas

9.6.10.1 Separate access points may be needed for particular types of ticket holders such as guests and VIPs, artists and their entourage, workers, officials and emergency services workers. Consider the location of the gates between these areas and the main arena to prevent any crowd build-up at such points. Clear identification of people permitted into such areas—using special passes or wristbands—will assist security in controlling admission and in minimizing delays in admission, which reduces queuing.

9.7 Entrance Searches

9.7.1 Searching at entrances using metal detector wands, walk through metal detectors, or pat-down searches along with bag checks, may be necessary to prevent prohibited items and weapons from being brought on site. A list of prohibited items should be posted at all entry gates to allow guests to return the items to their vehicles before the screening process. Arrange for the safe storage/disposal of confiscated items. Searching should only carried out by properly trained and supervised security personnel. The initial screening process can be streamlined by keeping the chutes moving and immediately sending persons with an issue to a secondary location so they do not obstruct the line moving through the chute.

9.8 Late Leavers

9.8.1 At the end of the event when most of the audience has left, if practical, stewards, event personnel and security can form a line in front of the stage and slowly walk to the furthermost exit, moving the remaining audience out of the area. Organizers should consider budgeting for the worst case scenario here so increased customer service requirements can be covered should they be required.

9.9 Crowd Sway/Surges

9.9.1 At large events it is sometimes effective to subdivide the audience into pens, which reduces the effects of sway and surge. If this method is used, put a system in place to prevent overcrowding. This is accomplished using a T-shaped barricade configuration or other audience control barriers.

9.9.2 Think carefully about where to position security and event personnel to monitor the audience for distress, crushing, sway, surges, or mosh pits as they all present a risk to the audience. Use of CCTV and/or the provision of raised viewing platforms, especially stage left and stage right, may help monitor the audience for signs of distress.

9.9.3 If people are at risk, you will need to take immediate action such as enlisting the assistance of performers by making an announcement. The performer's production and or security could be asked to alert you or the safety coordinator if they are concerned about a possible serious audience problem. It can then be investigated immediately.

9.10 Means of Escape

9.10.1 It is essential for effective crowd management during emergencies and evacuations that a well-planned and efficient means of escape exists for all occupants of the venue. As a guide, consider the following:

- For outdoor open field sites, a maximum 15 minute evacuation time is recommended;
- For stadiums, the maximum exit times should be between 2 minutes 30 seconds and 8 minutes;
- For indoor venues, the maximum evacuation time should be between 2 and 3 minutes depending on conditions and depending on the venue layout and arrangement;
- Calculate exit door flow rates at 40 persons per minute for a single, fully open door (36 inches or 0.9 m wide) and 60 persons per minute for wider passages;
- Use 24 inches (0.610 m) when calculating the width of a person over the width of wider passage ways. For example, if a passage way is 24 feet wide, 24 inches goes into 24 feet 12 times. Therefore, 12 is the maximum estimated number of people that can fit down that passage way at a time. Multiply the 12 persons wide by the 60 persons per minute mentioned above, and the result is a maximum exit capacity of 720 persons per minute. Note: This formula is intended only to acquaint one with the basics of egress space estimation and is not an official means by which egress capacity is measured. Chapter 4, *Fire Safety*, includes details more about means of egress. But, as always, work with the local fire and building authorities having jurisdiction to establish exactly what is required in terms of means of egress.

9.10.2 An emergency evacuation plan is essential, and it must be approved by the organizer, the venue, the emergency medical services, the fire department and any other relevant local authorities. This plan should provide arrangements for at least the following:

- Identification of key decision making personnel
- Location of a unified command center or point equipped with a communications network
- Arrangements for stopping the event in an emergency
- A gridded venue map
- Identification of dedicated sterile emergency routes
- Rendezvous points for emergency vehicles
- Identification of road closures and holding areas for the public and press
- Detail of the script of coded messages to initiate tasks to management, security and stewards e.g., to open gates, to stand down and so on.
- Detail of first aid casualty arrangements together with lists of hospitals in the area prepared for major catastrophes.

9.10.3 In an emergency at a concert, do not suddenly turn the music off. This can cause a major crowd disturbance and impede the work of first responders. Reduce the volume gradually or wait until the end of the song. If time allows, advise the artist's management and use them to calm the audience as the events unfold.

9.10.4 Emergency announcements should not be made by the performers, however their assistance may be necessary to calm the crowd. Artist tour management can be helpful here by having the artist help manage the situation. Persons making announcements should use a prepared script, with the most experienced person available making the announcements.

9.10.5 When stopping an event, the organizer, the venue management and the artist's management need to be unified in the actions to be taken.

9.11 Police Involvement

9.11.1 Any police presence in or at the event should be jointly coordinated and agreed to in advance with a clear vision of their functions and deployment. Uniformed police are effective deterrents to a variety of crowd issues, including loitering, vandalism and related crime on the perimeter, monitoring the security searches of prohibited items, and assisting with the ejections and or arrests of unruly patrons.

9.12 Aids to Crowd Management

9.12.1 Use of PA Systems and Video Screens

9.12.1.1 It may be helpful to arrange a safety announcement for the audience before the event starts. The announcement could give information about the location of exits, the identification of stewards, event personnel and security and procedures for evacuation. The use of video screens to provide entertainment before the event and during changeover periods can also help to inform the audience about safety arrangements, facilities on the site and transportation, etc. Screens may not be visible in all parts of the site so it may be necessary to plan supplementary means of giving information.

9.12.2 Security and Event Personnel

9.12.2.1 The main responsibility of security and event personnel is crowd management. They are also there to assist the police and other emergency services if necessary. Apart from the specialist workers provided for the protection of the performers, using separate teams for security and event staffing should not be considered without consulting all interested parties. The roles of these two groups are closely inter-linked and lack of communication can lead to ineffective crowd management

9.12.3 Deployment and Quantity of Event Personnel

9.12.3.1 The risk assessment will help you to establish the number of event personnel necessary to manage the audience safely. When preparing your risk assessment for crowd management, carry out a comprehensive survey to assess the various parts of the site and consider the size and profile of the audience along with the current trends in audience such as moshing, mud slides, bonfires, etc.

9.12.3.2 Security and event personnel staffing numbers and deployment should be determined primarily based on the size and scope of the venue, audience demographics and performers.

9.12.3.3 Examples to consider for the risk assessment include:
- Previous experience of specific behavior associated with the performers;
- Uneven ground, presence of obstacles, etc., within or around site, affecting flow rates;
- Length of perimeter fencing;
- Type of stage barrier and any secondary barriers;
- Provision of seating.

9.12.3.4 Further information regarding risk assessments for crowd management can be found in the UK Health Safety Executive's document titled "Research to Develop a Methodology for the Assessment of Risks to Crowd Safety in Public Venues" (1998, Parts 1 & 2, RM Consultants, LTD for the HSE). This document is available online at http://www.hse.gov.uk/research/crr_pdf/1998/crr98204.pdf.

9.12.4 Operations
9.12.4.1 There has to be an established chain of command and the use of the Incident Command System is recommended. In addition, consider appointing a security manager to oversee all security and event personnel contractors at the event. Also consider:
- A number of senior supervisors who are responsible for specific tasks and who report directly to the security manager; and
- A number of supervisors who report directly to a senior supervisor and who are normally in charge of six to ten personnel at specific areas in and around the venue.

9.12.4.2 All event personnel and security must participate in a pre-event briefing to receive a written statement of their duties as part of the incident action plan, a checklist (if appropriate), a plan showing key features, and credentials to be used for access control.

9.12.5 Conduct of Event Personnel
9.12.5.1 All personnel need to be ages 18 or older, need to be able to carry out their allocated duties, and while on duty must concentrate only on their duties and not on the performance. All event personnel must be capable of, and fully aware that, they must:
- Not leave their post without permission;
- Not consume or be under the influence of alcohol or drugs; and
- Remain calm and be courteous toward all members of the audience.

9.12.5.2 All event personnel should wear distinctive uniforms and be individually identifiable using a clearly visible number or ID badge.

9.12.6 Competency of Personnel
9.12.6.1 Duties and responsibilities of security and event personnel include:
- Understanding their general responsibilities toward the health and safety of all categories of audience (including those with special needs and children), others, event workers and themselves;

- Carrying out pre-event safety checks;
- Being familiar with the layout of the site and able to assist the audience by giving information about the available facilities including first aid, toilet, water, welfare and facilities for people with special needs, etc.;
- Staffing entrances, exits and other strategic points; e.g., exit doors or gates which are not continuously secured in the open position while the event is in progress; Controlling or directing the audience in and out, to help achieve an even flow of people into and from the various parts of the site;
- Recognizing crowd conditions to ensure the safe dispersal of audience and preventing overcrowding;
- Assisting in event safety by keeping gangways and exits clear and preventing standing on seats and furniture;
- Investigating any disturbances or incidents;
- Ensuring that combustible refuse does not accumulate;
- Responding to emergencies (such as the early stages of a fire), raising the alarm and taking the necessary immediate action;
- Being familiar with the audience evacuation procedures, including coded messages and undertaking specific duties in an emergency;
- Communicating with the incident control center if there is an emergency.

9.12.7 Event Personnel Training

9.12.7.1 Ensure that all personnel are trained to carry out their duties effectively. The level of training will depend on the type of functions to be performed. Keep a record of the training and instruction provided, including the:
- Date of the instruction or exercise;
- Duration;
- Name of the instructor;
- Name of the trainee; and
- Nature of the instruction or training.

9.12.7.2 All personnel need to be trained in fire safety procedures, emergency evacuation procedures and dealing with situations such as bomb threats. Those working in the pit area should be trained to lift distressed people out of the audience safely and without risk to themselves. They should also be trained to assist with lost children and vulnerable adults. For example, the UK "Guide to Safety at Sports Grounds" (2008, Fifth Edition, UK Department for Culture, Media and Sport) provides further information on specific training of personnel. This publication is available online at http://safetyatsportsgrounds.org.uk/pdf/GuidetoSafetyatSportsGrounds.pdf.

9.13 The Pit

9.13.1 In a concert environment, the security and medical teams in the pit area should be experienced and disciplined. During the event the two teams will work closely together. It is recommended a single supervisor manage this area and direct both the security and medical groups during event operations. The pit team should be trained to recognize distressed individuals in the crowd. The team must be physically capable of dealing with both disorder and fan extraction—and rescue, if necessary. To assist in the operations of the pit team, remote medical facilities and additional security resources should be located next to the stage to take the hand-off of fans pulled over the barricade. These individuals can then be escorted to the public re-entry point or an ejection point.

Fig. 9-2 - Large rear step on the pit barrier allows staff to monitor the crowd and manage the safe passage of Surfers and recover anyone who needs assistance. Note the well-disciplined and organized pit team – essential for large and highly dynamic audiences. Photo courtesy of The Event Safety Shop, LTD.

9.13.2 If your event is a concert, most artists allow for still photographers to shoot the first three songs and TV crews to film a portion of one song. Organizers should always confirm with the artist's tour management what the guidelines will be regardless of the comments in the contract rider.

9.13.3 All photographers working in the pit should be escorted by a publicist, tour manager, promoter or other non-pit crew staff member. During this period, the media in the pit serve as an obstruction and can be a distraction to those working there. Because the photographers are typically allowed in the pit at the start of the performance, security and medical teams must be prepared for the initial surge of enthusiasm by the crowd when the house lights go out, while sharing the space with the photographers. This means the pit crew must be even more alert during this period. The organizer should have earplugs on hand and ready for distribution to staff and media in the pit.

9.13.4 Videographers and TV crews usually shoot from another location further out into the house, like the mix location, and will also require an escort provided by the show or the promoter.

9.13.5 Many artists like to get as close to the fans as possible. Depending on the physical condition of the artist, they may attempt to get closer to the fans by whatever means are available, including: standing on anything they think will hold their weight; climbing onto structures they feel they can navigate; or jumping onto speakers, camera track risers even the crowd barricade to get closer to the fans and incite more enthusiasm. Any action like this can increase the workload on the pit crew.

Fig. 9-3 - Security, medical and photographers in confined space in the pit during a music festival.. Photo courtesy of The Event Safety Shop, LTD.

9.13.6 Tour staff should inform the organizer and security team whenever they feel the artist is likely to come off the stage and go into the pit.

9.13.7 The pit crew is busiest when the show is general admission or "standing." This means members of the audience are not in seats, they are on their feet and possibly standing for hours on end, sometimes in adverse weather conditions. This is the crowd with which the pit crew needs to be the most alert and attentive.

9.13.8 Considering all the people who may have access to the pit area, it is important this zone be as free as possible of trip hazards and other items that may cause injury. Sharp corners should be padded and marked with high visibility tape, cables should be contained in "cable-ramps" or other cable protector, etc. The more active the pit crew during a performance, the more "pull-overs" there are and the greater the risk of injuries.

9.13.9 Security and medical staff in the pit should always be focused on the needs of the fans and not watch the performers, especially when a performer has left the stage and entered the pit.

9.13.10 The pit can be hazardous even to a seasoned professional. It is even more dangerous for a member of the general public who ends up there. Care should be taken by the pit crew to supply a person to escort those members of the general public from the point of crossing into the pit to the nearest exit. Medical teams should check each person pulled over and exiting the pit in case medical attention is needed.

9.14 Personnel Welfare

9.14.1 Ensure that event personnel are not stationed for long periods near any loudspeakers and make sure they are provided with ear protection according to the OSHA Regulations if the sound levels warrant it (see Chapter 22, *Sound: Noise and Vibration*). Event personnel and stewards will need adequate rest breaks at reasonable intervals.

9.14.2 Here is a list of posts where security guards or stewards/ushers may need to be placed at an event:
- Parking direction and assistance
- Parking lot patrol

- Bus and truck parking lots
- Points where pedestrians and vehicles cross paths
- Monitoring of unauthorized vending of items
- Entry line control
- Soft ticket check
- ID check
- Line control
- Bag check tables
- Pat downs
- Magnetometers
- Secondary screening areas
- Rovers - both inside and outside the venue
- Cash locations
- Cash movements
- At the head of floor seating aisles and entrances to fixed seating sections
- Mix platform or control tower
- Camera platforms
- Projection towers
- Pyro and Special Effect locations
- Storage in public areas
- Merchandise booths
- Spotlight platforms
- Access points to back of house and secure areas like the venue rigging grid
- Pit in front of the stage
- Backstage access points
- Stairs leading to the stage
- VIP areas
- Beer gardens
- Points where access accreditation changes
- Entrances to hallways or compounds for offices and dressing rooms
- Office or dressing room doors
- Meet-n-Greet locations
- Talent escort
- Photographer/Media escort
- Catering/Parking, valuable protection and meal ticket collection
- Venue perimeter fence
- Late coverage after the event
- Overnight

9.14.3 Some posts are only required for short periods, others for the duration of the event. It is common for organizers and security companies to manage the available resources and redeploy event personnel and security staff to multiple posts during an event.

9.14.4 When planning the redeployment of security guards from posts at the entrance areas to other posts inside the venue, carefully consider the type of crowd, their anticipated arrival time to the venue and the pace with which the entrance team can get the ticket holders into the venue. For example, redeployment of security from the entrance to the pit in front of the stage is a common practice. However, if the entrance process is behind schedule with considerable crowd yet to enter, there is risk of understaffing the entrance and delaying the remaining crowd's entry into the venue. In this situation, if the music starts, especially the headline act, the crowd may resort to an act of civil disobedience. Even though there may be costs incurred from such a delay, safety must always take precedent over financial gain.

10. Children

10.0.1 This chapter is not intended to serve as a substitute for a comprehensive familiarity with the large amount of laws, regulations and information available on the care and handling of children at an event. The reader is cautioned not to underestimate the amount of preparation and work required to provide a safe and enjoyable children's activity or play area at an event. Know the requirements or hire someone who does. Care of others' children is not a subject to be taken lightly or pursued without extraordinary preparation.

10.0.2 Most parents know what it is like to "baby-proof" a house. At some point the adult eventually gets down on hands and knees and crawls through the house looking for hazards from the child's point-of-view. The exercise always reveals additional risks the average adult simply misses because of their perspective. The goal of this chapter is to inspire the reader to view the event through the eyes of a child and to remember an adult's perspective can differ significantly from that of a child.

10.0.3 In this chapter, the following terms will be used to describe age groups of minors (i.e., children) as defined by the U.S. Centers for Disease Control and Prevention (CDC), Child Development Section:
- Infants (not walking): ....0-1 year of age
- Toddlers (walking): ....1-3 years of age
- Preschoolers: ....4-5 years of age
- Middle Childhood: ....6-11 years of age
- Young Teens: ....12-14 years of age
- Teenagers: ....15-17 years of age

10.0.4 For clarity, the following terms will be added to the list:
- Young Adults: ....18-20 years of age
- Adults: ....21 and over

10.1 Planning Considerations

10.1.1 When integrating a children's activity area or event into a larger event, organizers should resist the temptation to plan and manage the children's area themselves. The manager for a children's activity or play area must be primarily, if not solely, dedicated to that area and event organizers with multiple other significant responsibilities are ill-suited for this task as they may often be pulled away or distracted.

10.1.2 Children's areas have specific guidelines and regulations that can be easily overlooked, even by an experienced event organizer. These areas require the attention to planning and management of a qualified and competent person or vendor. Children's activity or play areas are often mistakenly viewed as easy and uncomplicated additions to a larger event. This view completely underestimates the amount of work required to plan and operate such an area.

10.1.3 Event organizers should consider the following questions at the beginning of the planning phase to help estimate the number and age group(s) of children who may attend their event. Asking the right questions will help plan for the presence, needs and safety of children at an event. Here are some examples:

- Is it a children's event and if so, which age group is the target demographic? This answer identifies what age groups to plan for and what rides and activities to present.

- Is the event a children's event where several children may arrive with one parent, guardian, chaperone, e.g. a teacher? This answer may impact staffing and parking-arrivals plans.

- How are the children arriving? Generally it is advisable to plan for a dedicated group drop-off location near the venue's entry point. If vehicles transporting children drop-off out in the parking lot, other cars in the ingress and parking process increase the hazards to children as they walk across the parking lot to the entry. This may also slow the parking process, as children in groups can become disorganized during the transit from drop-off to the venue entry point. Event staffing may need to be increased to help chaperones monitor children.

- Are children going to be participants in your event where they or their organization may be a part of the event or a featured attraction or paying an entry fee, e.g. a parade, road race or competition? This answer may affect the infrastructure required, the food and beverages to be served, staffing needed and parking locations.

- Will children serve as talent or performers or as workers at the event as part of a volunteer group or paid staff? Child labor laws may come into play. If used as volunteers, you may need to communicate more information ahead of time with their organizations. All volunteers under the age of 18 will need a parent or guardian's signature on the appropriate release form. Children often do not carry the information necessary to fill out the paperwork required by the federal government, e.g. social security card, etc. For more information on the use of volunteers and the US Department of Labor – Fair Labor Standards Act (www.dol.gov/elaws/esa/flsa/docs/volunteers.asp).

- Will children be arriving with or working for a vendor who has arranged to sell at your event? Authorities are less likely to check for labor law compliance of a vendor if their child is accompanying them at an event while the parent is selling. Regardless, it is recommended that organizers clearly state in their vendor agreements any expectations regarding the vendor's liability and insurance requirements.

10.1.4 Once the needs of children at your event are examined, more questions arise. As the organizer, you want to know your exposure to risk and mitigate that exposure wherever possible. Ultimately, responsibility for safety rolls uphill to the organizer.

10.2 General Considerations for Children at Events

10.2.1 Children often accompany adults to events. Consider the presence of children in your risk assessment and emergency action plans even if they are not the target demographic of the event.

10.2.2 Event promotional materials should indicate if the event is suitable for children and what age groups, or if children under a certain age will not be allowed to enter.

10.2.3 When children attend events, there is always the potential for their separation from parents, guardians, or chaperones. Establishing lost child procedures is part of venue and event operations when the presence of children is reasonably foreseeable. See section 10.7, *Lost and Missing Children*, below for additional information.

10.2.4 Facilities should contain changing tables or other similar accommodation for infants and toddlers in restroom facilities.

10.2.5 Venue operating procedures should designate at what age a ticket and/or individual seat is required for a child attending an event. This generally is around 2 years of age but venue practices and policies vary. Check the customs in your region.

10.2.6 Review safety and code requirements for the safety of children, such as hand railing space/opening and guard rails. Conduct an inspection at least annually with the AHJ (fire marshal) or insurance inspector, or more frequently if there are changes and adjustments made to seating areas or structures. NFPA 102, *Standard for Grandstands, Folding and Telescopic Seating, Tents, and Membrane Structures*, provides useful information and may be a required standard in the jurisdiction in which you are operating.

10.2.7 Include children's needs in your emergency planning and hire or consider identifying and training a person to serve as the Children's Issues Coordinator; a qualified and competent person who knows what to do and can give direction when an incident involving a child occurs.

10.2.7.1 Develop policies and plans for the following:
- The ongoing care of children who are separated from their parents and are unable to be quickly reunited,
- Law enforcement and/or child social services notification,
- Transportation of children to a safe child-friendly place on site,
- Supervision for the child while in that child-friendly place,
- Medical care for children (if necessary),
- Nutrition for the child while in your care, and
- A location off site, e.g. a nearby school or multipurpose room, with facilities that can be used if there is an emergency and the attending parents and children need to be evacuated to an alternate location. You will need a check-in and check-out process, a system for leaving messages, and security for unattended/unsupervised children who are now in your care.

10.2.7.2 Communication to area managers, event staff, and the public, and the instructions they are to follow in an emergency situation. Keep in mind that there may be language barriers. Train event staff and crowd managers in the procedures and considerations for evacuating and/or assisting children in emergency situations.

10.2.7.3 In an emergency, families tend to stay together and the parents or guardians will make the decisions regarding their children's safety. Typically, the parents and guardians usually gather the family to make sure all are accounted for and then evacuate as a group or seek sanctuary nearby. Preserving family unity is an important step to provide for the physical safety and emotional stability of children in disasters.

10.2.7.4 Because family separation can occur even when well-prepared evacuation plans are carried out, it is important to identify protocols for reuniting families as quickly as possible.

10.2.7.5 If you are an organizer who also manages a venue, it is important to remember that during a state of an emergency, your venue may be called upon to serve as a rally or meeting point, temporary housing location for refugees, or emergency command center for public service agencies.

10.2.8 Events involving predominantly young teen audiences who attend without parents can present challenges of a different sort. Problems they experience may include becoming separated from their companions, missing the public transportation to get home, and losing items such as cell phones and money. Consider providing a "Solutions" or "Help" booth with at least one telephone so that these young teens can get help or contact their parents or guardians.

10.2.9 Parents sometimes have difficulties finding their children at the end of an event. Consider providing a specific, well labeled, meeting point that includes a staffed message facility where parents and teens can leave messages for each other. Alternately, offer a dedicated space with tables, chairs and other accommodations where parents can spend time at the venue waiting on site to collect their children at the end of the event.

10.2.10 The Americans with Disabilities Act (ADA) applies to the needs of all persons including children. Consider the requirements, such as those relating to staffing and equipment, and develop a policy for children with special needs to ensure that they can be fully engaged in all activities. See Chapter 11, *Facilities for People with Special Needs*, for more information on ADA.

10.2.11 Establish and maintain a temporary holding location and arrange for the safe care of children separated from their parents or guardian (a planning consideration for all events).

10.2.12 If your event has a children's play area, consider at what age you can allow a child to be left unattended in the area. There may be guidelines available from the AHJ or Child Protective Services (CPS). However, it is a good practice for organizers not to allow children to be left alone in their care or unattended by the parents at the event under any circumstances. This policy should be posted at all entrances in the applicable languages relevant to the event. For more information on laws and guidelines in your event's location, contact your local CPS agency.

10.2.13 Unaccompanied children should not leave the area unless accompanied by their parent or guardian. This is more easily monitored if the area is enclosed with limited entrances and exits and there is an effective accountability (i.e., sign-in/sign-out) policy.

10.2.14 Event staff must be familiar with the event's policy and procedures for discipline and dealing with uncooperative or abusive children or parents. These policies and procedures should be coordinated with—and ideally approved by—local law enforcement and the local child protective services agency.

10.2.15 Dangerous behavior by children should always be discouraged.

10.2.16 Post your policy regarding sick children at all play area entrances. Event staff should not undertake the care of sick children except in situations where a child is separated from his or her parent or guardian.

10.3 Children's Activity and Play Areas

10.3.1 Children's activity and play areas at events are very popular. Regarding the site and layout of these areas, consider, and plan appropriately for, the following:

10.3.1.1 When planning the location of the children's area, the overall event's throughput and the impact of the location of children's areas on that throughput must be considered.

10.3.1.2 Position the children's area near posts for medical and security/law enforcement, food and beverages, restrooms, and a plumbed source of water.

10.3.1.3 The maximum capacity of an activity and/or play area should not simply be based on the total size of the space. It should take into consideration the layout of space, the available area for queue lines, any open space areas, and the capabilities of the staff for each of the activities. The maximum capacity of the area should be calculated with input from (and often direction from) the AHJ, known by event staff working in the area, posted and enforced. It is usually acceptable to have people wait outside the area when the area is at maximum capacity as long as it does not obstruct exits or guest flow. If this is necessary, design and establish a contingency queue line space outside the entrance of the area.

10.3.1.4 For the safety and security of the children, the layout of the area should accommodate one main entry and exit point. Additional emergency exits may be required by the AHJ and those passageways may require staffing to monitor and control access. When designing the layout, be sure to meet any requirements of the AHJ and the Americans with Disabilities Act (ADA), and pay attention to the area's proximity to the primary venue exits in case evacuation is required.

10.3.1.5 Survey the play and activity areas for hazards. If hazards are cannot be removed, fence them off to prevent access by children. In addition, prevent young children's access to all open and standing water including pools, ponds, streams, fountains, and mud.

10.3.1.6 Children's ears are sensitive. Consider ambient noise levels, especially the noise from nearby entertainment stages, carnival rides, etc.

10.3.1.7 Provide perimeter fencing aimed at the containment of enthusiastic, running children and to provide a well-defined boundary for the area that prevents unmonitored access.

10.3.1.8 Provide dedicated vehicular access for emergency and service vehicles such as resupply trucks, ambulances, sanitation pump trucks, fire trucks, police vehicles, etc., and assure separation of children and vehicles.

10.3.1.9 Around the entire event site, post signage that directs guests to the activity and play area. In an open space event, consider placing large sandwich board maps around the anticipated main entry points that clearly identify the locations of activities, vendors and all other useful information such as where to find help. At entry points to the play area, post the area's hours of operation, schedules of activities and posters with the area's policies. Signage in the play area should be designed for, and visible to, children.

10.3.1.10 For security purposes, ensure that the entry and exit points are supervised and controlled by positioning an event staff member, such as a host, steward or security guard, near entrances and exits. If the budget allows, the presence of a uniformed law enforcement officer is often welcomed by parents and gives local law enforcement an outreach opportunity with the children.

10.3.1.11 Provide parking for strollers, wagons and bicycles (if applicable).

10.3.1.12 Consider whether the area has adequate lighting, heating/cooling and ventilation. If using an outdoor venue, make sure to plan for protection from wind, rain, snow (if applicable), and other environmental influences on both the structure and its occupants.

10.3.1.13 Plan and establish a location for diaper changes, breast feeding, and any other necessary family-provided care of children.

10.3.1.14 The provision of hand-washing facilities and tables for supervising adults is recommended.

10.3.1.15 Areas used for children's play and activities should be inspected and "cleaned" to remove or mitigate hazardous items such as broken glass, debris, exposed sprinkler heads, animal feces, cleaning solutions (poison), tools, holes, etc.

10.3.1.16 Toilets and running water must be available near the children's area. In addition, contractors must service and sanitize toilets in this area more often than adult facilities. Alert the cleaning contractor that there may be discarded diapers containing human waste and that special receptacles for the collection of soiled diapers may be needed.

10.3.1.17 In hot weather, ice, water, and sunscreen should be available in the play area; consider making these items available for purchase. If offering snacks, prepackaged items offer convenience. Check with your local health department about regulations regarding food sales.

10.3.1.18 Make sure there is plenty of suitable seating for both the expected age group of children and their parents/guardians.

10.3.1.19 Check that stair rails are at appropriate heights on stairs and steps. OSHA 29 CFR 1926.1052, *Stairways*, includes several requirements with which event organizers should be familiar. For example, when a stair rail is also a handrail, it must be not less than 36 inches (0.9144 m) and not more than 37 inches (0.9398 m) from the upper surface of the tread. Also, make sure a sphere 4 inches (0.1016 m) in diameter (the head of an infant) cannot fit between rail balusters or between the bottom of the rail and the flooring materials. For all construction and safety issues like this, consult with the AHJ to establish and apply all local requirements.

10.3.1.20 Check to determine whether the floors in the area are slippery or will be made slippery by moisture. If so, mitigate the hazard, and have a plan in place to quickly remove liquids from stairways and flooring.

10.3.1.21 Provide suitable food and drink for children in and around these areas. Avoid foods known to provoke allergic reactions (such as nuts) and, if these types of food items will be offered, post notices that clearly inform the public.

10.3.1.22 Smoking should be prohibited within or near children's areas.

10.3.2 Clearly indicate that parents and guardians are expected to watch and supervise their own children in play areas and that they are responsible for their behavior. Local parks may have already developed language you can borrow for this purpose.

10.3.3 The event's insurance provider is a good source of information regarding safety in general, but they can be particularly helpful with issues related to managing children and children's areas. Keep the insurance representative well informed of the intended activities and do not hesitate to ask him or her for advice and information. Insurers will likely have access to risk managers who are excellent sources of valuable information.

10.3.4 When planning and operating children's activity and play areas, your obligations as the organizer will vary depending on your location. Consult the local AHJ, Child Protective Services (CPS) and other child services agencies for a clear understanding of the requirements and responsibilities associated with such an endeavor.

10.4 Children's Activity Considerations

10.4.1 Activities should be age-appropriate. Ensure that a risk assessment for each activity is carried out prior to the event and that the staff involved in the activity is fully briefed. Clearly display information about potential hazards at all entrances to the area; pictures may be more effective for children.

10.4.2 If the activity is a craft such as tie-dying t-shirts or building a model (i.e., something requiring some dexterity and skill), expect the parents to engage in the activity along with their children. The presence of adults in the area may limit throughput and reduce the capacity for children. So, plan for more space if these types of activities are used.

10.4.3 Materials used in children's play activities should be clean, nontoxic and non-allergenic.

10.4.3.1 Before use, check materials from the scrap pile or recycle bin for cleanliness, suitability and any hidden sharp items such as slivers, staples, nails, sharp edges, etc.

10.4.4 For children ages four and under, face painters should only paint children's cheeks or hands and should always acquire parental permission before doing so. Ask face painters for their public liability insurance and proof that they are using a reputable brand of non-allergenic, water-based face paints.

10.4.4.1 Be prepared to provide information to parents on the safety of the products and about removing face paints.

10.4.5 Ensure there is constant supervision for activities that could be potentially dangerous. Consider these activities carefully. Some activities with inherent risks, such as woodworking and candle making, are not recommended.

10.4.6 The nature, size and operation of recreational equipment and rides to be placed, or already located, in the area will need to be evaluated to determine if each piece can remain, must be removed, or needs to be fenced off like any other hazard. Play equipment must be safe for the age group using it and children need to be closely supervised at all times when using the equipment. Post signage identifying the intended age group for the equipment.

10.4.7 Some rides and activities for older children may be best located outside the young children's area. This not only protects the smaller children from larger kids and moving rides, it increases foot traffic and customers for any vendors along the route.

10.4.8 Different age groups demand different levels of infrastructure, supervision and types of activity. These factors can all affect the design of the layout as well.

10.4.8.1 If mixed age group activities are planned, the planner should consider and plan the layout in such a way so as to avoid smaller children being knocked over or accidentally hurt by excited older and larger children.

10.4.8.2 For children of 3-5 years of age, knowledgeable and qualified workers should lead the activities. The entry point must be supervised at all times to ensure that children do not leave the event area unnoticed. There should be access to a nearby quiet rest area in the event that the parent/guardian needs time with the child away from the activity.

10.4.8.3 Children 6-14 years old should have more challenging activities that last for longer periods of time. Consider separating these kids into groups of 6-8, 9-11 and 12-14, as these groupings coincide with elementary and middle school age groups.

10.4.8.4 The 15-18 year old age group may arrive and not be accompanied by an adult. If your event will serve alcohol, consider that it may get into the hands of under aged individuals. Institute a method of easily identifying those who are underage such as wristbands for those of drinking age and the use of specific types or colors of cups for identifying alcoholic beverages. Where children are permitted, non-alcoholic beverages must be available.

10.4.9 Locate rides and amusements on your site in an environmentally and child-friendly space. Rides should be appropriate to the age and size of users. Ride operators must have relevant and up-to-date certificates for their equipment. Bouncy castle operators must provide proper supervision at all times and control the capacity of the units to prevent overcrowding, excessive excitement and possible injuries.

10.4.10 All items used by animals should be kept out of the reach of infants and toddlers. When an activity includes touching animals or items that come in contact with animals, children and parents should be encouraged to wash and/or sanitize their hands immediately after the activity. Provide hand sanitizer or hand washing facilities within the animal area and near the exit.

10.4.11 When an injury occurs in the children's area, the children's play area manager should notify the event command center to alert the event medical team any time a child sustains an injury more serious than a minor cut or scratch, and file a report.

10.4.12 Any sand boxes should be inspected regularly and kept free of potentially hazardous foreign materials such as glass, broken toys, hardware, twigs/sticks, animal feces, acorns, etc.

10.4.13 Areas beneath climbing equipment, swings, slides and similar equipment should be cushioned with material that absorbs energy during falls such as play sand, wood chips, pea gravel, shredded rubber, or rubber mats designed specifically for this purpose. Local parks may be a good source of suggestions and can help identify suitable materials for your circumstances.

10.4.14 If infant or toddler water activities are proposed, consult with the local authority having jurisdiction (AHJ) regarding any required staff-to-child ratios. The following are some examples included in the California Health and Safety Code:
- There shall be at least one adult present who has a valid water-safety certificate on file at the event command center.
- A ratio of one adult to two infants/toddlers shall be maintained during activities in or near any body of water.
- A ratio of one staff member to every four infants/toddler shall be maintained during activities in or near any container of water that a child can get into and get out of unassisted. This could include but is not limited to, wading pools, basins or water trays. This staff-child ratio may include authorized representatives of infants/toddlers in care and adult volunteers.

10.4.15 If water activities involving any age of child are proposed, consult with the local AHJ regarding any required staff-to-child ratios. The following are some examples included in the California Health and Safety Code.

10.4.15.1 There should be at least one adult present with a valid water-safety certificate on file at the event command center during water activities in or near any of the following bodies of water:

- Swimming pool;
- Any portable pool with sides so high that children using the pool cannot step out unassisted by a person or device (including a ladder); or
- Potentially dangerous natural bodies of water including, but not limited to, oceans, lakes, rivers and streams.

10.4.15.2 A ratio of not less than one adult, including teachers, to every six children, or fraction thereof, shall be maintained during water activities in or near any of the bodies of water specified above. Lifeguards or personnel supervising anyone other than children attending the event's water activity should not be included in this ratio.

10.4.16 All 50 states, the District of Columbia, Guam, the Northern Mariana Islands and the Virgin Islands require child safety seats for infants and children fitting specific criteria. If the event is offering transportation for infant, toddler or preschool children, the vehicles used must be equipped with the appropriate child seats and restraint devices. Drivers of these vehicles should also be qualified adult drivers and should be assisted by competent assistant(s) as necessary. There should be a well-stocked, age-appropriate first-aid kit in all vehicles, or carried by an attending event staff-member if public transportation is being used.

10.4.16.1 The Governor's Highway Safety Association (GHSA) has a web site that lists all child passenger safety laws by state (http://ghsa.org/html/stateinfo/laws/childsafety_laws.html) and the National Highway Traffic Safety Administration (NHTSA) has an informative web site that serves as a gateway to information and resources related to keeping kids safe when they are in a vehicle (http://www.safercar.gov/parents/).

10.5 Staffing

10.5.1 Organizers should consult the AHJ, the local child protection or social services agency, or a professional human resource consultant to learn about any requirements and solicit advice and suggestions regarding the staffing of children's areas. These organizations and agencies can also provide guidelines regarding adult-to-child ratios and square foot-per-child space standards, among other things, for premises where children and children's activities will be located.

10.5.2 It is best to start the event planning process with an experienced manager/coordinator to plan, operate and manage the children's area: the Children's Area Manager. Although the full scope of any such job will be determined by the size, circumstances and location of the event, below are some suggested job responsibilities for the Children's Area Manager (CAM).

- Children's Area Binder – A binder with all the relevant information about the children's area and its policies should be developed and maintained by the CAM. If the children's area is part of a larger event, a copy of this binder should be made a part of the overall event binder and filed in the event's command center.

- Risk Assessment and Emergency Planning – A risk assessment of the children's area should be conducted and included in the documentation of emergency and contingency plans for the event.
- Agency Communication – The CAM should liaise and communicate directly with the AHJ and the local law enforcement and child services agencies.
- Design and Activities – The CAM should lead the planning, design and layout of the children's area and the selection of children's activities and vendors.
- Equipment – The CAM should lead the selection, and supervise the installing and dismantling, of any play equipment used.
- Policies – The CAM should be responsible for developing and implementing all policies concerning the children's area, which can differ from the main event's policies. Often, the most difficult policy to define is the procedure for the release of custody of minors to parents and guardians. This process is seemingly simple but may be complicated by issues such as the adult's lack of identification or a parent being drunk or abusive to a child or event staff.
- Staff – The CAM should be responsible for the screening, selecting and training of all staff used in the children's area, as well as their credentialing and scheduling during the event.
- Shelter – The CAM is responsible for arranging a nearby and secure shelter for the children in the event of an evacuation.
- Infrastructure – The CAM is responsible for the planning, installation, operation and dismantling of any required infrastructure such as tents, power, food and beverage, sanitation, solid waste collection and removal, etc.
- Staff Communication – Making provisions for communication between the various area managers, event management staff and the event command center is the responsibility of the CAM. The CAM assumes responsibility for the system and ensures that radios are issued to key staff (using cell phone service only as a back-up) to ensure rapid communication ability between the managers of children's areas and other relevant on-site services such as event management, the event command center (or other dispatch services for anything needed), security, other area managers. etc. The CAM and the other relevant area managers should be introduced to one another during the pre-event briefing.

10.5.3 General recommendations for staff working in children's areas, and all event staff who may come in contact with children, include the following:
- The Children's Area Manager (CAM) for the children's play area should be competent, qualified and experienced in managing such areas. An assistant manager should be fully qualified to serve as the manager in the absence of the manager.
- Although relative youth can be helpful in a CAM and assistant CAM, the CAM should be at least 18 years of age and preferably 21 years of age.
- Photos of all staff working the children's area should be taken and filed for reference.
- Criminal background checks are recommended for all persons who will work in this area or could come into contact with children, and some states and/or jurisdictions may require it. No persons with a known history of child- or violence-related offenses should

be permitted to work in or near the area. It may be helpful to review the National Center for Missing and Exploited Children's publication entitled "What You Need to Know About Background Screening" (http://ric-zai-inc.com/Publications/cops-p260-pub.pdf).

- All staff working in the children's area should be medically fit for the assignment and not under the influence of alcohol or drugs.

- Ideally, all children's area staff should be certified in first aid and CPR, but any time the area is operating at least one adult working in the area should have these qualifications. It is also a good idea to position an event first aid team either in or next to each children's activity area.

- Required staff training should include use of radios, emergency scenarios, event safety policy, child protection issues, and the reporting policy and procedures. All event staff should be trained on the appropriate handling and/or touching of children and how to report violations and questionable situations.

- Event staff must never employ any practices that threaten, frighten or humiliate children.

- Event staff must never engage in corporal punishment (i.e., spanking, slapping, shaking, etc.). Corporal punishment is a criminal offense in some states. It is advisable for organizers to require event staff witnessing corporal punishment of any child, even by the child's parents or guardians, to report it.

10.5.4 Records' requirements for those who work with children will vary based on state and local jurisdiction. Contact the AHJ, local child protective services, or professional human resource consultant for information regarding required documentation for employees and volunteers who will work with children. It is always a good idea to collect and maintain employee and volunteer records related to position within the organization, job description/duties, and number of hours worked or volunteered. In addition, it is recommended that the following information be collected and retained for each employee or volunteer:

- Employee's full name;
- Driver's license number and type, especially if the employee is to transport children;
- Date(s) of employment;
- A statement signed by the employee that he/she is at least 18 years of age;
- Employee's current home address, phone number(s), and email address;
- Documentation of employee's educational background, training and/or experience;
- Employee's experience, including types of employment and former employers;
- A health screening and tuberculosis test of the employee, if required by the AHJ; and
- Documentation of the employee's criminal record check and clearance.

10.5.4.1 The maintenance of certain personnel records may be required by some authorities having jurisdiction (AHJ) for volunteers working in the children's area. Some examples of these requirements include the following:

- A health statement such as the one required by Section 101216(g) of California's DSS/CCL Evaluators Manual Procedure;
- Tuberculosis test documents such as those specified in Section 101216(g)(3) of California's DSS/CCL Evaluators Manual Procedure;
- Fingerprints of volunteers who are required to be fingerprinted by the AHJ;
- A signed statement regarding their criminal record history if required by the AHJ;

- Documentation of either a criminal record clearance or a criminal record exemption if required by the AHJ.

10.5.5 All event workers and volunteers working around children should be clearly identifiable through the use of credentials, uniform clothing (e.g., t-shirts, caps, vests), and other unique and difficult to duplicate methods. This combination of unique and easily recognizable attire and credentials should quickly identify each individual to both the children and the parents as a trusted adult. If funding for uniforms and/or signage is an issue, consider finding a sponsor.

10.6 Detention and Release of Minors

10.6.1 Event organizers should have a written policy for handling security issues involving minors. These issues should be addressed during the pre-event planning meeting by the event organizer, a representative of the venue and the AHJ.

10.6.2 When handling minors under the influence of alcohol or drugs, it is inadvisable for the crowd management team to simply eject the underage guests from the venue. Event organizers should consult with the AHJ before establishing their minor detention policy. The venue and/or AHJ will likely already have a policy regarding this issue of detaining minors; if not, they will still be able to help guide organizers and crowd management contractors in proven methods and best practices for that location.

10.6.3 If minors are detained by event staff, they should be held in a supervised area separate from adult detainees until they can be passed to law enforcement, a child protective services agency or are released to their parent/guardian.

10.6.4 Any person, especially a minor, requiring detention should be immediately handed over to law enforcement by event staff. For this reason, effective radio communications are important between area managers, crowd management and the event command center, all of which should have direct access to law enforcement.

10.6.5 The terms of release of a minor should be determined in advance by the AHJ and the venue. These terms should be included in the event policy with at least specific guidelines on the following:
- The policy regarding which entity is responsible for the minor's detainment and welfare while on site;
- The procedure regarding notification of parents/guardians and law enforcement;
- The procedure for a minor requiring medical attention or with a condition that requires transport to a medical facility;
- The policy regarding who is authorized to take custody of a minor and the conditions under which they are allowed to do so; and
- Reporting procedures.

10.7 Lost and Missing Children

10.7.1 A child found separated from his or her parents or guardian and the report of a missing child by parents or guardians are two of the most distressful and emotional situations an event organizer may experience. However, they are extraordinarily common and occur at virtually all events. Thus, understanding and planning for such situations is necessary and important.

10.7.2 Event organizers should be familiar with the vernacular used by experts who deal with lost and missing children. The National Center for Missing and Exploited Children® (www.missingkids.com) generally defines a "missing child" as a person less than 18 years of age, "whose whereabouts are unknown to his or her custodial parent, guardian, or responsible party" (p. 2, *Law Enforcement Policy and Procedures for Reports of Missing and Exploited Children: A Model*, 2011). The NCMEC goes on to define an "at risk" missing child as being 13 years of age or younger or believed or determined to be experiencing one or more of the following circumstances (pp. 2-3):

* Is out of the zone of safety for his or her age and developmental stage;
* Has mental or behavioral disabilities;
* Is drug dependent, including prescribed medication and/or illegal substances, and the dependency is potentially life-threatening;
* Has been absent from home for more than 24 hours before being reported to law enforcement as missing;
* Is in a life-threatening situation;
* Is in the company of others who could endanger his or her welfare;
* Is absent in a way inconsistent with established patterns of behavior and the deviation cannot be readily explained; or
* Is involved in a situation causing a reasonable person to conclude the child should be considered at risk.

10.7.3 In contrast to the too common "missing child" scenario most familiar to law enforcement, search and rescue personnel refer to a person as "lost" when he or she is disoriented as to location and unable to find a preferred location. Children who have been temporarily separated from their parent or guardian at an event likely fit this definition. But, combine this with the often young age of the child (i.e., "at risk" is 13 years of age or younger) and it becomes clear that the situation should be addressed quickly to keep it from escalating to a tragedy.

10.7.4 Discovering a child separated from his or her parent or guardian is a common occurrence and can usually be resolved quickly. However, during that time of separation, the event staff will play a key role in maintaining the child's physical security and emotional stability. Event staff involved in the handling of lost children should be fully briefed and trained in exactly what to do and how to do it.

10.7.5 When a child is discovered who appears to be lost, the National Center for Missing and Exploited Children® (NCMEC) recommends the following (which can be found online at www.missingkids.com/en_US/publications/PDF15A.pdf):

What Should You Do If You See A Child Who Appears To Be Lost?

The National Center for Missing & Exploited Children® (NCMEC) encourages people to be alert and report situations regarding children who appear to be lost.

NCMEC teaches children to seek the assistance of adults who may be sources of help if they become lost. NCMEC encourages children to ask for help from adults such as uniformed law-enforcement or security officers and business/store personnel wearing nametags.

But what should you do when you see a child who appears to be lost? The steps noted below are ways you can provide assistance to a lost child.

1. **Get involved if you see a child who appears to be lost.**

2. **Comfort the child** but avoid physically touching him or her.

3. **Ask the child if he or she is lost or knows the location of his or her parent/guardian.**

4. **Refrain from requesting too much personal information** since children are taught not to give out this information to people they do not know.

5. **Contact law-enforcement authorities to report the incident.**

6. **Ask other adults in the area for assistance in reporting the incident to a person** in a position of authority in the area while waiting for law enforcement's arrival.

7. **Remain in the immediate location, and do not take the child elsewhere.** Do not place the child in your vehicle and drive to a different location to seek help.

8. **Wait with the child until help arrives.**

Note: If you see a child who you recognize as one who is missing, based on information from sources such as media reports, missing-child alerts, or fliers issued by law-enforcement authorities, immediately call law enforcement and follow their instructions.

This project was supported by Grant No. 2010-MC-CX-K001 awarded by the Office of Juvenile Justice and Delinquency Prevention, Office of Justice Programs, U.S. Department of Justice. Points of view or opinions in this document are those of the author and do not necessarily represent the official position or policies of the U.S. Department of Justice. National Center for Missing & Exploited Children® is a registered trademark of the National Center for Missing & Exploited Children. NCMEC Order PDF-15A.

10.7.5.1 Although this procedure is a good one for the public in general, an event organizer may want to direct event staff to notify law enforcement thorough the event command when the incident is reported so the relevant event procedure can also be implemented immediately.

10.7.6 When a child is discovered missing, the National Center for Missing and Exploited Children® (NCMEC) recommends the following (which can be found online at www.missingkids.com/MissingChild):

What to do if your child is missing

1. Immediately **call your local law enforcement agency**.

2. After you have reported your child missing to law enforcement, call the National Center for Missing & Exploited Children at **1-800-THE-LOST (1-800-843-5678)**.

3. If your child is missing from home, search through:
   - Closets.
   - Piles of laundry.
   - In and under beds.
   - Inside large appliances.
   - Vehicles – including trunks.
   - Anywhere else that a child may crawl or hide.

4. Notify the store manager or security office if your child cannot be found when in a store. Then **immediately call your local law enforcement agency**. Many stores have a Code Adam plan of action in place.

When you call law enforcement:
   - Provide law enforcement with your child's name, date of birth, height, weight and descriptions of any other unique identifiers such as eyeglasses and braces. Tell them when you noticed your child was missing and what clothing he or she was wearing.
   - Request law enforcement authorities immediately enter your child's name and identifying information into the FBI's **National Crime Information Center Missing Person File**.

10.7.6.1 This NCMEC procedure can be adapted for use at an event by borrowing from the well tested principles and applying them to the event environment. An example of an adaptation of this procedure for use by event staff might look something like this:

What to do if a child is reported missing at our event:

1. Keep the informant (the person who reported the child missing) with you and quickly get a detailed description of the child to pass to the command center. This should include the child's name, height, weight, hair color, eye color, date of birth, and any other unique identifiers including clothing worn, shoes, shoe treads, hats, glasses, braces, and any details that might help identify the child. Stay in the area where the missing child was reported if the child was last seen or known to be in the area.

2. Immediately contact the event command center (*insert phone number as back up*) who will (a) notify all staff including those at entrances and exits of the physical features and clothing of the child, and (b) contact local law enforcement if the child is not found within 10 minutes.
    - See if the informant can provide a photo of the child, which has been made easier with the wide use of smart phones and camera phones.

3. If the child is not located quickly, do not hesitate to call the National Center for Missing & Exploited Children at 1-800-THE-LOST (1-800-843-5678) to inform them of the situation and ask for advice.

4. If the child was last seen, or last known to be, in your immediate vicinity, quickly check the following:
    - Closest toilet facilities, then all toilet facilities.
    - Nearby children's games and activities that might attract the child.
    - Inside buildings, sheds, booths, trailers, large boxes, and any place a child could fit.
    - Beneath straw, in sandboxes, in ball pits, etc. – consider the child's low perspective.
    - Inside large appliances.
    - Vehicles – including trunks, spare tire storage areas, and underneath them.
    - Anywhere else that a child may visit, crawl or hide.

5. Initiate the Code Adam plan (have one in place!).
    - Page/announce "Code Adam" with the child's description to everyone at the event.
    - Designated employees will immediately stop working, look for the child and monitor entrances/exits to ensure the child does not leave the premises.
    - Call law enforcement (if not already done) if the child is not found within 10 minutes.
    - If the child is found and appears to have been lost and unharmed, reunite the child with the searching family member, guardian or responsible party.
    - If the child is found accompanied by someone other than a parent or legal guardian, make reasonable efforts to delay their departure without putting the child, staff or visitors at risk. Immediately notify law enforcement and give details about the person accompanying the child.
    - Cancel the Code Adam page after the child is found or law enforcement arrives.

6. Request law enforcement authorities enter the child's name and identifying information into the FBI's National Crime Information Center (NCIC) Missing Person File.

10.7.7 Relevant staff training and scenario planning discussions related to lost and missing children should be conducted in advance of an event. A procedural checklist that described the exactly what should be done and by whom should be created, kept in the event binder, and distributed to everyone with responsibilities on the checklist as well as the AHJ. The AHJ or local child protective services agency may have helpful information on procedures that will work well in their region. This checklist should contain, and the event supervisory staff should be familiar with, the following:

- How to identify a child who may be separated or lost.
- How to approach and make verbal contact with a child and listen carefully (remember the child may not speak English); refrain from touching the child until a second staff member arrives.
- Instructions directing the staff member who makes the first contact with the child to remain with the child through the entire process, until custody of the child is assumed by the parent/guardian, law enforcement or other agency.
- How to calm a frightened child (or parent/guardian).
- Steps to be taken to minimize the duration of the separation such as immediately scanning the area for a parent/guardian who might be looking for their child. Keep in mind the parent may not be able to see the child because the child may be short.
- The process of alerting one's supervisor and/or the event command center of the situation, and the requirement that they wait for a second event staff member or law enforcement to join them before relocating the child to a holding area.
- The implementation of the event's policy regarding lost child announcements.
- The policy regarding relocating the child to a secure, controlled, child friendly place while assuring the child that he or she is safe and going to be okay.
- How to entertain children while being cared for and waiting to be retrieved.
- How to determine an adult is the true parent or guardian of the child for whom they are seeking custody.
- The procedures to follow in the event that the collecting parent or guardian appears unfit to take custody of the child (e.g., drunk, angry, violent). Consult the local child protection agency for guidelines and advice.
- When it comes to handing over a child to the parent or guardian who may be unfit to take custody, event staff should avoid making judgment calls on their own. Instead, they should contact another event staff member to witness the proceedings and discretely summon law enforcement to assist with the situation.
- Staff may be faced with the rare and delicate situation of a found child resisting the person claiming to be their parent/guardian. The staff member should contact the event command center and request immediate back-up assistance from another staff member. The command center should immediately notify the children's area manager, law enforcement and/or the crowd management team. The staff member should remain calm and stall if possible until authorities arrive or try to keep the parent and child in sight until law enforcement arrives. The staff member should consider offering to take a picture of the couple and sending it to them as a goodwill gesture. This could come in handy later on.
- Dealing with an excited or angry parent/guardian. If the discussion with an adult over a child should become heated, no staff member should place themselves in danger. Instead, they should report the incident immediately and turn it over to law enforcement.

- Completing and filing a lost child report with the event office. This should be done for every reported or discovered lost child, even if the child is quickly reunited with his or her parents.

10.7.8 Documenting a reported or discovered lost child is important, even if the situation is quickly resolved. Insist that a report is filed with the event office in all situations involving a missing or lost child.

10.7.9 A clearly labeled collection point (holding area) for lost children should be established and supervised at all times by qualified and well briefed staff. No less than two staff members must be present when a child is being supervised.

10.7.10 A "two person minimum" approach should be the standard of care when supervising children for any reason at an event, including when taking a child to the toilet. A good practice in this delicate situation is for both adults to stay in visual contact with each other while one is assisting the child.

10.7.11 Any announcements made regarding a child whose whereabouts are unknown should not include the child's name, age, personal details or description. Rather, these announcements should inform the listening public of where to go if they are missing a child. These announcements should also be made in (multiple) languages relevant to the event.

10.7.11.1 If a child cannot be located, some descriptive details will have to be announced during the implementation of the Code Adam plan. However, only physical features and clothing should be announced.

10.7.12 When the child is retrieved, proof of identity and a signature should be obtained from the parent or guardian and a release form should be completed and filed in the children's area binder. Once a child has been reunited with his or her parent or guardian, inform security, staff and law enforcement immediately if they have been involved.

10.7.13 There are number of helpful children's protection programs and resources available through The National Center for Missing and Exploited Children® (www.missingkids.com) and venue managers and event organizers should become familiar with them. One that may be of immediate use is the handout "Know the Rules…For Child Safety in Amusement or Theme Parks," which can be found at: http://www.missingkids.com/en_US/publications/NC33.pdf

10.7.14 There are two particular programs in use throughout the United States with which event organizers should be familiar: Code Adam (www.missingkids.com/CodeAdam) and the AMBER Alert™ program (www.amberalert.gov). Code Adam is a tool for lost and potentially abducted children that involves a specific protocol for immediate use in public venues when a child is reported missing. The AMBER Alert program is a Federally-sponsored (Department of Justice), voluntary partnership between law-enforcement agencies, broadcasters, transportation agencies, and the wireless industry, to activate an urgent bulletin in the most serious child-abduction cases.

10.7.14.1 Code Adam

10.7.14.1.1 Code Adam is one of the country's largest child safety programs and is currently used in tens of thousands of establishments across the nation, including all Federal facilities where it is required by law to be used. The Federal "Code Adam Act of 2003" (Public Law 108-21) requires the designated authority for a public building to establish procedures for a child missing in a federal facility (www.gsa.gov).

10.7.14.1.2 According to the NCMEC (www.missingkids.com/CodeAdam), Code Adam works as follows:

Code Adam decals (navy blue background with CODE★ADAM in white letters across the front and white diagonal stripes at the top and bottom) are posted at the entrances of participating businesses and facilities. Employees are trained to take the following steps when Code Adam is activated:

1. Obtain a detailed description of the child, including what he or she is wearing.

2. Page "Code Adam" Describe the child's physical features and clothing.

3. Designated employees will immediately stop working, look for the child and monitor front entrances to ensure the child does not leave the premises.

4. Call law enforcement if the child is not found within 10 minutes.

5. If the child is found and appears to have been lost and unharmed, reunite the child with the searching family member.

6. If the child is found accompanied by someone other than a parent or legal guardian, make reasonable efforts to delay their departure without putting the child, staff or visitors at risk. Immediately notify law enforcement and give details about the person accompanying the child.

7. Cancel the Code Adam page after the child is found or law enforcement arrives.

10.7.14.1.3 Code Adam is named in honor of Adam Walsh, a 6-year old abducted in 1981 from a Florida department store and later found murdered.

10.7.14.2 AMBER Alert™

10.7.14.2.1 The AMBER Alert System began in 1996 when Dallas-Fort Worth broadcasters teamed with local police to develop an early warning system to help find abducted children. AMBER stands for "America's Missing: Broadcast Emergency Response" and was created as a legacy to 9-year-old Amber Hagerman, who was kidnapped while riding her bicycle in Arlington, Texas, and then brutally murdered. Other states and communities soon set up their own AMBER plans as the idea was adopted across the nation. The goal of an AMBER ALERT is to instantly galvanize the entire community to assist in the search for and the safe recovery of the child.

10.7.14.2.2 AMBER Alerts are broadcast through radio, television, road signs and all available technology referred to as the AMBER Alert Secondary Distribution Program. These broadcasts let law enforcement use the eyes and ears of the public to help quickly locate an abducted child. The U.S. Department of Justice coordinates the AMBER Alert program on a national basis.

10.7.14.2.3 Once law enforcement has determined that a child has been abducted and the abduction meets AMBER Alert criteria, law enforcement notifies broadcasters and state transportation officials. AMBER Alerts interrupt regular programming and are broadcast on radio and television and on highway signs. AMBER Alerts can also be issued on lottery tickets, to wireless devices such as mobile phones, and over the Internet. Through the coordination of local, state and regional plans, the U.S. Department of Justice is working towards the creation of a seamless national network.

10.7.14.2.4 The Department of Justice's guidance on criteria for issuing AMBER Alerts is as follows:
- Law enforcement must confirm that an abduction has taken place.
- The child is at risk of serious injury or death.
- There is sufficient descriptive information of child, captor or captor's vehicle to issue an Alert.
- The child must be 17 years old or younger.
- It is recommended that immediate entry of AMBER Alert data be entered into the FBI's National Crime Information Center. Text information describing the circumstances surrounding the abduction of the child should be entered, and the case flagged as Child Abduction.

10.7.14.2.4.1 Most states' guidelines adhere closely to the Department of Justice's recommended guidelines. The program is used in all 50 states, the District of Columbia, Puerto Rico and the U.S. Virgin Islands.

10.8 Emergencies and Children

10.8.1 There are few sources available that offer specific information related to children and emergencies at events. However, there is a wealth of general information on children and emergencies that can be applied to events and that offer valuable insights for event organizers.

10.8.2 The Federal Emergency Management Agency (FEMA) offers two free, online, independent study courses designed to instruct persons charged with the responsibility for children. The IS-366 course, *Planning for the Needs of Children in Disasters*, provides guidance for those leading children's programs for meeting the unique needs that arise among children as a result of disasters and emergency situations. The IS-36 course, *Multi-hazard Planning for Child Care*, covers the steps to help childcare providers prepare for incidents to ensure the safety of the children at their site. Both can be found online at http://training.fema.gov/IS/.

10.8.3 The local government's Hazard Vulnerability Assessment (HVA) and the local, county and state's Emergency Operations Plans (EOP) described in Chapter 2, *Planning and Management*, will also contain valuable information and guidance regarding children and their needs during an emergency.

10.8.4 The National Center for Missing and Exploited Children® (NCMEC) provides The Unaccompanied Minors Registry (http://umr.missingkids.com/), a tool for reporting children displaced during a disaster such as a hurricane, tornado or terrorist attack. Through the Unaccompanied Minors Registry, the NCMEC is able to assist emergency management personnel on the ground in their efforts to reunite families. The NCMEC also provides the National Emergency Child Locator Center, a call center that, when a national disaster is declared by the President, can be activated to assist in the location of children and the reunification of families resulting from the disaster or subsequent evacuations.

10.8.5 The following web sites also offer advice for dealing with children in emergencies. Although none directly relate to entertainment events, some offer valuable insights that might be useful to event organizers.
- http://www.savethechildren.org/site/c.8rKLIXMGIpI4E/b.6192515/k.319F/Protecting_C hildren_in_Emergencies.htm
- http://www.redcross.org/prepare/location/home-family/children
- http://yosemite.epa.gov/ochp/ochpweb.nsf/content/Emergencies.htm
- http://children.webmd.com/guide/when-call-911-7-emergencies-children

11. Facilities for People with Special Needs

11.0.1 Consider suitable arrangements, wherever possible, to ensure that all people with special needs are able to attend. It is also recommended that a complete access strategy be prepared which includes the technical issues as well as factors which will encourage and attract persons with special needs to your event.

11.0.2 Organizers should always keep in mind that not all disabled persons are in wheelchairs or mobility impaired.

11.1 Americans with Disabilities Act (ADA)

11.1.1 The Americans with Disabilities Act of 1990 (ADA) is a U.S. Federal law that was enacted by the U.S. Congress in 1990 and later amended with changes (ADA Amendments Act of 2008 or ADAAA) effective January 1, 2009. The ADA is a wide-ranging civil rights law that prohibits, under certain circumstances, discrimination based on disability. It affords similar protections against discrimination to Americans with disabilities as the Civil Rights Act of 1964, which made discrimination based on race, religion, sex, national origin, and other characteristics illegal. Disability is defined by the ADA as, "…a physical or mental impairment that substantially limits a major life activity." The Equal Employment Opportunity Commission (EEOC) later interpreted the phrase "substantially limits" more narrowly to mean "significantly or severely restricts." Certain conditions are excluded as disabilities, such as current substance abuse and visual impairment that is correctable by prescription lenses.

11.1.2 There are five major sections or Titles in the ADA. Titles II and III are most likely to be of interest to those constructing or modifying entertainment venues:
- Title I – Employment - The ADA states that a covered entity shall not discriminate against a qualified individual with a disability. This applies to job application procedures, hiring, advancement and discharge of employees, workers' compensation, job training, and other terms, conditions, and privileges of employment.
- Title II – Public Entities (and public transportation) - Title II prohibits disability discrimination by all public entities at the local (i.e. school district, municipal, city, county) and state level and applies to public transportation provided by public entities through regulations by the U.S. Department of Transportation. It includes all commuter authorities.
- Title III – Public Accommodations (and commercial facilities) - Under Title III, no individual may be discriminated against on the basis of disability concerning the full and equal enjoyment of the goods, services, facilities, or accommodations of any place of public accommodation by any person who owns, leases (or leases to), or operates a place of public accommodation. "Public accommodations" include most places of lodging (such as inns and hotels), recreation, transportation, education, and dining, along with

stores, care providers, and places of public displays, among other things. Under Title III of the ADA, all "new construction" (construction, modification or alterations) after the effective date of the ADA (approximately July 1992) must be fully compliant with the Americans With Disabilities Act Accessibility Guidelines (ADAAG) found in the Code of Federal Regulations at 28 CFR, Part 36, Appendix A. Title III also has application to existing facilities.

- <u>Title IV – Telecommunications</u> - Title IV of the ADA amended the landmark Communications Act of 1934 and requires that all telecommunications companies in the U.S. take steps to ensure functionally equivalent services for consumers with disabilities, notably those who are deaf or hard of hearing and those with speech impairments.
- <u>Title V – Miscellaneous Provisions</u> - Title V includes technical provisions. It discusses, for example, the fact that nothing in the ADA amends, overrides or cancels anything in Section 504 of the Rehabilitation Act of 1974.

11.1.3 In the United States, compliance with the Americans with Disabilities Act by event organizers is not optional, it is mandatory. The Americans with Disabilities Act is readily available online and is recommended reading for all event organizers (http://www.ada.gov/).

11.1.4 ADA Title III mandates the immediate removal of architectural and communications barriers from existing public accommodations if such action is "readily achievable." Changes must be made whenever it is possible to remove barriers, "without much difficulty or expense." Unfortunately, there are no solid guidelines for when changes must be made. The determination is made on a case-by-case basis.

11.1.5 No numerical formula or threshold is provided to indicate whether a given action is readily achievable. Many non-compliant offenders of the ADA choose to gamble that they will not have a problem at their event. This approach is not recommended.

11.2 General Accommodation

11.2.1 The kinds of accommodations to consider at an event include:
- Accessibility ramps with an incline less than 1:12, e.g., 1 unit in vertical height over a 12 unit distance;
- Curb cuts at access points;
- Accessible parking spaces;
- Widening of doors;
- Lowering of drinking fountains and phones if applicable;
- Accessible counters at select food and beverage outlets;
- Adding braille to elevator control buttons and signage and informational literature;
- Replacing door knobs with handles;
- Widening toilet seats and installing grab bars in restrooms; and
- Using a signer to sign lyrics or spoken word, use of a picture in picture (PIP) in the live image magnification using cameras (IMAG) video can help here also.

11.2.2 Consider provision for people with:
- Mobility problems (including wheelchair users);
- Difficulty in walking; and
- Impaired vision and/or hearing.

11.2.3 Event publicity should provide a contact number and website where people with special needs can obtain information on arrangements the event has made to accommodate persons with special needs.

11.2.4 When designing your event site or venue consider how people with special needs can best be accommodated. This includes easy access and adequate means of escape in an emergency. The number of wheelchair users who can be admitted will be dependent upon a number of factors including the structural and internal layout of the venue.

11.2.5 In open field events, creation and/or compaction of pathways or creating a road for wheelchairs may need to be considered.

11.2.6 Wheelchair spaces in parts of a seated area should allow for adequate room for maneuvering a wheelchair. Generally, a manual wheelchair needs approximately 3 feet (1 m) in width and about 5 feet (1.5 m) in depth including an attendant. Electric wheelchairs may need more space.

11.2.7 Avoid placing accommodations in dense crowded areas.

11.3 Access

11.3.1 Place parking facilities for people with special needs at the most directly accessible point to those areas set apart for wheelchair users. Spaces allocated should be wider than normal (about 12 feet or 3.6 m) to allow room to maneuver. At outdoor events parking for people with special needs should also be placed at the most directly accessible point to the allocated seating areas, as well as the most directly accessible point to designated and accessible campsites. Thought should be given to the means of having direct and safe access links between the designated parking, camping and seating areas. Use flat surfaces or ramps to provide access from parking or drop-off areas to designated areas.

11.3.2 Ramps for wheelchairs should comply with the ADA. The ramp should not be steeper than 1 in 12. Ramps should begin with a 6 feet (1.8 m) long level landing and have a five foot (1.5 m) long level resting space landing every 30 feet (9.1 m). They should also have raised safety edges and handrails 2 feet 8 inches (0.81 m) high.

11.4 Viewing Areas

11.4.1 As standing audiences can cause surging movements, all people attending the event that have any mobility difficulties should be located in an area where they will not be affected. When setting aside viewing areas for people with special needs, the area should have a clear view of the stage, often beside the "mixing" tower or area. The area should be constructed using non-slip materials with direct access to an exit. Accessible toilets should be located nearby.

11.4.2 At outdoor concerts wheelchair users can be accommodated either on an open area or on a flat terrace with direct access to toilet facilities and concessions. The eye level of a wheelchair user is estimated to be between 43 and 49 inches (0.9144 m and 1.2446 m) high.

11.4.3 Many wheelchair users will be accompanied by an able-bodied companion. Make sure that space in the wheelchair users area can accommodate these companions, preferably with chairs provided which do not block the view of other wheelchair users in the area.

11.4.4 When someone transfers from a wheelchair to a seat, provision needs to be made for the wheelchair to be readily accessible without it causing an obstruction in any aisle or exit route. Where a person remains in their wheelchair, the wheelchair should be placed in a position where there is a ready means of escape and it will not create an obstruction for other people.

11.5 Facilities

11.5.1 Concession stands should also be encouraged to have either varied level of serving counter space or an access ramp in front of the serving counter. When using temporary sanitation facilities, ADA toilets should be unisex with wheelchair access and it is suggested that one unit per 50 wheelchair users should be provided, along with additional provision for the use of attendants, etc. (see Chapter 14, *Sanitary Facilities*).

11.6 Support

11.6.1 Guest services personnel, other trained event staff, or special needs assistants should be in attendance to ensure that facilities which are provided for people with special needs are available for the intended purpose.

11.6.2 Consider providing designated "ground support" staff. They could be people with special skills (e.g., signers, medics, etc.) who can provide on-site support for people with special needs. These workers should be easily recognizable by the use of an easy-to-read emblem or logo. Stewards operating in and near the area set aside for people with special needs require specialized training in the evacuation and exit procedures. Also consider using safe sites for people with special needs in the event of an evacuation.

11.7 People with Impaired Vision

11.7.1 People with impaired vision or color perception may have difficulty in recognizing information signs including those used for fire safety. Signs therefore need to be designed and positioned so that they can easily be seen and are distinguishable. Good lighting and the simple use of color contrasts can also help visually impaired people find their way around. Where practical, consider admitting guide dogs.

11.8 Evacuation

11.8.1 People in the audience may be affected by a range of disabilities that influence their ability to evacuate an event site, such as restricted mobility, epilepsy, impaired hearing, or mental health issues. So, reasonable accommodations should be included in major incident and contingency plans. Where they exist, electronic display systems should be used to give information, including evacuation messages, particularly for people with impaired hearing.

11.9 Publicity

11.9.1 It would be helpful to potential visitors if the facilities that are available are publicized. This can be achieved by contacting a local disability association, access groups and local clubs or organizations for people with disabilities. These organizations are often willing to help event organizers provide suitable accommodations at the event site.

12. Transportation Management

12.0.1 This chapter discusses the management of traffic and various modes of transportation approaching the site and inside the venue perimeter, including pedestrians.

12.0.2 When created in cooperation with local police and highway authorities, traffic management plans help ensure safe and convenient site access and minimize off-site traffic disruption.

12.1 Traffic Signs and Highway Department road Closures

12.1.1 Identify the need for temporary traffic signs before the event. If temporary signs are needed, prepare and agree on detailed traffic sign plans and schedules with the police and local highway authorities before the event. It may be necessary for people living in the area to be consulted over route changes and to be advised of the impact, once agreement has been reached. Consider using a traffic sign contractor for events where the majority of people will be arriving by cars or buses.

12.1.2 Consider the need for temporary traffic regulation orders to provide for road closures, banned turns, lane closures, parking restrictions, temporary speed limits and rest stop closures. For large events and particularly if special traffic management arrangements and temporary traffic regulation orders are required, consultation with the local highway authority is essential. Highway authorities include the highways agency for all thoroughfares and highways, and the local authority for all other roads.

12.1.3 Consult the local highway authority as to the best way of carrying out traffic orders and allow sufficient time for any temporary traffic regulation orders to be processed.

12.2 Traffic Marshaling

12.2.1 Only law enforcement, a similar official authority (e.g., authorized road crew, emergency service personnel, etc.) or someone under their direction or authority can legally regulate traffic on the public highway. Consultation is essential to secure the appropriate provision of resources. Authorized personnel or stewards directing traffic on site should have suitable personnel protective equipment such as high-visibility clothing and weather protection. Stewards should receive traffic marshaling training, e.g., safe positioning of the marshal and awareness of visibility problems for drivers of reversing vehicles.

12.2.2 Make sure there is sufficient communication between on-site and off-site traffic marshaling and provide adequate numbers of stewards to manage the traffic flows and deal with the parking of vehicles.

12.3 Public Transportation

12.3.1 A comprehensive attendee transportation management plan should consider the availability of public transportation options in addition to private vehicles parking at or near the event site.

12.3.2 Trains and Underground Trains

12.3.2.1 If appropriate, consult with rail authorities about introducing additional trains or enhancing existing services to accommodate the demands of the event and to limit the demand for on-site and off-site parking.

12.3.2.2 It may also be worth investigating the use of combined event/rail package tickets. Consideration, however, needs to be given to the distance between railway stations and the venue and the availability of connecting bus and coach services to and from the event. Advertising and additional signage may be carried out on trains and at stations before the event, stating any additional service (or lack of) being provided.

12.3.2.3 It is also important to consult the rail authorities concerning the maximum number of people that a station can accommodate at any one time. Most railway stations will have contingency plans, which identify the safe number of people allowed on the platform at any one time. These contingency plans can be used at the event planning meetings between the relevant authorities.

12.3.2.4 Train-operating companies have responsibility for the queuing of large numbers of people at their stations. Plan how you are going to communicate with the train-operating companies and law enforcement in the event of a major incident to ensure that the stations receive advance information in case the event finishes earlier than planned or emergency evacuation is necessary.

12.4 Public Transportation Management

12.4.1 Advice to Train-Operating Companies

12.4.1.1 Train-operating companies need to consider their own planning procedures to ensure that they can safely manage the potential increased 'throughput' of passengers associated with the event, e.g., ensuring suitable entrances and exits, control of passenger numbers on platforms, footbridges and tunnels, crowd flow plans and temporary queuing system and communicating travel information by PA systems.

12.4.1.2 It is important for transportation providers to have their own contingency plans for dealing with train delays or incidents on the track and to consider the suitability of the rolling stock, provision of first-aid points and first aiders, additional toilets and additional workers.

12.4.2 Coaches/Buses

12.4.2.1 Planning the arrival and departure of coaches can greatly reduce congestion at the beginning and end of a large music event. Careful consideration has to be given to the routing of such vehicles. Bus "loops," parking areas and access roads should be provided to reduce as far as possible the need for coaches/buses to reverse, e.g., creating one-way systems.

12.4.2.2 Coaches and buses need a wide turning radius and easily accessible entrance and exit points, as well as large turning areas into allocated parking areas. Consider specific arrangements to ensure the free flow of coach/bus routes in consultation with the police, and ensure that this is documented in the transportation management plan. Coach and bus parking areas may need to contain toilet facilities.

12.4.2.3 Private bus/coach operators are often prepared to provide special shuttle bus services between local rail and/or bus stations. However, shuttle bus systems may not be appropriate for all events. Congestion caused by a natural mass exodus at the end of an event is likely to prohibit free flow of traffic routes and consequently shuttle buses become unable to operate effectively. Consider the potential for dedicated shuttle bus routes or consult local bus operators about enhancing or extending their established services to serve any proposed event.

12.4.3 Vehicle Parking and Management

12.4.3.1 Include proposals in the transportation management plan for the management of vehicle parking which identify the likely resources required (space necessary, traffic marshals and equipment) and methods to be used for parking management.

12.4.3.2 Make sure the traffic management team and law enforcement can communicate with the vehicle parking management team, so that resources can be directed quickly to deal with any incidents within the parking lots or at the various site accesses. It is good practice to have a tow truck on-site or on call to assist where needed.

12.4.3.3 For large events, consider the appointment of a traffic management coordinator who will liaise with the police, parking lot management, traffic signs contractor, local highway authority and local authority.

12.4.4 Vehicular Access

12.4.4.1 Ensure that the road signs are appropriate and easily visible, the capacities of the parking areas are adequate and the surface is capable of withstanding the anticipated traffic volume. Consider using a suitable temporary surface which can prevent damage to the ground and prove invaluable in wet ground conditions.

12.4.4.2 Detailed capacity assessments may be needed to ensure access entry capacity is adequate. Queuing on entry into the site can cause blocking of traffic flows leading, potentially, to severe congestion. Exit capacity is less problematic. If congestion occurs on exit, it is contained within the site and will not adversely affect off-site conditions. However, the risks associated with poor vehicle exit management should not be underestimated. Methods for ensuring the safe exit of vehicles from the site need just as much careful planning. Consider planning alternative routes and accesses. These can be used if main access points or routes become blocked.

12.4.4.3 Consider vehicle access for service vehicles before, during and after the event, e.g., waste collection vehicles and sanitary service vehicles.

12.4.5 Parking

12.4.5.1 Consider separate parking areas for the general audience traffic, vehicles for people with special needs (close to event site), coaches, shuttle buses, guests/VIPs, artists, emergency service workers and event workers. Facilities either on site or at a convenient location off-site to accommodate the potential for excess visitors (e.g., overflow parking) may also need to be planned. This may take the form of a vehicular circulation/holding area as a temporary measure. When parking event workers off-site in a remote location consider a shuttle system to transport them to the staff check-in area. They will likely have tools and work boxes which they need to perform their duties so head counts on shuttles may be reduced and increased numbers of screening security, if applicable, should be on hand.

12.4.5.2 Car and bus parking areas need to be adequately lit, signposted and labeled with reflective numerals or letters so that vehicles can be easily located at the end of an event or in any other emergency. Ideally, separate buses from parking lots. For large outdoor events, position signs at exit gates leading from the parking area to the venue to assist in identifying where cars have been parked and consider clear signs for exiting vehicles showing route direction.

12.4.5.3 Ground conditions within parking and walkway areas should be closely monitored. Conditions that may present a safety hazard to those entering or exiting the parking area should be immediately corrected or visibly flagged, such as broken concrete, standing water or deep mud. For the safety and security of guests, adequately mark walkways for the patrons to go safely to and from the event.

12.4.6 Emergency Access

12.4.6.1 Plan to provide for the entry and exit of emergency service vehicles. Many existing facilities will have established entry and exit points for emergency vehicles. Incorporate them into your event plan. Ideally these routes should be separate and safeguarded from routine vehicle and pedestrian traffic. The routes and access chosen must allow for access by fire department apparatus (generally, 20-26 feet [6-8 m] wide and capable of supporting at least a 75,000 pound [34 050 kg] vehicle) and may be required to be approved by the local fire department. Even if it is not required, requesting fire department approval for these routes is always prudent and an opportunity to demonstrate one's interest in safety. Be mindful that these routes may include access through gates, and must include access to every structure and fuel storage facility. These routes should also be signposted as a "no parking - fire lane" or whatever is required locally.

12.4.6.2 Get advice from the fire department concerning access route specification and incorporate this into the transportation management plan. In this respect, early application for road closures and temporary traffic regulation orders may be necessary. It is also important to identify allocated emergency vehicle rendezvous points in the transportation management plan.

12.5 Pedestrians

12.5.1 Generally speaking, pedestrians sharing the same space as any mode of transportation with wheels should be avoided where alternative routes are available.

12.5.2 Identify safe means of entry and exit for pedestrians, ideally segregated from vehicular access. Where pedestrian access is difficult, consider the provision of alternative means of access, e.g., shuttle buses to collect pedestrians en route. Consider making specific arrangements for those attending who have a physical disability and may not be able to walk long distances. Avoid entry and exit routes crossing parking lots and traffic routes. Where the latter is unavoidable, plan for adequate traffic control measures.

12.6 On-Site Vehicle Management and Temporary Roadways

12.6.1 It is important to minimize traffic movement within the site and conflicts between vehicles and pedestrians. Consider moving vehicles into the parking areas as efficiently as possible and having a dedicated access to parking areas with no ticket checks on entry. In some circumstances, ticket checks can be undertaken on pedestrian exits from the parking into the event area. This may, however, not be practical for camping events.

12.6.2 Restrict traffic movement in the event arena to emergency service vehicles and other essential services. Consider speed restrictions on site and plan separate access for production vehicles.

12.6.3 Temporary roadways are useful to allow suitable hard-surfaced access for pedestrians and service vehicles. Plan temporary access roads, ideally to provide for two-way emergency access or one-way with passing places and working space as appropriate. All on-site vehicles must display adequate lighting during darkness and remember to keep pedestrian and vehicle conflict points to a minimum. Plan how vehicles delivering equipment and provisions enter and exit the site safely during the load-in and load-out of the event.

12.6.4 Where vehicle routes change from those arranged at planning stage, due to heavy rain or some other unforeseen circumstance, make sure that arrangements are in place for reinforcing the alternative route. Vehicle recovery from soft ground should be planned.

12.7 Vehicles

12.7.1 Forklifts

12.7.1.1 No one should be permitted to operate a forklift unless they have been trained and authorized to do so.

12.7.1.2 Trained operators will have a certificate from an accredited organization indicating the type of forklift for which they have received training. A certificate to drive one forklift does not qualify an operator to drive other types of forklifts. Do not allow workers to operate forklifts without checking that they are fully trained for the type of truck they are to use. Confirm that your event labor provider is supplying trained and certified operators.

12.7.1.3 If the event is renting forklifts, check that the equipment is delivered in a safe working condition with a manual and adequate fuel. Each should be marked with its safe working load and that load should never be exceeded. Ensure that each forklift comes with a current service record which adequately covers the period it will be used on the site.

12.7.2 Other Vehicles Used on Site

12.7.2.1 As well as forklifts, there is likely to be the need for other types of vehicles to operate on site such as:

- Other specialty lifting vehicles (e.g., boom lifts, man lifts or personnel lifts and scissor lifts);
- Hand powered conveyances like pallet jacks and hand trucks;
- Vehicles used to deliver equipment around the site or venue (e.g., golf carts, four-wheel drive flatbed carts and electric carts); and
- Other vehicles (e.g., tractors, trailers and waste-collection vehicles).

12.7.2.2 The use of all vehicles on site should be carefully planned and monitored to ensure that accidents do not result from the incorrect use of the vehicle or that pedestrians are not injured as a result of their use. Consider a motor pool check-out and check-in system to track drivers, damage to vehicles as well as fueling or charging as necessary.

12.7.2.3 Vehicular traffic at loading docks and in trailers can present a real possibility of injuries. Trailers left in the dock without a tractor attached should have their front wheels fully in contact with the dock surface, wheels chocked to prevent accidental movement while workers are in the trailer, and a warning cone placed in front of the trailer to remind drivers to check the status of the workers before moving the trailer.

13. Food, Drink and Water

13.1 Catering Operations

13.1.1 To the extent possible, ensure that the delivery, storage, preparation and sale of food complies with the relevant food safety regulations and relevant industry standards. This will include mobile catering units, catering stalls and tents, crew catering outlets, hospitality catering, bars and ice cream vendors, etc.

13.1.2 In an effort to ensure that food businesses carry out their work in a safe and hygienic way, the event organizer can request evidence from each caterer regarding:
- The identification and control of potential food hazards by all catering operations;
- The identification and control of potential health and safety hazards by all catering operations;
- Provision of appropriate fire extinguishers;
- Proper training of all food handlers, including obtaining a food handler's permit where required;
- The suitability of all premises used for the production or sale of food;
- The suitability of the equipment being used;
- Transporting food safely and separate from any potential source of contamination;
- Storing and disposing of food waste (solid and liquid) properly;
- The maintenance of high standards of the personal hygiene of food handlers;
- The proper storing, handling and preparation of food;
- The provision of a drinking water supply (see section 13.7);
- Insurance of all food businesses including public, product and employers liabilities;
- The possession of electrical and gas installation compliance certificates by all food businesses; and
- The possession of a properly equipped first-aid box by each operating unit.

13.1.3 Where needed, contact the local authority environmental health officers (EHOs) for advice on food safety and hygiene. EHOs may wish to carry out an inspection of the catering facilities provided at the event. They may also require you to provide them with a list of caterers who will attend the event.

13.1.4 Additional requirements may be necessary in certain types of catering operations, e.g., barbecues and spit roasting. Such operations may present an increased risk of fire, contamination or food poisoning. Ensure that a suitable risk assessment is carried out, taking into account the particular factors of the operation.

13.2 Positioning

13.2.1 The site plan of the event must include a detailed layout of all catering operations (see Chapter 8, *Venue and Site Design*), bearing in mind the need to:

- Prevent any obstruction that may affect the health and safety of people attending or working at the event;
- Prevent, as far as is possible, access to the rear of catering operations by the audience;
- Allow entry and exit for emergency vehicles;
- Take into account suitable spacing between individual operations;
- Provide readily accessible and preferably lockable facilities for the storage and Disposal of solid and liquid waste;
- Allow for the efficient removal of refuse (see Chapter 15, *Waste Management*);
- Position catering operations within close proximity to a supply of drinking water, foul drainage and within a safe minimum distance from any source of possible contamination, i.e. fuel, waste or refuse storage;
- Consider manual-handling issues involved in the disposal of water, the delivery of supplies, etc.;
- Provide separate toilet facilities for the exclusive use of food-handlers, with hot and cold hand-washing facilities;
- Provide suitable facilities for parking and access of support vehicles;
- Position mobile sleeping accommodation away from the catering operations.

13.3 Propane

13.3.1 Propane, a liquefied petroleum gas, is the main source of fuel for outside catering operations. It does present a substantial fire/explosion risk, therefore ensure that:
- All operators using propane can demonstrate a basic understanding of its safe use, its characteristics and emergency procedures;
- All propane tanks are properly anchored and secured;
- Storage at each catering operation does not exceed that which is required for a 24-hour period;
- All propane should be handled, stored and used in accordance with *The International Fuel Gas Code*, *The International Mechanical Code*, and NFPA 58, *Liquefied Petroleum Gas Code*, which may be required by the local authorities having jurisdiction; and
- All supplies of propane whether in compounds or within catering operations are protected and isolated from audience interference and vehicular traffic.

13.4 Electrical Installations

13.4.1 Electrical power to catering operations should, wherever possible, be provided by the site electrical supply (see Chapter 17, *Electrical Installations and Lighting*). If portable generators are used, preference should be given to propane or diesel-fueled types.

13.4.2 Ensure that:
- They are of a suitable rated power output for the intended use;
- They have been tested and certified by a competent person;
- They are sited in a well-ventilated place away from propane cylinders, combustible and flammable materials;

- They are adequately guarded to avoid accidental contact by people or combustible material;
- Cables and sockets are appropriate for their intended use;
- The electrical installation is protected by a ground fault circuit interrupter (GFCI);
- Cables do not create a trip hazard;
- Fueling and refueling are carried out in a safe manner; and
- Fuel is stored in a safe manner in suitable containers.

13.5 Fire-Fighting Equipment

13.5.1 Suitable fire-fighting equipment should be provided at the catering operation dependent on the activity type. The equipment must conform to the relevant U.S. standards (see Chapter 4, *Fire Safety*). No combustible materials should be allowed to accumulate next to any catering outlets.

13.5.2 It is important to use the correct type of fire extinguisher for the particular type of fire. For information on selecting the appropriate type of extinguisher for various situations (ordinary combustibles, flammable liquids, electrical equipment, combustible metals, combustible cooking) see Chapter 4, *Fire Safety*. The U.S. Fire Administration also publishes a guide that can be found online at http://www.usfa.fema.gov/citizens/home_fire_prev/extinguishers.shtm.

13.5.3 Generally, at least one portable fire extinguishers with a minimum rating of 2-A:20-B:C should be installed within 30 feet (9.144 m) travel distance of commercial cooking operations. However, if solid fuel or deep fat fryers are in use, a Class K fire extinguishing system may be required and at least one 2.5 gallon (9 L) or two 1.5 gallon (6 L) Class K portable fire extinguishers should be installed and may be required.

13.6 Alcohol and Bar Areas

13.6.1 Alcohol comes under the definition of food and should meet the requirements of the relevant food safety regulations, associated industry guides and codes of practice. Ensure that:
- The structure used for the sale of alcohol, usually tents, complies with the structural requirements (see Chapter 19, *Structures*);
- The operation is designed to allow the free flow of people to and from the bar server areas to prevent congestion and crushing hazards (this may involve the use of suitable barriers, providing consideration has been given to the barriers becoming a hazard in themselves);
- The electrical installation complies with the requirements detailed in Chapter 17, *Electrical Installations and Lighting*;
- Suitable and sufficient lighting is provided;
- Alcohol storage tanks are positioned on stable, even ground allowing suitable access for delivery vehicles, particularly in bad weather;
- Risk assessments for both food and health and safety, have been carried out;
- Carbon dioxide cylinders are suitably secured;
- Chemicals to clean pipelines are properly handled and stored;

- The type of containers that drinks are served in conform to any site/event specifications (e.g., no glass policy);
- There is a suitable means of disposal for glass bottles, used to decant drinks before serving;
- Bar areas are kept free of litter and the floors are cleared of spillages; and
- If a "token system" is used instead of cash, the "change areas" need to be separate from the bar service area.

13.7 Drinking Water

13.7.1 The provision of free drinking water is important at all events, especially open-air concerts and dance events, due to the volume of people, confined conditions and the weather.

13.7.2 Generally all water should be provided from a water main supply, but if this is not possible then water tanks are permissible provided they are suitable for the purpose. All water dispensing equipment should be clean, well maintained and suitable. It is considered good practice to sample and test temporary water supplies for bacterial safety, especially those provided at outdoor events.

13.7.3 Pit Area

13.7.3.1 There should be an adequate supply of drinking water points in the pit area, together with an adequate supply of paper or plastic cups. The number of drinking water points will be determined by the risk assessment.

13.7.3.2 If storage containers are used to supply the water, they should be of sufficient capacity and number for the anticipated needs of the people within the first 16 feet (4.9 m) of the pit barrier. Pit area water points should not be within the reach of the audience.

13.7.4 General Area

13.7.4.1 There must be a supply of drinking water within easy reach of the audience and all catering operations. At outdoor sites (one-day events) a general guideline is one water outlet per 3,000 people and one outlet per 10 caterers provided they are in the same area.

13.7.4.2 All water points should:
- Have unobstructed access;
- Be clearly marked;
- Be clearly lit at night if the event continues after dark; and
- Have self-closing taps.

13.7.4.3 The ground surrounding all water points should be well drained or provision made to "bridge" any flooded areas.

14. Sanitary Facilities

14.0.1 Ensure that adequate sanitary provision is made for the number of people expected to attend the event. Consideration should be given to location, access, construction, type of temporary facilities, lighting and signage.

14.0.2 The American Restroom Association provides some guidelines online at http://www.americanrestroom.org/pr/policy and FEMA offers their "FEMA Special Events Contingency Planning- Toilets" pamphlet, which can also be found online at http://www.americanrestroom.org/pr/policy/#fema.

14.0.3 Construct and locate toilets so that people are protected from bad weather and trip hazards. The floors, ramps and steps of the units should be stable and of a non-slip surface construction. Protect connecting pipe work to avoid damage.

14.0.4 Toilets should be readily visible, lit and clearly signed from all parts of the venue. The areas and, where appropriate, the individual units, should be adequately lit at night and during the day, if required.

14.1 Maintenance

14.1.1 Regularly maintain, repair and service toilets using suitably experienced competent workers throughout the event to ensure that they are kept safe, clean and hygienic. Toilets need to be supplied with toilet paper, in a holder or dispenser at all times. Some modern units come equipped with hand sanitizer dispensers which can relieve some pressure from your hand-wash units. Arrangements should be made for the rapid clearance of any blockages.

14.2 Location

14.2.1 Where possible, locate toilets at different points around the venue rather than concentrating in one small area, to minimize crowding and queuing problems. Consider placing toilets outside the perimeter fenced venue area (e.g., parking lots, box office queue lines, event campsites, etc.). Attention should be given to access requirements for servicing and emptying. This may include the need for temporary roadways and dedicated access routes for pumper trucks, subject to the layout of the site.

14.3 Type

14.3.1 Where temporary toilets are required, an assessment should be made of the suitability of each of the available types of temporary unit, for the nature and duration of the event being organized. Consider the perceived peak usage of any toilet units and the time taken for cisterns to fill. Rapid and constant use of any toilet can cause the bowls to become unsanitary and prone to blockages.

14.3.2 Temporary units can be used if a sewer, drain, septic tank or sewage holding tank is available, and provided an adequate water supply and adequate water pressure are available. Recirculating self-contained units do not require drains or water services. Provision must be made for service vehicles and safe access to all units requiring regular maintenance.

14.3.3 Single self-contained units are versatile and easily relocated during events but are limited to a maximum number of uses before requiring service.

14.3.4 Wherever field toilets are used, provision for safe and hygienic waste removal must be arranged with holding tank facilities, if required.

14.4 Quantity

14.4.1 Recommendations as to the minimum scale of toilet provision for buildings of public entertainment are available online and from the various vendors who specialize in this service.

14.4.2 In all circumstances, the sanitary accommodation will depend on the nature of the event, the audience profile and the type of venue. To calculate sanitary provision requires knowing the audience size and then estimating the anticipated male to female ratio. When there is insufficient information to assess this ratio, a split of male to female (1:1) should be assumed.

14.4.3 Consider the following when determining the minimum number of units:
- The duration of the event;
- Perceived audience food and fluid consumption and whether alcohol is to be sold;
- Adequate provision during intervals and breaks in performance;
- Requirements for event-related temporary campsites;
- Provision of suitable facilities for children, elderly or infirm people attending who may take longer to use a facility;
- Facilities inside a fenced venue at a 'no-readmission' event; and
- Weather conditions and temperature.

14.4.4 The experience of a competent consultant or responsible contractor could prove invaluable when determining numbers of sanitary conveniences.

14.4.5 Table 14-1 (below) shows a general guideline for a music event, though these figures may be too high for short duration/'non-peak' period events such as fairs and parties, or too low for events with high levels of fluid consumption or where extended use or camping will occur.

14.4.6 Special event organizers should refer to Table 14-1 to determine the approximate number of portable restrooms needed at a special event. Factors that may skew this chart:
- Does not allow for excessive consumption of beer.
- Does not consider existing permanent venue toilet facilities.
- The space available. If space is tight, increasing the frequency of service might be an alternative which will allow your event to use fewer units.

Table 14-1.

Number of Portable Restrooms Required for Special Events.

		NUMBER OF HOURS FOR EVENT PER DAY (ASSUMES SERVICING PER DAY)									
		1	2	3	4	5	6	7	8	9	10
	250	2	2	2	2	2	3	3	3	3	3
N	500	2	3	4	4	4	4	4	4	4	4
U	1000	4	5	6	7	7	8	8	8	8	8
M	2000	6	10	12	13	14	14	14	15	15	15
B	3000	9	14	17	19	20	21	21	21	21	22
E	4000	12	19	23	25	28	28	28	30	30	30
R	5000	15	23	30	32	34	36	36	36	36	36
O	6000	17	28	34	38	40	42	42	42	44	44
F	7000	20	32	40	44	46	48	50	50	50	50
P	8000	23	38	46	50	54	57	57	58	58	58
E	9000	26	42	52	56	60	62	62	62	64	64
O	10000	30	46	57	63	66	70	70	72	72	72
P	12500	36	58	72	80	84	88	88	88	88	92
L	15000	44	70	84	96	100	105	105	110	110	110
E	17500	50	80	100	110	115	120	125	125	126	126
PER	20000	57	92	115	125	132	138	138	144	144	150
DAY	25000	72	115	144	154	168	175	175	176	176	184
	30000	88	138	168	192	200	208	208	216	216	216

14.5 Washing Facilities

14.5.1 An experienced vendor will guide you best when it comes to all sanitation needs. In the event you do not have access to an experienced and knowledgeable vendor or consultant, the following is a good rule of thumb. Where possible, provide hand-washing facilities in the ratio of one per five toilets with no less than one hand-washing facility per 10 toilets provided. Provide suitable hand-drying facilities. If paper towels are supplied, arrange for regular waste disposal and restocking.

14.5.2 Where warm water hand-washing facilities are available, provide adequate supplies of suitable soap. Hand sanitizer, antiseptic hand wipes or antibacterial soap should be provided where warm water is not available.

14.5.3 On sites where hand-washing facilities are supplied in the open air, consider the management of the facility to ensure that the surrounding ground does not become waterlogged leading to localized pooling or flooding.

14.6 Long Duration Events

14.6.1 Hand-washing facilities alone may not provide adequate provision for events longer than one day, or when overnight camping is available. In these instances, consider whether it may be appropriate to supply shower facilities on site, subject to the availability of adequate water supply and water pressure.

14.7 Sanitary Provision for People with Special Needs

14.7.1 Provide appropriate sanitary accommodation for wheelchair users and other people with special needs attending the event. The Americans with Disabilities Act will apply regarding sanitary accommodation for people with special needs.

14.7.2 Also consider access to toilets for people with special needs. Supply fixed and stable ramps where appropriate. Position facilities close to any area set aside for people with special needs such as viewing platforms, and ensure they are designed to comply with the ADA.

14.7.3 The provision of facilities should relate to the expected numbers of people with special needs attending the event. It is suggested that one toilet with hand-washing facilities should be provided per 50 people with special needs.

14.7.4 If there is any possibility that tampons or sanitary napkins may block sanitation facilities, supply suitable and clearly identified designated containers and arrange for regular emptying of those containers.

14.7.5 If infants are expected at an event, provide appropriate baby-changing facilities including receptacles for the hygienic disposal of disposable diapers and baby wipes. Provide prominent signage within the baby-changing cubicle to ensure the receptacles are used.

14.7.6 If for any reason sewage needs to be stored on site until offsite disposal facilities are open or available, it is essential that adequate holding tanks are provided on site in a safe and secure location. Seek advice on safe sewage disposal from the appropriate authorities and ensure that a licensed contractor is employed to remove and dispose of that sewage. Arrangements should be documented and agreed with the contractor before the beginning of the event.

14.8 Facilities for Employees and Event Staff

14.8.1 OSHA requires that suitable and sufficient toilets and washing facilities must be provided at workplaces. Guidance should be available from your service provider. When in doubt, use the same formulas as for the general public and consider increasing the number of hand wash stations to accommodate 25% more uses than the public units.

14.8.2 Sanitary accommodation for use by event workers should be located near work areas and, in particular, behind the stage, near the mixer tower, next to the catering areas and parking lots, the first-aid areas, welfare, and children's areas. Specific dedicated toilets with hot and cold hand-washing facilities should be provided for food handlers.

14.9 Contractors Providing or Servicing the Sanitary Facilities

14.9.1 Discuss requirements for the type, numbers, positioning, servicing and maintenance of sanitary facilities with the contractor before the event. It is advisable to provide contractors with a plan of the site, showing the proposed location of the facilities along with a copy of the site safety rules and information concerning any significant risks highlighted in the overall event risk assessment.

14.9.2 Examine contractors' safety policies and risk assessments. Contractors should ensure that their workers are provided with and wearing the correct personal protective equipment. Protective overalls, boots or shoes, gloves and eye protection are needed to ensure that workers are protected from accidental splashes of the disinfecting and odorizing chemicals as well as accidental contamination by sewage.

15. Waste Management

15.0.1 Large quantities of waste materials will be generated by the concessionaires and the audience at most music events. A waste management plan for a music event must account for waste removal both during and after the event, to minimize the risks associated with accumulated waste.

15.1 Types of Waste

15.1.1 Types of waste generated include the following:
- Paper and cardboard packaging;
- Food and drink containers;
- Food waste from attendees and vendors;
- Glass;
- Plastics;
- Metal cans;
- Construction materials (e.g., scrap metal, steel, aluminum, etc.)
- Clothing;
- Human waste products (e.g., vomit, urine and feces, sanitary towels and tampons often placed in miscellaneous containers);
- Medical waste such as needles and bandages;
- Remains of camp fires;
- Fireworks and pyrotechnics;
- Waste water from toilets, showers and hand-washing basins;
- Waste water from food concessions; and
- Needles used by intravenous (IV) drug users.

15.2 Hazards Posed by Waste

15.2.1 Hazards posed by waste include the following:
- Injury to workers during collection and removal of waste from the site. Examples include cuts and grazes, needle stick injuries; back strains due to manual handling difficulties and possible infection;
- Blocking emergency access routes, hampering movement around the site, creating tripping hazards to the audience;
- Smoke and fire hazards when waste is accidentally or purposely ignited;
- The misuse of waste by the audience (e.g., throwing bottles, cans, etc.);
- Vehicle movements associated with the collection of waste materials; and
- Waste which attracts insects and vermin.

15.3 Areas Where Waste is Generated

15.3.1 Waste and the type of waste products will not be generated evenly across the venue or site. The buildup of waste and the need to collect it promptly will vary in different areas over time. A competent waste contractor will need to manage their workers and equipment to ensure that there are suitable and adequate resources directed to the appropriate areas at appropriate times. Each area of the venue or site may need to be managed differently.

15.3.2 Pay special attention to the following areas:
- Access routes to music event (e.g., surrounding streets or land);
- Entrances and exits;
- Arenas and stages;
- Sanitary areas;
- First-aid areas;
- Food service areas; and
- Camping areas.

15.4 Information to be Exchanged with Waste Contractor

15.4.1 Ensure that details are given to the waste contractor concerning audience size, arena size, site boundaries, numbers of campers, food concessions and other relevant factors. The waste contractor cannot accurately plan working methods or employ the correct number of workers without this information. Insufficient information could have serious consequences for the audience and employee health, safety and welfare, and the overall success of the event.

15.5 Methods of Collection

15.5.1 Waste collection from the site or arena usually involves a combination of the following:
- Contractors' workers specifically trained to pick the waste up (litter pickers), and/or empty the receptacles placed around the site or venue;
- The use of sweeper vehicles and vacuum suction vehicles;
- Vacuum tankers for collection of waste water temporarily held in smaller tanks; and
- Other vehicles, trailers and towing vehicles.

15.5.2 Discuss arrangements with the waste contractor before the event so that any special requirements regarding access or height restrictions, storage space for vehicles or accommodation for the litter pickers can be incorporated into your overall event planning.

15.6 Receptacles

15.6.1 Waste receptacles can be positioned around the perimeter of the venue or site, and they can also be positioned within the venue or site or other areas as appropriate. Care must be exercised in choice, size and location of receptacles. Wheeled containers or similar receptacles appear to be the most versatile at present as they can be obtained in a variety of sizes, are equipped with lids and are easily positioned as required. Also consider providing tamper-proof bins for sharp objects.

15.6.2 Steel drums are difficult to maneuver and empty when full so assess their use.

15.6.3 Large onsite compactors can reduce the bulk of the refuse. Compactors requiring a power source should only be operated by trained personnel. Compactors, dumpsters and other front-end-loader containers should be separated from the audience for reasons of safety, access for loading and to prevent the audience placing non-compacting, hazardous or other inappropriate waste into these units.

15.3.4 The collection company must be a licensed waste hauler that dumps in a professionally managed landfill that complies with federal regulations. The collection company's trucks must be well maintained to prevent leakage of the waste materials being hauled.

15.7 Times of Collection

15.7.1 Discuss with the waste contractor the strategy for waste collection for the duration of the event, including pre- and post-event collections. Different collection methods may be planned for each of these phases.

15.8 Methods of Removal

15.8.1 Discuss with the waste contractor the methods of waste removal from the venue or site. There may be areas that are subject to a ban on vehicle movements during the event to protect the audience. The sites chosen for the bulk collection must have a suitable access route capable of taking the weight of heavy collection vehicles. Consider local noise ordinances (and common courtesy) when scheduling late night and early morning trash hauls.

15.9 Health, Safety and Welfare of Employees and Event Workers

15.9.1 Waste contractors have a legal duty to ensure that the health, safety and welfare of their employees are protected on site. OSHA requires the use of personal protective equipment (PPE) to reduce employee exposure to hazards when engineering and administrative controls are not feasible or effective in reducing these exposures to acceptable levels. Employers are required to determine if PPE should be used to protect their workers. If PPE is to be used, a PPE program should be implemented. This program should address the hazards present; the selection, maintenance, and use of PPE; the training of employees; and monitoring of the program to ensure its ongoing effectiveness. See Chapter 16, *Personal Protective Equipment*, for more information.

15.9.2 Examples of suitable clothing and personal protective equipment for event personnel who are handling waste include:
- Protective boots or shoes with metal toe caps;
- Trousers and jackets;
- Waterproof suits, as appropriate;
- Fluorescent vests;
- Hard hats;
- Goggles; and

- Different types of gloves for different tasks.

15.9.3 Ensure that hand washing stations are available throughout the waste collection process. Those handling waste need access to hot and cold running water, soap and nail brushes to wash their hands and bodies if they become contaminated. Toilets and washing facilities must be available, particularly at the final waste collection process and in some circumstances showers will be necessary.

15.9.4 Brief workers before beginning work to explain site hazards and risks, hours of work and meal breaks and estimated completion time.

15.10 Recycling

15.10.1 Two commonly used methods of recycling during public assembly events are single stream and multiple stream recycling.

15.10.2 Single stream (also known as "fully commingled" or "single-sort") recycling systems refer to a system in which all paper fibers, glass, plastics, and metals are commingled in one container at the collection point. These recyclable materials are hauled to a recycling facility for sorting. A benefit of single stream recycling is that only two types of containers have to be deployed: trash and recycling. Attendees are likely to sort waste at this level.

15.10.3 Multiple stream recycling systems require separate containers for the different recyclable materials: paper, glass, plastics, steel, aluminum, etc. When recyclable materials are sorted at the site, many haulers will pay for the materials or waive hauling fees. However, more containers are required for multiple stream recycling and event attendees are less likely to sort their waste materials at this level.

15.10.4 The effectiveness of the segregation systems for recycling will depend upon the cooperation of the event attendees, adequate supervision, suitable clear labeling and the location of the containers.

15.11 Planning Guide

15.12.4 A planning guide titled *Developing Trash-Free Special Events* was put together by the Eno River Association in North Carolina in 2000, in conjunction with the North Carolina Division of Pollution Prevention and Environmental Assistance, Department of Environmental and Natural Resources. This thorough guide discusses recycling programs, waste producers, placement, etc. and can be found online by doing an Internet search for "Developing Trash-Free Special Events."

16. Personal Protective Equipment (PPE)

16.0.1 Personal protective equipment, commonly referred to as "PPE," is equipment worn to minimize exposure to a variety of hazards.

16.0.2 Hazards exist in most workplaces, including sites and venues of live entertainment events. Sharp edges, falling objects, flying sparks, noise and a myriad of other potentially dangerous situations present risks to workers at event sites and venues. Commonly used PPE for the concert and live event industry includes gloves, foot and eye protection, protective hearing devices and hard hats.

16.0.3 The U.S. Occupational Safety and Health Administration (OSHA)(http://www.osha.gov/) requires that employers protect their employees from workplace hazards that can cause injury. The information in this chapter is general in nature and does not address all workplace hazards or PPE requirements.

16.0.4 The information, methods and procedures in this chapter are based on the OSHA requirements for PPE as set forth in Title 29, *Labor*, of the U.S. Code of Federal Regulations (a.k.a., 29 CFR) at Part:
- 1910.132 (General requirements);
- 1910.133 (Eye and face protection);
- 1910.135 (Head protection);
- 1910.136 (Foot protection);
- 1910.137 (Electrical protective equipment); and
- 1910.138 (Hand protection).

16.0.5 The information, methods and procedures in this chapter are also based on the OSHA requirements that cover the construction (Part 1926) industries as set forth in 29 CFR at Part:
- 1926.95 (Criteria for personal protective equipment);
- 1926.96 (Occupational foot protection);
- 1926.100 (Head protection);
- 1926.101 (Hearing protection); and
- 1926.102 (Eye and face protection).

16.0.6 This guide does not address PPE requirements related to respiratory protection (29 CFR 1910.134) as this information is covered in detail in OSHA Publication 3079, *Respiratory Protection.*

16.1 The Hazard/Risk Assessment

16.1.1 A first critical step in developing a comprehensive safety and health program is to identify physical and health hazards at the work site. This process is known as a "hazard/risk assessment." Potential hazards may be physical or health-related and a comprehensive assessment should identify hazards in both categories. Examples of physical hazards include moving objects, fluctuating temperatures, high intensity lighting, rolling or pinching objects, electrical connections and sharp edges. Examples of health hazards include overexposure to harmful dusts, chemicals or radiation.

16.1.2 The hazard/risk assessment should begin with a walk-through survey of the facility to develop a list of potential hazards in the following basic hazard categories:
- Impact;
- Penetration;
- Compression (roll-over);
- Chemical;
- Heat/cold;
- Harmful dust;
- Light (optical) radiation; and
- Biologic.

16.1.3 In addition to noting the basic layout of the work site and reviewing any history of occupational illnesses or injuries, things to look for during the walk-through survey include:
- Sources of electricity
- Sources of motion such as machines or processes where movement may exist that could result in an impact between personnel and equipment
- Sources of high temperatures that could result in burns, eye injuries or fire
- Types of chemicals used in the workplace
- Sources of harmful dusts
- Sources of light radiation, such as welding, brazing, cutting, furnaces, heat treating, high intensity lights, etc.
- The potential for falling or dropping objects
- Sharp objects that could poke, cut, stab or puncture
- Biologic hazards such as blood or other potentially infected material.

16.1.4 When the walk-through is complete, the employer should organize and analyze the data so that they may be efficiently used to determine the proper types of PPE needed at the work site. The employer should become aware of the different types of PPE available and the levels of protection offered. It is recommended that employers select PPE that will provide a level of protection greater than the minimum required to protect employees from hazards.

16.1.5 Documentation of the hazard/risk assessment is required through a written certification that includes the following information:
- Identification of the workplace evaluated;
- Name of the person conducting the assessment;

- Date of the assessment; and
- Identification of the document certifying completion of the hazard assessment.

16.2 Training Employees in the Proper Use of PPE

16.2.1 Employers are required to provide the following information to each employee who must use PPE:
- When PPE is necessary;
- What PPE is necessary;
- How to properly put on, take off, adjust and wear the PPE;
- The limitations of the PPE; and
- Proper care, maintenance, useful life and disposal of PPE.

16.2.2 The employer must document the training of each employee required to wear or use PPE by preparing a certification containing the name of each employee trained, the date of training and a clear identification of the subject of the certification.

16.3 Eye and Face Protection

16.3.1 OSHA requires employers to ensure that employees have appropriate eye or face protection if they are exposed to eye or face hazards from flying particles, molten metal, liquid chemicals, acids or caustic liquids, chemical gases or vapors, potentially infected material or potentially harmful light radiation.

16.3.2 Everyday use of prescription corrective lenses will not provide adequate protection against most occupational eye and face hazards, so employers must make sure that employees with corrective lenses either wear eye protection that incorporates the prescription into the design or wear additional eye protection over their prescription lenses. It is important to ensure that the protective eyewear does not disturb the proper positioning of the prescription lenses so that the employee's vision will not be inhibited or limited. Also, employees who wear contact lenses must wear eye or face PPE when working in hazardous conditions.

16.4 Eye Protection for Exposed Workers

16.4.1 OSHA suggests that eye protection be routinely considered for use by carpenters, electricians, pipefitters, sheet metal workers and tinsmiths, grinding machine operators, welders, and laborers.

16.4.2 Examples of potential eye or face injuries include:
- Dust, dirt, metal or wood chips entering the eye from activities such as chipping, grinding, sawing, hammering, the use of power tools or even strong wind forces.
- Chemical splashes from corrosive substances, hot liquids, solvents or other hazardous solutions.
- Objects swinging into the eye or face, such as tree limbs, chains, tools or ropes.
- Radiant energy from welding, harmful rays from the use of lasers or other radiant light (as well as heat, glare, sparks, splash and flying particles).

16.4.3 Some of the most common types of eye and face protection include:

16.4.3.1 Safety spectacles. These protective eyeglasses have safety frames constructed of metal or plastic and impact-resistant lenses. Side shields are available on some models.

16.4.3.2 Goggles. These are tight-fitting eye protection that completely cover the eyes, eye sockets and the facial area immediately surrounding the eyes and provide protection from impact, dust and splashes. Some goggles will fit over corrective lenses.

16.4.3.3 Welding shields. Constructed of vulcanized fiber or fiberglass and fitted with a filtered lens, welding shields protect eyes from burns caused by infrared or intense radiant light; they also protect both the eyes and face from flying sparks, metal spatter and slag chips produced during welding, brazing, soldering and cutting operations. OSHA requires filter lenses to have a shade number appropriate to protect against the specific hazards of the work being performed in order to protect against harmful light radiation.

16.4.3.4 Laser safety goggles. These specialty goggles protect against intense concentrations of light produced by lasers. The type of laser safety goggles an employer chooses will depend upon the equipment and operating conditions in the workplace.

16.4.3.5 Face shields. These transparent sheets of plastic extend from the eyebrows to below the chin and across the entire width of the employee's head. Some are polarized for glare protection. Face shields protect against nuisance dusts and potential splashes or sprays of hazardous liquids but will not provide adequate protection against impact hazards. Face shields used in combination with goggles or safety spectacles will provide additional protection against impact hazards.

16.4.4 An employer may choose to provide one pair of protective eyewear for each position rather than individual eyewear for each employee. If this is done, the employer must make sure that employees disinfect shared protective eyewear after each use.

16.4.5 Protective eyewear with corrective lenses may only be used by the employee for whom the corrective prescription was issued and may not be shared among employees.

16.4.6 Each type of protective eyewear is designed to protect against specific hazards. Employers can identify the specific workplace hazards that threaten employees' eyes and faces by completing a hazard assessment as outlined in the earlier section.

16.5 Head Protection

16.5.1 Protecting employees from potential head injuries is a key element of any safety program. A head injury can impair an employee for life or it can be fatal. Wearing a safety helmet or hard hat is one of the easiest ways to protect an employee's head from impact and penetration hazards as well as from electrical shock and burn hazards.

16.5.2 Hard hats must have a hard outer shell and a shock-absorbing lining that incorporates a headband and straps that suspend the shell from 1 to 1-1/4 inches (2.54 cm to 3.18 cm) away from the head. This type of design provides shock absorption during an impact and ventilation during normal wear. Optional brims may provide additional protection from the sun and some hats have channels that guide rainwater away from the face.

16.5.3 Construction workers, carpenters, scaffold erectors, stagehands, ground riggers, electricians, welders, among others, should be required to wear head protection on an event site.

16.5.4 Types of Hard Hats

16.5.4.1 There are many types of hard hats available in the marketplace today. In addition to selecting protective headgear that meets ANSI Z 89.1 (2009 is the latest) standard requirements, employers should ensure that employees wear hard hats that provide appropriate protection against potential workplace-specific hazards. It is important for employers to understand all potential hazards when making this selection, including electrical hazards. This can be done through a comprehensive hazard analysis and an awareness of the different types of protective headgear available.

16.5.4.2 Hard hats (helmets) are divided into three industrial classes:
- Class G (General, formerly Class A) hard hats provide impact and penetration resistance along with limited voltage protection (up to 2,200 volts)
- Class E (Electrical, formerly Class B) hard hats provide the highest level of protection against electrical hazards, with high-voltage shock and burn protection (up to 20,000 volts). They also provide protection from impact and penetration hazards by flying/falling objects.
- Class C (Conductive) hard hats provide lightweight comfort and impact protection but offer no protection from electrical hazards.

They are also broken into two "types:"

- Type I helmets offer protection from blows to the top of the head; and
- Type II helmets offer protection from blows to both the top and sides of the head.

16.5.4.3 Another class of protective headgear on the market is called a "bump hat," which is designed for use in areas with low head clearance. They are recommended for areas where protection is needed from head bumps and lacerations. These are not designed to protect against falling or flying objects and are not ANSI approved. It is essential to check the type of hard hat employees are using to ensure that the equipment provides appropriate protection. Each hat should bear a label inside the shell that lists the manufacturer, the ANSI designation and the class of the hat.

16.5.4.4 A daily inspection of the hard hat shell, suspension system and other accessories for holes, cracks, tears or other damage that might compromise the protective value of the hat is essential. Paints, paint thinners and some cleaning agents can weaken the shells of hard hats and may eliminate electrical resistance. Consult the helmet manufacturer for information on the effects of paint and cleaning materials on their hard hats. Never drill holes, paint or apply labels to protective headgear as this may reduce the integrity of the protection. Do not store protective headgear in direct sunlight, such as on the rear window shelf of a car, since sunlight and extreme heat can damage them.

16.5.4.5 Hard hats with any of the following defects should be removed from service and replaced:
- Perforation, cracking, or deformity of the brim or shell; or
- Indication of exposure of the brim or shell to heat, chemicals or ultraviolet light and other radiation (in addition to a loss of surface gloss, such signs include chalking or flaking).

16.5.4.6 Always replace a hard hat if it sustains an impact, even if damage is not noticeable. Suspension systems are offered as replacement parts and should be replaced when damaged or when excessive wear is noticed. It is not necessary to replace the entire hard hat when deterioration or tears of the suspension systems are noticed.

16.6 Foot and Leg Protection

16.6.1 Employees who face possible foot or leg injuries from falling or rolling objects, or from crushing or penetrating materials, should wear protective footwear.

16.6.2 If an employee's feet may be exposed to electrical hazards, non-conductive footwear should be worn. On the other hand, workplace exposure to static electricity may necessitate the use of conductive footwear.

16.6.3 Examples of situations in which an employee should wear foot and/or leg protection include:
- When heavy objects such as barrels or tools might roll onto or fall on the employee's feet;
- Working with sharp objects such as nails or spikes that could pierce the soles or uppers of ordinary shoes;
- Exposure to molten metal that might splash on feet or legs;
- Working on or around hot, wet or slippery surfaces; and
- Working when electrical hazards are present.

16.6.4 All ANSI-approved footwear have a protective toe and offer impact and compression protection, but the type and amount of protection is not always the same. Check the product's labeling or consult the manufacturer to make sure the footwear will protect the user from the hazards they face.

16.6.5 Foot and leg protection choices include the following:
- Leggings protect the lower legs and feet from heat hazards such as molten metal or sparks from sawing or welding. Safety snaps allow leggings to be removed quickly.
- Metatarsal guards protect the instep area of the foot from impact and compression. Made of aluminum, steel, fiber or plastic, these guards may be strapped to the outside of shoes.
- Toe guards fit over the toes of regular shoes to protect the toes from impact and compression hazards. They may be made of steel, aluminum or plastic.

- Combination foot and shin guards protect the lower legs and feet, and may be used in combination with toe guards when greater protection is needed.
- Safety shoes have impact-resistant toes and heat-resistant soles that protect the feet against hot work surfaces common in scaffold erection, and work on asphalt.

16.6.6 The metal insoles of some safety shoes protect against puncture wounds. Safety shoes may also be designed to be electrically conductive to prevent the buildup of static electricity in areas with the potential for explosive atmospheres or nonconductive to protect workers from workplace electrical hazards.

16.7 Special Purpose Shoes

16.7.1 Electrically conductive shoes provide protection against the buildup of static electricity. Event site workers in explosive and hazardous locations such as and pyrotechnics assembly areas must wear conductive shoes to reduce the risk of static electricity buildup on the body that could produce a spark and cause an explosion or fire. Foot powder should not be used in conjunction with protective conductive footwear because it provides insulation, reducing the conductive ability of the shoes. Silk, wool and nylon socks can produce static electricity and should not be worn with conductive footwear. Conductive shoes must be removed when the task requiring their use is completed.

16.7.2 Note: Employees exposed to electrical hazards must never wear conductive shoes.

16.7.3 Electrical hazard, safety-toe shoes are nonconductive and will prevent the wearers' feet from completing an electrical circuit to the ground. These shoes can protect against open circuits of up to 600 volts in dry conditions and should be used in conjunction with other insulating equipment and additional precautions to reduce the risk of a worker becoming a path for hazardous electrical energy. The insulating protection of electrical hazard, safety-toe shoes may be compromised if the shoes become wet, the soles are worn through, metal particles become embedded in the sole or heel, or workers touch conductive, grounded items.

16.7.4 Note: Nonconductive footwear must not be used in explosive or hazardous locations.

16.8 Care of Protective Footwear

16.8.1 As with all protective equipment, safety footwear should be inspected prior to each use. Shoes and leggings should be checked for wear and tear at reasonable intervals. This includes looking for cracks or holes, separation of materials, broken buckles or laces. The soles of shoes should be checked for pieces of metal or other embedded items that could present electrical or tripping hazards. Employees should follow the manufacturers' recommendations for cleaning and maintenance of protective footwear.

16.9 Hand and Arm Protection

16.9.1 If a workplace hazard assessment reveals that employees face potential injury to hands and arms that cannot be eliminated through engineering and work practice controls, employers must ensure that employees wear appropriate protection. Common potential hazards on event work sites include bruises, fractures, abrasions, cuts, punctures, and electrical or thermal burns.

16.9.2 Protective equipment includes gloves, finger guards and arm coverings or elbow-length gloves.

16.9.3 Employers should explore all possible engineering and work practice controls to eliminate hazards and use PPE to provide additional protection against hazards that cannot be completely eliminated through other means. For example, machine guards may eliminate a hazard. Installing a barrier to prevent workers from placing their hands at the point of contact between a table saw blade and the item being cut is another method.

16.10 Types of Protective Gloves

16.10.1 There are many types of gloves available today to protect against a wide variety of hazards. The nature of the hazard and the operation involved will affect the selection of gloves. The variety of potential occupational hand injuries makes selecting the right pair of gloves challenging. It is essential that employees use gloves specifically designed for the hazards and tasks found in their workplace because gloves designed for one function may not protect against a different function even though they may appear to be an appropriate protective device.

16.10.2 The following are examples of some factors that may influence the selection of protective gloves for a workplace:
- Area requiring protection (hand only, forearm, arm);
- Grip requirements (dry, wet, oily);
- Thermal protection;
- Size and comfort; and
- Abrasion/resistance requirements.

16.10.3 Gloves made from a wide variety of materials are designed for many types of workplace hazards. In general, gloves fall into four general groups: (1) leather, canvas or metal mesh gloves, (2) fabric and coated fabric gloves, (3) chemical and liquid-resistant gloves, and (4) insulating rubber gloves. For detailed requirements on the selection, use and care of insulating rubber gloves, see 29 CFR 1910.137. The other three groups will be described briefly here.

16.10.4 Leather, Canvas or Metal Mesh Gloves
16.10.4.1 Sturdy gloves made from metal mesh, leather or canvas protect against cuts and burns. Leather or canvass gloves also protect against sustained heat.
- Leather gloves protect against sparks, moderate heat, blows, chips and rough objects.
- Aluminized gloves provide reflective and insulating protection against heat and require an insert made of synthetic materials to protect against heat and cold.
- Aramid fiber gloves protect against heat and cold, are cut- and abrasive-resistant and wear well.
- Synthetic gloves of various materials offer protection against heat and cold, are cut- and abrasive-resistant and may withstand some diluted acids. These materials do not stand up against alkalis and solvents.

16.10.5 Fabric and Coated Fabric Gloves

16.10.5.1 Fabric and coated fabric gloves are made of cotton or other fabric to provide varying degrees of protection.

16.10.5.2 Fabric gloves protect against dirt, slivers, chafing and abrasions. They do not provide sufficient protection for use with rough, sharp or heavy materials. Adding a plastic coating will strengthen some fabric gloves.

16.10.5.3 Coated fabric gloves are normally made from cotton flannel with napping on one side. By coating the un-napped side with plastic, fabric gloves are transformed into general-purpose hand protection offering slip-resistant qualities. These gloves are used for tasks ranging from handling bricks and wire to chemical laboratory containers.

16.10.5.4 When selecting gloves to protect against chemical exposure hazards, always check with the manufacturer or review the manufacturer's product literature to determine the gloves' effectiveness against specific workplace chemicals and conditions.

16.10.6 Chemical- and Liquid-Resistant Gloves

16.10.6.1 Chemical-resistant gloves are made with different kinds of rubber: natural, butyl, neoprene, nitrile and fluorocarbon (viton); or various kinds of plastic: polyvinyl chloride (PVC), polyvinyl alcohol and polyethylene. These materials can be blended or laminated for better performance. As a general rule, the thicker the glove material, the greater the chemical resistance but thick gloves may impair grip and dexterity, having a negative impact on safety.

16.10.6.2 Some examples of chemical-resistant gloves include:
- Butyl gloves are made of a synthetic rubber and protect against a wide variety of chemicals, such as peroxide, rocket fuels, highly corrosive acids (nitric acid, sulfuric acid, hydrofluoric acid and red-fuming nitric acid), strong bases, alcohols, aldehydes, ketones, esters and nitro compounds. Butyl gloves also resist oxidation, ozone corrosion and abrasion, and remain flexible at low temperatures. Butyl rubber does not perform well with aliphatic and aromatic hydrocarbons and halogenated solvents.
- Natural (latex) rubber gloves are comfortable to wear, which makes them a popular general purpose glove. They feature outstanding tensile strength, elasticity and temperature resistance. In addition to resisting abrasions caused by grinding and polishing, these gloves protect workers' hands from most water solutions of acids, alkalis, salts and ketones. Latex gloves have caused allergic reactions in some individuals and may not be appropriate for all employees. Hypoallergenic gloves, glove liners and powderless gloves are possible alternatives for workers who are allergic to latex gloves.
- Neoprene gloves are made of synthetic rubber and offer good pliability, finger dexterity, high density and tear resistance. They protect against hydraulic fluids, gasoline, alcohols, organic acids and alkalis. They generally have chemical and wear resistance properties superior to those made of natural rubber.
- Nitrile gloves are made of a copolymer and provide protection from chlorinated solvents such as trichloroethylene and perchloroethylene. Although intended for jobs requiring dexterity and sensitivity, nitrile gloves stand up to heavy use even after prolonged exposure to substances that cause other gloves to deteriorate. They offer protection when

working with oils, greases, acids, caustics and alcohols but are generally not recommended for use with strong oxidizing agents, aromatic solvents, ketones and acetates.

16.11 Hearing Protection

16.11.1 Determining the need to provide hearing protection for employees can be challenging. Employee exposure to excessive noise depends upon a number of factors, including:
- The loudness of the noise as measured in decibels (dB);
- The duration of each employee's exposure to the noise;
- Whether employees move between work areas with different noise levels; and
- Whether noise is generated from one or multiple sources.

16.11.2 Generally, the louder the noise, the shorter the exposure time before hearing protection is required. For instance, employees may be exposed to a noise level of 90 dB for 8 hours per day (unless they experience a Standard Threshold Shift) before hearing protection is required. On the other hand, if the noise level reaches 115 dB hearing protection is required if the anticipated exposure exceeds 15 minutes.

16.11.3 Table 22-1 in Chapter 22, *Sound: Noise and Vibration*, shows the permissible noise exposures that require hearing protection for employees exposed to occupational noise at specific decibel levels for specific time periods. Noises are considered continuous if the interval between occurrences of the maximum noise level is one second or less. Noises not meeting this definition are considered impact or impulse noises (loud momentary explosions of sound) and exposures to this type of noise must not exceed 140 dB. Examples of situations or tools that may result in impact or impulse noises are powder-actuated nail guns, a punch press or drop hammers.

16.11.4 If engineering and work practice controls do not lower employee exposure to workplace noise to acceptable levels, employees must wear appropriate hearing protection. It is important to understand that hearing protectors reduce only the amount of noise that gets through to the ears. The amount of this reduction is referred to as attenuation, which differs according to the type of hearing protection used and how well it fits. Hearing protectors worn by employees must reduce an employee's noise exposure to within the acceptable limits noted in Table 22-1. Refer to Appendix B of 29 CFR 1910.95, *Occupational Noise Exposure*, for detailed information on methods to estimate the attenuation effectiveness of hearing protectors based on the device's noise reduction rating (NRR).

16.11.5 Manufacturers of hearing protection devices must display the device's NRR on the product packaging. If employees are exposed to occupational noise at or above 85 dB averaged over an eight hour period, the employer is required to institute a hearing conservation program that includes regular testing of employees' hearing by qualified professionals. Refer to 29 CFR 1910.95(c) for a description of the requirements for a hearing conservation program.

16.11.6 Some types of hearing protection include:
- Single-use earplugs are made of waxed cotton, foam, silicone rubber or fiberglass wool. They are self-forming and, when properly inserted, they work as well as most molded earplugs.
- Pre-formed or molded earplugs must be individually fitted by a professional and can be disposable or reusable. Reusable plugs should be cleaned after each use.
- Earmuffs require a perfect seal around the ear. Glasses, facial hair, long hair or facial movements such as chewing may reduce the protective value of earmuffs.

16.11.7 Refer to OSHA Publication 3074, *Hearing Conservation*, for more detailed information on the requirements to protect employees' hearing in the workplace.

17. Electrical Installations and Lighting

17.0.1 Electricity can cause death or serious injury to performers, workers or members of the public if the installation is faulty or not properly planned. This chapter gives some general guidance limited to installations of 600 volts or less. Very large loads and high-voltage systems are not covered as they require special considerations. In many circumstances the electrical supply may be of a temporary or portable nature, but this does not mean that it can be sub-standard or of an inferior quality to a permanent installation. Only qualified personnel should plan and carry out electrical work as well as operate electrical equipment.

17.0.2 All electrical installations and equipment must comply with the general requirements of local, state, provincial, and national regulations. For the United States, this would include NFPA 70, National Electrical Code (NEC), and NFPA 101, Life Safety Code. The Occupational Safety and Health Administration (OSHA) 29 CFR 1910 Subpart S, *Electrical*, addresses electrical safety requirements that are necessary for the practical safeguarding of "employees" (including unpaid individuals) in their workplaces. Another important OSHA regulation, 29 CFR 1910.147, *The Control of Hazardous Energy (Lockout/Tagout)*, sets forth regulations for the servicing and maintenance of machines and equipment in which the unexpected energizing or startup of the machines or equipment, or release of stored energy could harm employees. It is the responsibility of the organizer to understand and implement the relevant local requirements.

17.0.3 This chapter provides a general overview of some of the matters to be considered by an event organizer when planning electrical installations within the overall venue design. Technicians working on site should refer to the specific guidance mentioned above, the guidance of qualified personnel and should not rely only on the general overview offered herein.

17.1 Planning

17.1.1 Factors to consider when planning the electrical installation include:
- The location of any existing overhead power lines or buried cables and other buried utilities;
- Positioning of temporary overhead or underground cables;
- The total power requirements for the site;
- The use and location of generators (separately derived systems) and/or transformers;
- The main service disconnects controlling the electrical supplies to the stage lighting, sound, special effects, lifting equipment, concessions, and wardrobe;
- Grounding and bonding;
- The location of the stages;
- The location of mixer positions, delay towers, video towers, etc.;

- Special power supplies for some equipment, e.g., equipment from other countries which operates on voltage and frequency other than 120/208V, 60 Hz, e.g., European equipment with power requirements of 230/398 volts, 50Hz;
- Power supplies required for hoists, portable tools, etc.;
- The electrical requirements for emergency power, emergency lighting and exit signs;
- High amperage areas such as cooking areas, coffee pots, clothes dryers, hair dryers, etc.;
- Power supplies for catering equipment, first-aid points, incident control room, CCTV cameras, etc.; and,
- Power supplies for heating or air conditioning.

17.2 Installation

17.2.1 The main electrical supply source(s) and disconnects should be located where it is accessible for normal operations and emergencies, but segregated from public areas of the venue. Display danger warning signs around the enclosure(s) of the source(s).

17.2.2 All electrical equipment which could be exposed to the weather and is not rated as weatherproof, e.g., consumer electronic units, distribution boards, etc., should be protected by means of suitable and sufficient covers, enclosures or shelters. As far as is practical, all electrical equipment should be located so that it cannot be touched by members of the public or unauthorized workers. Refer to ANSI E1.19-2009, *Recommended Practice for the Use of Class A Ground-Fault Circuit Interrupters in the Entertainment Industry*. It is the responsibility of the organizer to understand and implement the relevant local requirements.

17.2.3 The electrical installation should be inspected and tested upon completion according to the applicable regulations. A written record of compliance and testing should be kept by the organizer.

17.3 Cabling

17.3.1 All cables selected for any event must be well maintained and suitable for their purpose and application.

17.3.2 Temporary overhead cables, whether they are carrying electricity, communication, or television signals, etc., must be installed according to code/regulation. Consideration needs to be given to the minimum height above the ground (final grade) for proper clearance of vehicles, structures, personnel and any other obstacle or object that could pose a risk. Advisory notices should be clearly displayed and effectively warn of the location of the overhead cables and the voltage being carried.

17.3.3 Wherever possible, segregate vehicular and pedestrian traffic and cable routes. If this is not possible, cable height must be of minimum clearance as specified in the NEC or other prevailing local code. Fences should be considered to segregate roadways from overhead cables running parallel to the roadway to prevent inadvertent contact.

17.3.4 If it is necessary to run cables underground, reference all applicable regulations. Consideration must be given to guard against:
- Crushing by vehicles;
- Damage by machinery, equipment or tools;
- Other mechanical damage;
- Damage to buried cables, gas lines, water lines or any other buried utilities by the driving of grounding rods, trenching, or other activity.

17.3.5 Cables run on the surface should be protected against sharp edges or crushing by heavy loads by approved cable protection devices such as cable ramps. Ramps should be conspicuously marked to avoid tripping hazards. In the United States, the use of ADA compliant cable ramps must be used in areas where the general public will cross cable runs.

Fig. 17-1 – Commercial, ADA compliant cable protector in use. Photo courtesy of Cross-Guard.

17.4 Electricity Utility Cables

17.4.1 Overhead or underground electrical supply cables belonging to an electrical supply company may cross the site, or its access roads. If so, precautions must be taken to avoid danger from these cables.

17.5 Access to Electrical Equipment

17.5.1 Appropriate minimum clear working space must be established and maintained around electrical equipment to allow access for operation or maintenance. Fire codes also require open space around electrical equipment (e.g., no storage) so that fire fighters can readily access it in an emergency. In the United States, the NEC and OSHA establish these requirements. Other equipment should also be given consideration, including but not limited to:
- Control switches and equipment;
- Amplification equipment;
- Special effects equipment;
- Follow spots;
- Dimmers;
- Ballasts; and,
- High-voltage discharge lighting.

17.5.2 The main control equipment and items specified above should be clearly identified, and their locations marked on a plan located in the command center.

17.5.3 Protect all electrical service, supply and distribution equipment to prevent access by unauthorized people. Where this equipment is installed in a locked enclosure, specific key holders should be given responsibility for operating the equipment to comply safely with any request made by the emergency services. Note: multiple people should be given keys and authorization to operate the equipment.

17.6 Generators

17.6.1 If generators are to be used, consider their location and accessibility for refueling. Allow for fuel storage and accessibility for the purposes of refueling. The generator and its fuel source must be protected from unauthorized access (including members of the public) and may require fencing.

17.6.2 If generators are located close to occupied spaces, consider noise nuisance and the risks from exhaust. Generators specifically designed for the entertainment industry should be used.

17.7 Electricity to the Stage Area and Effects Lighting

17.7.1 The electrical supply to the stage should be controlled by a switch or switches and installed in a position accessible at all times to authorized people in the stage area.

17.7.2 If there is a structural collapse of the stage, canopy, or an outlying sound / video / lighting tower, then there is the possibility that electrical service cables can be stretched and/or sheared. This can potentially cause the structure to come in contact with live electrical wires which could energize the structure. In the event that electrical service cables are stretched or sheared and could energize equipment or an occupied area, there should be a protocol in place that has numerous personnel trained in the shut-down / disconnect procedure for power sources that serve the event structure(s). Multiple persons are needed so that if any of the trained crew is incapacitated or unavailable then there will still be someone qualified to perform the task.

17.7.3 Follow established local codes/regulations for proper grounding and bonding of electrical systems and of electrically conductive material that is likely to become energized if a fault were to occur. In the United States, the NEC establishes this regulation.

17.7.4 Distribution systems must consist of equipment specifically suitable for their purpose. In the United States, all distribution system equipment must be listed with a Nationally Recognized Testing Laboratory for the purpose for which it is being used. It is the responsibility of the organizer to understand and implement the relevant local requirements.

17.7.5 Use of Ground Fault Circuit Interrupters. Reference ANSI E1.19-2009, *Recommended Practice for the Use of Class A Ground Fault Circuit Interrupters*, gives guidance on the use of GFCIs in the entertainment industry.

17.8 Emergency Lighting Systems

17.8.1 In the United States, Article 700 of the NEC (NFPA 70, 2011) establishes requirements for emergency electrical systems including lighting. Emergency lighting systems are legally required, are classified as emergency systems and should be marked as such. It is the responsibility of the organizer to understand and implement the relevant local requirements.

17.8.2 In addition to the normal lighting arrangement, emergency lighting should be provided as determined by the risk assessment and fire-risk assessment. These assessments should cover all possible hazards associated with the venue, including but not limited to emergency evacuation routes, pits, holes, trenches, ditches, etc. Also consider the provision of emergency lighting within generator enclosures and the main electrical service(s).

17.8.3 The emergency lighting power supply should come from a source independent of the normal lighting. The emergency lighting should be of a maintained type (continuously lit), which includes the exit signs located around the venue for directional purposes, and located above the final exit doors. In the United States, emergency lighting should be installed, operated and maintained in accordance with NFPA 101, *Life Safety Code* (2012), section 7.8 and 7.9 and applicable sections of NFPA 70 (NEC). It is the responsibility of the organizer to understand and implement the relevant local requirements.

17.8.4 Any source of supply used for providing emergency lighting should be capable of maintaining the full light load as determined by the event risk assessment and the major incident plans prepared for the event, in case of a mains failure. It is important to keep any battery used for this purpose in a fully charged condition whenever the venue is in use. Ground fault circuit interrupters (GFCIs) must not be used on emergency lighting circuits.

17.8.5 The normal and emergency lighting systems should be installed independently of one another so that a fault or accident arising with one system cannot jeopardize the other. Suitable provision should be made to enable repairs to be undertaken if one or more parts of these lighting systems fail. Both the normal lighting circuits and emergency lighting circuits, including generators, should be protected from acts of vandalism.

17.9 Lighting Levels for Means of Egress

17.9.1 All public areas of the venue should be provided with normal and emergency lighting capable of giving sufficient light for people to leave safely as determined by the risk assessment. Consider providing additional lighting, operating in a maintained mode, to the gangways passing through temporary seating structures. In the United States, lighting for stairways, gangways/corridors, exit doorways, gates, emergency lighting, etc., should be installed, operated and maintained in accordance with NFPA 101, *Life Safety Code* (2012). It is the responsibility of the organizer to understand and implement the relevant local requirements.

17.10 Portable Electrical Appliances

17.10.1 Portable electrical equipment is defined as equipment fed with listed portable cords or cables intended to be moved from one place to another. Portable equipment should be used by a competent person and should be appropriately inspected and tested prior to use. In the United States, all portable equipment must be listed with a Nationally Recognized Testing Laboratory for the purpose for which it is being used. It is the responsibility of the organizer to understand and implement the relevant local requirements.

18. Rigging

18.0.1 Rigging is the system of devices used to support or manipulate suspended objects. Rigging system equipment may include wires, ropes, winches, counterweights, sandbags, electric chain hoists and other such elements. For events, suspended objects typically include lighting, sound, scenery, video, special effects, banners and soft goods.

18.0.2 This is not a tutorial or list of specifications. It is a selection of recommendations and best practices to be considered when planning or using a rigging system to hang, raise or lower anything for an event, indoors or out. Specialized conditions of permanent installations are not discussed here.

18.0.3 A typical event has at least three legal entities with substantial authority and responsibility for the event's execution: the venue, the producer, and the subcontractors who provide goods and services to the producer in order to execute the event. The relationships between these entities, and the decision-making authority regarding rigging safety, may be expressly set forth by contract, impliedly, or agreed upon by custom, practice, and prior dealings. From a risk management standpoint, it is preferable to have lines of authority and responsibility stated in clear language in contracts executed by each of the affected parties. This last statement is intended to underscore two important points:

18.0.3.1 <u>Authority Must Match Responsibility</u>. Regarding rigging and any other operational aspect, there is no universally established amount of responsibility that any one entity must have. For example, some rigging service providers will use all their own equipment and personnel, providing supervision from load-in through load-out, ensuring all aspects of worker safety in between. Some venues or producers may seek to insert their own supervisors, workers, or equipment. The key is that however much responsibility one entity has for its work and safety, the entity MUST have the corresponding authority to bear that responsibility. The worst possible scenario is to be responsible for conditions at an event that you are powerless to change.

18.0.3.2 <u>Written Contracts Prevent Disputes</u>. The days immediately before an event can be very hectic; this is no time to be making spontaneous, important agreements. Instead, to the extent possible, lines of authority and responsibility should be memorialized in advance in written contracts that the agreeing parties sign and date. Memories often differ after an event, but recorded communications always remember. While a signed contract is always preferred, use whatever means of recording an agreement you can to clearly document your agreements for posterity.

18.0.4 While reading this chapter, consider your legal and functional responsibility with regard to rigging. Your situation may be unique and such responsibilities may contractually fall outside your scope. If this is the case, this chapter should at least help you understand the perspectives of the entities involved who are responsible.

18.1 Individuals Involved with Rigging

18.1.1 Rigging creates life safety hazards that affect almost every person associated with an event. Anything dropped or lowered from above can cause serious injury to someone below. Large forces and complex structures are often involved, even in small events. The possibility of a catastrophic rigging failure must be considered and planned for by all responsible parties.

18.1.1.1 In rigging there can be no compromise in the planning, engineering, equipment, maintenance, or qualifications of rigging supervisors or persons performing the work. No person should engage in the organization, planning and supervision of work at height or equipment for use in such work, unless he or she is competent to do so, or, if being trained, is under the direct supervision of a competent person.

18.1.1.2 Rigging is not found exclusively in stage or performance areas. All structures with hanging signs, banners, delay speakers or other items are potentially hazardous. All event staff, especially supervisors and managers, should be aware of any rigging work in progress, familiar with safe site practices, and consider how he or she can contribute to a safe rigging system.

18.1.2 The persons and roles involved with rigging—and who may be personally responsible and liable for the rigging—include the following:

18.1.2.1 <u>Event Organizer or Producer</u>. As the general contractor for the event, this person or entity is responsible for:
- Ensuring a professional rigging consultant and/or engineer is involved in all phases of the event including a final evaluation of the system prior to commencing operations;
- Procuring reputable and competent rigging equipment and service suppliers;
- Verifying that suppliers and individuals involved with rigging are competent to do so and have qualifications, training, certifications, insurance and experience commensurate with the size and complexity of the event; and
- Ensuring that appropriate method statements, safety policy documents, contracts and insurance are in place before work begins onsite.

18.1.2.2 <u>Engineer</u>. A qualified engineer establishes and approves the capacities and limitations of the entire rigging system and how it applies to the event's structures. Calculations should always take into account and include the operation of any show elements that move or other forces that may cause loads to change during the load in, event or load-out. This includes but is not limited to:
- The locations and weights of all rigging attachment points;
- The application of loads to event structures including static and dynamic (moving) loading;
- The effects of weather related loading (e.g., snow or wind); and
- The effects of any other loading that is applied to the structures (e.g., additional lighting, scenery, scrims, soft goods, score boards);
- Design and specification of detailed lateral system requirements.

18.1.2.3 <u>Production Supervisor, Manager, Technical Director, or Operations Supervisor (titles vary)</u>. This role is generally responsible for the following:

- Overseeing the installation, operation and dismantle of production elements;
- Integration of production elements and the coordination of production vendors and suppliers;
- Communications with the various vendors and suppliers relative to the production;
- Representing the production or the event organizer as described above in production related matters;
- The management of information and documents that apply to the event's rigging in conjunction with the event's production rigger and/or the engineer described above;
- Serving as the event organizer's representative and liaison between the Producer, the Venue, the Subcontractors and the local authorities with regard to the production of the event;
- Obtaining the applicable and approved documents from the engineer and storing those documents in a readily available safe location e.g. the event binder for the duration of the event. These documents should be in hand and verified prior to the commencement of rigging work;
- At the conclusion of the rigging installation, obtaining signed statements from the responsible rigging/installer that the rigging system is safe for use.

18.1.2.3.1 Please note that on occasion the head rigger or rigging vendor may assume some or all of these duties. This does not relieve the organizer of their obligation to verify that the above work has been completed.

18.1.2.4 <u>Production Rigger</u>. The person in this role is generally responsible for the following:

- Participating in the development of the rigging risk assessment, fall protection plan, rigging rescue plan and the overall event emergency response plan;
- Reviewing the event safety policy with all rigging workers immediately prior to starting work;
- The integrity of the event related rigging system and its operation;
- Working with and guiding the subcontractors, rigging team supervisors and rigging service suppliers during the installation, operation and removal of the event related equipment.
- Usage, maintenance and inventory of event related rigging equipment;
- Monitoring the event related rigging installation to ensure the rigging plan is properly executed per the approved documents and drawings;
- Maintaining applicable documentation including relevant engineering information and an accurate as-built rigging plan;
- Verification of the event's actual rigging loads, particularly dynamic loads; Inspection of event related production elements for integrity before the equipment is raised and again once those elements have been locked off and secured;
- Oversight of the operation and working crew for all show elements that move, including any automation;
- Monitoring all rigging equipment operators for competence and reporting incompetent workers for removal from the rigging crew;

- Contacting the design engineer to review any variations and/or discrepancies from the approved engineering documents and drawings; and
- Receiving written approval from the design engineer for any variations and/or discrepancies from the approved engineering documents and drawings.

18.1.2.4.1 A production rigger should meet the definition of a "qualified person" as defined by OSHA, which is "One who, by possession of a recognized degree, certificate, or professional standing, or who by extensive knowledge, training and experience, has successfully demonstrated his ability to solve or resolve problems relating to the subject matter, the work or the project" (29 CFR 1926.32(m)).

18.1.2.5 Riggers. Riggers install and operate the rigging components. Competency of the rigging staff is central to safety. Riggers should be familiar with the applicable regulations, industry standards and practices, have an understanding of the specifics of the venue and supporting structures, either temporary or permanent, and be trained in the proper use of personal protection equipment (PPE)(see Chapter 16, *Personal Protective Equipment*).

18.1.2.5.1 Rigging teams consist of high riggers and ground riggers and may be supplied through a labor contractor or a subcontractor. High riggers install rigging points with the assistance of the ground riggers.

18.1.2.5.2 All members of the rigging team(s) touch and inspect their respective elements of the rigging and should vigilantly look for questionable (i.e., worn or compromised) rigging equipment. They must know and use proper equipment handling procedures and rigging terminology. Describing the full responsibilities of a qualified entertainment rigger is beyond the scope of this document.

18.1.2.5.3 At a minimum, a rigger should meet the definition of a "competent person" as defined by OSHA, which is "One who is capable of identifying existing and predictable hazards in the surroundings or working conditions which are unsanitary, hazardous or dangerous to employees, and who has authorization to take prompt corrective measures to eliminate them" (29 CFR 1926.32(f)).

18.1.2.5.4 In addition to the safe installation, operation and dismantling of the rigging equipment, all riggers are expected to observe and monitor the event's environment for issues in support of the production rigger and his or her duties as described above in 18.1.2.4 and to report those issues to the production rigger.

18.2 Safe Rigging Environment

18.2.1 Everyone associated with the execution of an event is expected to work together to create a safe working environment. As this relates to rigging, factors to consider include:

18.2.1.1 Chain of command. It is essential that a clear chain of command and control is established and understood by everyone on the job site prior to the start of work.

18.2.1.2 <u>Area control</u>. All event workers must be aware at all times of where rigging teams are working overhead and avoid the area directly below them. A ground rigger or another designated person should be tasked with the job of monitoring the zone under where rigging work is occurring and keeping all non-rigging crew personnel out of immediate harm's way. Only competent ground riggers should be in the work zone directly below the high riggers, with all others outside the perimeter of that work zone. It is important that everyone is aware of the rigging work occurring in the area. Ground riggers should keep all persons in the vicinity informed of the rigging process, especially those persons near the work zone under the rigging teams above. If possible, warning signage or cone markers should be utilized to discourage persons from entering the work zone under where rigging is occurring overhead.

18.2.1.3 <u>PPE</u>. All persons in the area below active overhead work should have personal protective equipment (PPE) suitable to the working environment, according to the event safety policy and the local authority (see Chapter 16, *Personal Protection Equipment*).

18.2.1.4 <u>Loud and clear communication</u>. Due to ambient noise typically present during an installation, radios are often used by the rigging crew to communicate. However, due to the nature of rigging work, simple voice commands are often more efficient. Communication should be limited to clearly understood terms.

18.2.1.5 <u>Lifting or moving operations</u>. When lifting, moving or lowering rigging equipment, always make a loud and clear announcement such as "truss moving," "grid moving" or "line set #24 – backdrop coming in" to inform all persons in the area of the intended action. Work, especially noisy operations, such as construction, gas operated blowers, audio checks and loud conversations must be suspended immediately before and during the lifting operation. This is a critical issue because the first sign of trouble will often be heard before being seen, such as a motor problem or a lift chain running from a chain bag. Prior to the move, all persons in the area should move to a safe position.

> ## NEVER stand directly under a
> ## lifting or lowering operation!

18.2.1.5.1 Event workers should closely watch the lifting operation and be ready to call out the word "STOP" if they see a problem. This is the only acceptable term to use in an urgent situation when something may be wrong and the operation must temporarily halt. While preparing to make the lift, it is a good idea to reiterate to those workers monitoring the movement that the universally-recognized word to halt operations is **"STOP"** not "WHOA," "HO," "HEADS UP" or any other term.

18.2.1.5.2 Only designated competent persons should observe and monitor the lifting operation. In doing so, they should pay close attention to all hoist locations, watching for uneven conditions, fouled rigging, stopped motors, or other problems. These persons must have a direct line-of-sight to the objects involved in the operation as well as, for flying objects, the paths the objects travel. One very serious and common problem is the occurrence of a hoist chain not feeding chain into the chain bag during a lift, causing the chain to run free and descend at an increasing rate of speed toward the floor—an episode that inevitably ends with the tail of the chain freefalling to the floor, potentially causing serious injury to people or damage to equipment below.

18.2.1.6 <u>Fall protection</u>. During pre-production, a personnel risk assessment must be conducted in the rigging planning stage. Rigging staff MUST BE TRAINED in proper safety practices prior to coming to the jobsite. On the job under stressful load-in conditions is not the best time or place to train a new rigging crew on any aspect of the job especially fall protection.

18.2.1.6.1 All workers on the job site who meet the criteria for working at height (six feet or more over a lower level) during the course of providing their services must use the appropriate fall protection equipment as required by law, the event safety policy, the venue's safety policy and accepted industry best practices. This includes workers involved with the operation or use of equipment such as articulated boom lifts, man lifts, scissor lifts or similar high reach equipment.

18.2.1.6.2 OSHA, in its definition of "unprotected sides and edges" requires that, "Each employee on a walking/working surface (horizontal and vertical surface) with an unprotected side or edge which is 6 feet (1.8 m) or more above a lower level shall be protected from falling by the use of guardrail systems, safety net systems, or personal fall arrest systems" (29 CFR 1926.501(b)(1)).

18.2.1.6.3 The event organizer or their agent should contact the venue in advance to determine if specialized equipment is necessary to access the venue's rigging or if any specific regulations exist for rigging in that venue. If there is a fall protection system installed in the venue, all riggers must be trained in its proper use prior to the commencement of work.

18.3 Rigging Rescue

18.3.1 Fall protection equipment does not prevent falls; it prevents the worker from falling to his or her death by arresting the fall. After an arrested fall, an injured or unconscious worker may hang in a safety harness for an extended period of time requiring rescue. Time is often critical during a rigging rescue operation. A worker hanging in a harness, especially an injured one, may experience orthostatic hypotension (abnormally low blood pressure that can lead to shock, a life-threatening condition, due to being in a position that does not allow proper blood flow) or suspension trauma (injury from hanging or falling in a harness). Either of these conditions can be fatal.

18.3.2 For all work that is performed at height a rescue plan must be created. This plan outlines the procedures to rescue a worker from an elevated location. The Rigging Rescue Plan must be understood by all pertinent parties, not just the rigging crew and communicated verbally during the preload-in safety meeting. This meeting typically includes venue representatives, production and stage management, touring representatives, local responders, event EMT services, security supervisors and labor supervisors. All workers performing work at heights must review and be knowledgeable of the rescue plan. Rescue plans must include methods of self-rescue when possible and safe, assisted recovery performed by trained personnel, and include the process of how and when to contact local emergency responders. There must be a designated rescue leader assigned the responsibility of implementing the rescue plan. In the event of a rescue operation, the rescue leader must have the authority and means to stop all other work in progress. Other qualified persons should act in support of the rescue leader until the situation is resolved, or until emergency responders relieve the rescue leader and the support team of their responsibilities and involvement in the rescue.

18.3.3 A kit of rescue equipment must be available in a known and quickly accessible location. It must be suitably packaged and should be clearly labeled "RIGGING RESCUE." This equipment must NOT be used for any other purpose and may include both general rigging gear and specialized equipment designed for the specific event or venue. Do not depend on borrowing rigging rescue equipment from the riggers on the job. Their personal equipment will likely be in use when needed for rescue and/or not be appropriate to the rescue operation.

18.3.4 During the pre-production phase when local emergency responders are briefed (i.e., fire fighters, police officers, emergency medical personnel), everyone should be informed of and familiarized with the venue and the planned rigging operations. In an emergency, the designated rigging rescue leader should be involved in the decision to contact emergency responders. However, it is recommended that any time an injured or incapacitated worker is suspended in a harness, public safety assistance should be immediately summoned.

18.3.5 A rescue plan cannot depend solely on public safety emergency responders for rescue. Because time is critical in these situations, the activation of a rescue plan should first involve trained and qualified personnel located on site in addition to the notification of local emergency responders. The rescue plan must be created well in advance, specific to the anticipated event's rigging situations and made part of the event's overall emergency action plan. Where feasible, such as in a fixed venue, rescue operations training and drills should be exercised periodically to minimize the time required to activate and perform a rigging rescue.

> **Effective rigging rescue requires a defined plan, proper equipment, a well-trained team and frequent practice.**

18.4 Riggers' Involvement in the Emergency Action Plan

18.4.1 Riggers are integral to several elements of an event's emergency action plan. In such plans, riggers are often required to lower wind walls or backdrops during high winds, rain or other emergency conditions. Thus, the supervising rigger should be included in all relevant meetings regarding event emergency planning.

18.4.2 No employer can require workers to work in hazardous conditions so emergency action plans must consider the safety of all event staff as well as the general public. Demobilizing an event in progress can take more time than the full evacuation of the audience, which is why organizers must prepare and talk through scenarios well in advance of the event. Tough decisions are never easy, but they are even harder when they have not been considered in advance. Event organizers and safety coordinators must be kept informed of all developing conditions that may require execution of any part of the emergency action plan. Scenario planning and tabletop discussions as described in Chapter 2, *Planning and Management*, will help streamline the crisis management process and build a team that is better prepared to face unexpected challenges and reduce the time required to effectively demobilize an event.

18.4.3 The production rigger or his agent should be present on the job site at all times. In the event of severe weather, high winds or any other scenario requiring action by the rigging crew, it is essential to have a production rigger capable of making decisions on behalf of the event and a suitable number of riggers standing by to carry out the emergency action plan or any other non-emergency task involving rigging.

18.5 The Rigging Plan

18.5.1 The rigging plan is a document (typically drawings) that identifies the location and magnitude of the loads to be applied to the event's support structures. The accuracy of this data is the responsibility of the production, act, rigging vendor or event coming into the venue. The plan should be submitted to the venue in advance so the venue can verify that all locations and loads can be accommodated. This verification must be done by a qualified person (see 18.1.2.4.1), preferably a licensed professional engineer familiar with the venue. The complete plan must be available to the venue supervising rigger or other supervising person for review before starting work.

18.5.2 Rigging Plan Requirements

18.5.2.1 Rigging plans can be generic in nature, with only the production elements drawn and no indication of the overhead supporting structure, or, they can be detailed and specific to the production and venue. Generic load plans, such as for touring productions, require a reference point or datum (usually indicated by 0' – 0' on the drawings where the downstage edge of the stage intersects with the stage's center line) to locate rigging. With this information, the organizer and the venue can determine where they would like the show to hang in the structure. With that information the structural engineer can then determine if the venue's structure can support the rigging loads and how the loads must be distributed.

18.5.2.2 The rigging load plan must include and clearly communicate the location and maximum magnitude of all loads to be attached to the supporting structure. All markings on the rigging plan should be clear and obvious as to their maximum loading, hoist capacity and any other information pertinent to the loads to be applied.

18.5.2.3 All loads must be accurately identified for each rigging point. Load figures should include the weight of hoists and include an allowance for typical rigging cable lengths.

18.5.2.4 A dynamic (changing) load changes in the direction or degree of force during operation. If any loads are dynamic, the maximum potential load must be indicated. Typically these loads are associated with a video screen, scenic element or lighting element that moves during the presentation. This causes the total load to change as the loading elements move in relation to the rigging points. These loads can produce significant changes in the structure's overall loading and can lead to an unsafe (overloaded) condition. If accurate information regarding these loads cannot be obtained through pre-production documentation, load measurement equipment, such as dynamometers or load cells, must be employed. Any dynamic rigging loads should be analyzed and monitored in use by a qualified person.

18.5.2.5 Identification of suspended elements: The plan should also indicate by illustration and label what specific equipment is to be suspended at each point, such as lighting, sound, performers, or scenic.

18.5.2.6 Elevation of suspended elements: The final elevation or "high trim" of all elements should be indicated.

18.5.2.7 Each diagram contained in the rigging plan should carry information in its title plate regarding its drawing number or other identifier, date, source, author and the author's qualifications and/or licensure.

18.5.2.8 Design wise, rigging plans should be adaptable to local venue conditions.

18.6 Flying People

18.6.1 At this time, it is beyond the scope of this document to discuss "flying" or aerial suspension of performers. Organizers should plan to suspend persons from rigging only after expert guidance from a qualified professional consultant and a qualified vendor who specializes in aerial suspension of human performers.

18.7 Rigging Installation

18.7.1 Always consider the possibility of a pre-rig (scheduling the rigging work prior to the general overall mass crew call). The rigging process requires a clear floor and quiet environment and can be limiting to other vendors trying to install their equipment. Scheduling rigging a few hours earlier than the load in time for other production elements or even on a prior day contributes to safety and in some cases can be cost effective. A pre-rig allows the rigging work to be carried out without the added stress of other workers unwittingly entering the area beneath the rigging teams working at height or the added pressure of other departments waiting to work.

18.7.2 A rigging team usually includes at least two high riggers and one ground rigger. Generally at least two rigging teams are required for an efficient load-in. Bridle points can easily require three high riggers and can take many minutes to rig if there are building obstructions such as catwalks and handrails, ductwork, conduits, lighting or dropped ceilings.

18.7.3 Large scale productions often require four or more rigging crews for efficient installation and removal. The tempo for the day is usually set by the pace the riggers install the rigging. In an industry where it is often said that, "the one thing you cannot buy is time," savvy organizers and producers know that by properly staffing the rigging department, stress is ultimately reduced because time is gained at the end of the load in process. For example, if a rigging install is going to take 60 man-hours, in theory it will take 6 riggers 10 hours to complete the work, while it will only take 10 riggers 6 hours to complete the same amount of work, a gain of 4 hours for approximately the same amount of money. This theory only works if there are enough qualified and competent rigging staff to field that many in the locality you are working. Many cities and towns simply do not have the resources in the environment to field a large rigging crew without importing resources from a distance.

18.7.3 Onsite changes: Even though the rigging plan has been submitted to the venue in advance for proper assessment and application, changes often take place during installation. If changes are required in the rigging plan during installation, the production rigger and the responsible venue representative must be consulted. This situation may also require consultation with the structural engineer of record and the designer/author of the original rigging plan. Some examples of changes due to venue specific rigging issues include the following:

- If a suitable directly vertical point (a "dead hang") is not available, rigging loads are often hung with a bridle to connect to two or more points in the structure. Bridle legs will exert horizontal forces on the support structure. These horizontal forces need to be considered and allowed for in the structural analysis.

- Other materials can also be used in conjunction with the primary overhead structure to provide supplemental support for rigging points; spanner beams and sub grids are some examples. When these secondary elements are used, the loads transferred from them to the primary overhead structure must be clearly identified, analyzed and approved.

- Rigging in non-standard venue locations, such as over auditorium seats or in areas not normally used, may require specific analysis.

18.8 Real Time Load Monitoring

18.8.1 While loads can be calculated from adding up the specifications of supported elements, this is often inaccurate. Only by actually weighing the loads at the rigging points can an exact figure be determined. This is often done during equipment preparation in the shop, in pre-production or rehearsals, or during the initial dates of a tour. As event rigging evolves and becomes more complicated, the need for accurate rigging information increases.

18.8.2 A simple straight truss supported at two points can be considered a "determinate" load. That is, as long as the truss is level, we can determine with some confidence the point loads at each support. "Indeterminate" loads are created when additional support points are added such as a center hoist on a long continuous truss. Any single element supported by three or more rigging points can have significant differences between point loads, even with careful analysis and leveling. This is because it is very difficult to balance the loads between multiple points on a single object. The only sure way to verify indeterminate loads to a structure is with load measuring devices, which require real time load monitoring.

18.8.3 Guy line systems on temporary outdoor structures can create static loads that reduce a structure's overall load capacity. Guy line loads also increase or decrease in windy conditions. Load measuring devices on guy lines can identify overloading due to winds. This information can be essential to the implementation of the Event Safety Plan. See Chapter 19, *Structures*, for additional guy line information.

18.8.4 Automation or high speed winch loads: Many productions use specialized winches for moving effects. This type of rigging can increase point loading dramatically and should be monitored with measuring devices. When moving elements start and stop they create momentary increases in loading. For a high speed winch in operation, or especially in an emergency stop situation, the rapid increase in point loading can be well over twice the static load. Never use automated winches without qualified professional consultation.

18.8.5 "Dynamometers" can be electronic or mechanical scales that display the applied load with a needle and a dial. These devices need to be read up close where they are installed, which may be impractical. Electronic dynamometers or "load cells" are devices inserted in the suspension system that can produce an electronic readout of load data which can be monitored from a convenient, remote location. This enables the production rigger to monitor loads during lifting and show conditions. Some load cell systems also have the ability to stop entire groups of motors should one fail. They can also monitor dynamic loads and record data.

18.8.6 Any truss system that moves during the event should be actively monitored by a load monitoring system. Indeterminate loads should be monitored, as should loads generated during load-in and load-out. In time, it may be possible to monitor every rigging point with a load cell system. Load cells are particularly valuable where suspended elements are subject to external sources such as wind loading.

18.8.7 Currently there are a number of load cell systems available, specifically designed for entertainment rigging systems. As the market for these systems grows they will have additional features and become more economically feasible. Rigging equipment vendors will be able to incorporate them into each rigging package as a standard practice and productions will be able to specify them as a condition of use. This is generally viewed as a positive development and the use of load cells in many event rigging applications is encouraged.

18.9 Rigging Training and Certification

18.9.1 Entertainment rigging knowledge is generally acquired through experience. There are also various nationally recognized programs to advance rigging knowledge and provide certification. All workers are encouraged to research and further their education. Many rigging suppliers have created training programs for their employees.

18.9.2 The Entertainment Technician Certification Program (ETCP; http://etcp.plasa.org/) offers certification programs for both theatre and arena riggers. This program has become widely accepted in the United States for the certification of entertainment riggers. Many venues have adopted the requirement that riggers be ETCP certified so organizers are encouraged to verify the venue's requirements on rigger certification well in advance during the planning phase

18.10 Integration of Moving and Flying Units into a Performance

18.10.1 Following the load in, all persons, (i.e., performers, crew, front of house personnel, etc.) who will be in areas where the flying of any object will occur, need to be briefed on and familiar with what elements will fly and when.

18.10.2 Under the leadership of the person in charge of performer integration, an event organizer, promoter, or stage manager, all persons should attend a rehearsal where:

- Under full white work lights and in silence, all moving and flying units should be demonstrated, in the exact sequence in which they will be executed during the performance, first at half speed (if possible) and then in real time.
- The crew should conduct simulations demonstrating how they can stop any unit at any time.
- Under full white work lights and in silence, all flown units are then run with talent and personnel in the places where they will be during the performance, and each person is given the time to ask questions, express concerns, and taught the process to STOP the motion of any flown unit at any time—including during the performance.
- Next, under show conditions (i.e., show lighting, other moving units such as scenery actually moving, full audio, full effects, etc.), a final run through of the moving and flying units occurs.

18.10.2.1 Following this rehearsal, a meeting should be held where questions can be asked and answered by anyone involved in the event.

18.11 Summary

18.11.1 It is in everyone's best interest to account for rigging early in the planning stages of a project, especially considering the potential safety and budget ramifications of doing otherwise. A producer or event organizer planning to incorporate rigging into their event is advised to consult the rigging team (if applicable), venue representatives, and rigging subcontractors. These resources are often the best sources of reliable rigging information and expertise.

18.11.2 Every event has many applications of rigging hardware in use at any given moment—from the laborers pulling out a stump with a wire rope to the collateral installer hanging a tent banner. You will find ropes, pulleys, cables, clamps, winches and more in use by many working at an event site. If you or any of your event team sees a questionable use of equipment by anyone on the job site, the riggers are a good source of information on the safe application of this type of equipment.

19. Structures

19.0.1 Many events require temporary structures, such as grandstands, stages, scaffold, tent and roof structures. Managing the hazards connected with these structures is just as important as managing other hazards. This can only be achieved if all those responsible for these structures undertake their duties conscientiously.

19.0.2 The failure of any temporary structure, no matter how small, in a crowded, confined space could have devastating effects. It is therefore essential to design and erect structures to suit the specific intended purpose and to recognize that the key to the safety of these structures is largely in the:

- Choice of appropriate design and materials;
- Correct positioning;
- Proper planning and control of work practices; and
- Careful inspection of the finished product.

19.0.3 This chapter gives guidance on safe temporary structures. It starts with the preliminary decisions that need to be made—choosing the site and the supplier—and continues to give general guidance on:

- The safety requirements for temporary structures;
- The documentation required (for example, guidance on temporary structure safety requirement listed in Appendix C, *Requirements for Outdoor Event Structures, Preparation Checklist*) to ensure that the essential safety requirements are provided; and,
- Advice on operations management of temporary structures.

19.0.4 This document provides general guidelines for structures. It is not meant to replace the advice of a professional engineer or the requirements dictated by the authorities having jurisdiction for the event site.

19.1 Scope

19.1.1 This section addresses the kinds of structure usually found at events including stages, sets, barriers, fencing, tents and roof structures, seating, lighting and special effect towers, platforms and masts, video screens, TV platforms and crane jibs, dance platforms and structures erected indoors and outdoors. Temporary structures erected outdoors need to meet all the requirements of indoor structures plus the additional factors created by the effects of the weather and site conditions.

19.1.2 Any structure erected in conjunction with an event, or on the land over which the organizer has control, should be subject to an equal degree of scrutiny and due diligence. Structures in parking lots and peripheral areas may present similar life safety risks as previous examples and the organizer needs to ensure that all such elements are installed and operated safely.

19.2 How the Law Applies

19.2.1 It is the organizer's responsibility to understand and implement the relevant local, state and national requirements as required by the authorities having jurisdiction.

19.2.2 Americans with Disabilities Act of 1990

19.2.2.1 The extent to which the Americans with Disabilities Act (ADA) Standards for Accessible Design (http://www.ada.gov/ 2010ADAstandards_index.htm) and accessible stadiums (http://www.ada.gov/stadium.pdf) may apply to event buildings, structures, equipment and devices is beyond the scope of this document and must be established by competent legal counsel. However, organizers and venue managers would be prudent to familiarize themselves with these requirements and be constantly mindful of their intent as they relate to event venues: to improve accessibility (to all aspects of the venue, including safety features) and enjoyment of those with disabilities and certain physical limitations who wish to attend.

19.2.2.2 See Chapter 11, *Facilities for People with Special Needs*, for more information about the Americans with Disabilities Act and its applicability.

19.3 Preliminaries

19.3.1 Choosing the Location

19.3.1.1 Many factors may influence the choice of location for temporary structures:
- Is the site adequately drained? If the site is liable to flooding, this could cause either the load-bearing capacity of the ground to be reduced or wash away the ground under the supports. Take measures to control these effects.
- Is the site flat or can it be made flat? Where there is a gradient or the ground is uneven, the structure needs to be capable of being modified to accommodate such variations.
- Are there overhead power cables, and if so are they sufficiently clear of the upper part of the structure? Are there cranes, which may be employed in the assembly of temporary structures)?
- Does the proximity of surrounding buildings, structures and vegetation create the possible spread of fire?
- Are there prevailing winds or weather patterns that could affect the performance of the stage? The stage should be located to lessen the hazards of potential weather events.
- Do you need to reduce or mitigate sound levels due to current neighborhood zoning?
- Is there enough space for trucks, power generators and other equipment or will it cause a hazard?
- Will the placement of lighting, sound and video equipment affect the safety?
- What about evacuation procedures? Is there enough exit space?
- Does the safety policy address it?
- Is it in the engineering drawings?

19.3.1.2 Obtain information from appropriate authorities or those with jurisdiction about the load-bearing capacity of the ground or stage floor. For outdoor events, ensure that the ground load-bearing capacity is capable of supporting the imposed loadings in all weather conditions.

Some venues may have space or basements under the stage floor surface. High point-loads may be created by the use of cranes or lift trucks utilized to install sections of structures or equipment and need to be evaluated by an engineer or other qualified person.

19.3.2 Choosing the Structure

19.3.2.1 It should be the responsibility of the event organizer to specify the requirements of the structure including structural performance, site limitations, equipment access, design and event duration. To establish these requirements, the organizer will likely have to consult with all the relevant parties including the artist, venue, production suppliers, labor provider and authorities having jurisdiction.

19.3.3 Choosing the Supplier

19.3.3.1 Choose a competent supplier for all temporary structures to be erected and used on site. A competent supplier will be able to demonstrate:

- A knowledge and understanding of the work involved;
- That they can manage/mitigate the risks involved in constructing these types of structure;
- That they can supply suitable engineering documentation—drawings and calculations— for their structures as detailed in section 9.11, below, Requirements for Outdoor Events Structures.
- That they comply with the relevant guidelines in section 9.11, below, Requirements for Outdoor Events Structures; and,
- That they employ a suitably trained workforce.

19.3.3.2 It is important to note that the design and engineering of temporary structures is generally outside of mainstream civil and structural engineering and is considered a specialized field. The design and engineering of temporary structures should be carried out only by qualified individuals with documented experience in the design of said structures.

19.4 Essential Requirements

19.4.1 Design

19.4.1.1 All temporary structures must possess strength and stability, in service and during construction, consistent with the intended purpose and requirements of the structure.

19.4.1.2 The design of a temporary structure should provide protection against falls during all phases of use including erection, operation and dismantle. Consider guardrails for all stage areas, platforms and access ways. The height of the guardrails as required by the local, state or national regulations must be referenced.

19.4.1.3 Consideration should be given to the surface of any ramp or tread, particularly those which could become wet.

19.4.2 Erection

19.4.2.1 To prevent the incorrect erection and subsequent use of temporary structures, attention should be paid to the following:

- The assembly of temporary structures should be carried out according to calculations, plans and specifications provided by a qualified designer or engineer. It is recommended that the final design documents bear the seal and signature of a licensed professional engineer.

- Apparent similarities between proprietary systems used for temporary structures may only be cosmetic. Be sure to obtain specific design information for the system you are utilizing. When components from different manufacturers are utilized together in a single system, approval should be obtained from each manufacturer and a qualified engineer.

- Erection should take place in a way that ensures stability during the erection process.

- Many temporary structures cannot be assembled except by climbing the framework as it is erected. These conditions should be addressed in the risk assessment and safety method statement. The use of personal protective equipment should be mandatory any time the working surface is greater than 6 feet (1.8 m) from the floor.

- Equipment should be checked to ensure it is fit for the purpose intended and fully meets any specification which has been provided; for example, steel, aluminum or other metal structural elements with cracked welds, bent or buckled members, or with large amounts of corrosion should be rejected and replaced where necessary.

- Components should not be bent, distorted or otherwise altered to force them to fit.

- Particular attention should be given to fastenings and connections. It is essential to provide suitable protection for bolts and fittings which project into or adjoin audience areas. Structures should not pose a hazard or risk to the audience.

19.5 Guy Line and Anchoring Stabilization Systems

19.5.1 This section discusses guy line and anchoring stabilization systems for temporary structures, including movable ballast and fixed point anchors.

19.5.2 When temporary structures depend on guy line and ballast systems to resist gravity and lateral loads from self-weight in addition to wind and seismic forces, their design is critical to the integrity of the structure. These components must be used in a manner consistent with the engineering calculations. Specifically, the location of ballast and angle of guy lines should not deviate from the site plan drawings without the approval of a qualified engineer.

19.5.3 Site conditions are variable for each application (venue and performer) and should be independently engineered. The engineering specification for the guying system must be included in the overall engineering documentation for each specific application.

19.5.4 The specification for the guying system should consider a worst-case condition, which should be clearly identified by the qualified engineer.

19.5.5 All fittings, cables and attachments should be specified by a qualified person and have sufficient safe working load. Guy line assemblies should be designed and rated based on their weakest component.

19.5.6 Guy lines must not be slack and should be pre-tensioned as specified by the qualified engineer. Any slack will be exacerbated by wind and other loading and can result in instability of the towers. Wind loading will deflect a structure with slack guy wires until the guy becomes taut. This condition greatly compromises strength and stability and can lead to unpredictable behavior and catastrophic collapse.

19.5.7 Tension on guy lines will increase the loads imposed on the structure by pulling horizontally on the trusses and by pulling down on the towers. Such tension includes pretensioning and tension induced by externally applied lateral forces (often environmental). These forces can reduce total structure capacity and should be considered in the overall design of the structure.

19.5.8 Fixed point anchors, such as earth anchors (Fig. 19-1) or permanently installed venue anchor points, are preferable to movable ballast, such as "Jersey" barriers. If fixed point anchors are used they must be tested for the anticipated maximum loading before structure erection.

19.5.9 Movable ballast systems must be designed to resist uplift, sliding, twisting and rolling resulting from guy wire tension where it is applied to the ballast, as well as environmental forces on the ballast (such as wind, earthquake). Sliding depends on the amount of weight bearing on the ground to create frictional resistance. The frictional resistance should be determined by a qualified engineer for the conditions of use. For ballast engaged by the guy through its center of gravity, sliding will

Fig. 19-1 – Installing an Earth Anchor. Earth anchors can be an effective and space efficient alternative to using bulk weight to ballast structures and rigging.

always occur before the ballast lifts off the ground. Any ballast movement can lead to structural collapse.

19.5.10 Guy line attachment to movable ballast must be designed to engage the entire ballast weight. Do not attach guy lines to non-structural embeds such as loops, eyes or bales at ends of ballast (such as Jersey barriers, concrete blocks, water tanks, or steel plates). Unless tested, these points are not reliable. Attachment to ballast should wrap the entire ballast in such a way that the attachment cannot slide and the loading is not biased to one end or the other. Note that the force to pick up one end of the ballast is less than the total weight of the ballast.

19.5.11 Where guying is used, care should be taken to ensure that the guy lines and their anchors do not cause an obstruction or hazard. The guy line anchor locations within public access must be barricaded to maintain the integrity of the guy lines and to prevent a hazard from occurring due to impact with a vehicle or spectator.

19.5.12 The resistance provided by movable ballast is NOT equal to the weight/mass of that ballast and is typically far less. Where a guy line "load" or "capacity" is provided on erection drawings without a specific quantity/size of ballast, a qualified engineer must be consulted to determine the equivalent amount of ballast mass/weight that is required to resist the load generated by the structure.

19.6 Fall Protection

19.6.1 The U.S. Bureau of Labor Statistics reported that in 2010, 751 construction workers died on the job, with 35 percent of those fatalities (263) resulting from falls (http://www.osha.gov). This is why 29 CFR 1926.501 (a)(2) states that,

> Each employee on a walking/working surface (horizontal and vertical surface) with an unprotected side or edge which is 6 feet (1.8 m) or more above a lower level shall be protected from falling by the use of guardrail systems, safety net systems, or personal fall arrest systems.

19.6.2 Fall protection is addressed in OSHA's standards for the construction industry. The following are some of the OSHA standards, Federal Registers (rules, proposed rules, and notices) preambles to final rules (background to final rules), directives (instructions for compliance officers), standard interpretations (official letters of interpretation of the standards), example cases, and national consensus standards related to fall protection. All can be found in 29 CFR 1926, *Safety and Health Regulations for Construction*.

- 1926.451, General requirements (Scaffolding)
- 1926.452, Additional requirements applicable to specific types of scaffolds
- 1926.454, Training requirements (Scaffolding)
- 1926.501, Duty to have fall protection
- 1926.502, Fall protection systems criteria and practices
- 1926.503, Training requirements (Fall protection)
- 1926.760, Steel erection (Fall protection)
- 1926.800, Underground construction
- 1926.1051, General requirements (Stairways and ladders)
- 1926.1052, Stairways
- 1926.1053, Ladders
- 1926.1060, Training requirements (Stairways and ladders)
- 1926.1423, Cranes and derricks in construction (Fall protection)
- 1926.1501, Cranes and derricks used in demolition and underground construction

19.6.3 Employers must issue personal protective equipment (PPE) in situations where the wearing of PPE, which includes hard hats and harnesses, is determined to be an effective means of preventing injury. Employers are required to train employees on the proper use of PPE and ensure that both the training and the equipment meet the requirements of the applicable local, state and/or federal regulations. It is the employer's responsibility to understand and implement the relevant local requirements.

Fig. 19-2 – Example of a climber clipped in on a wet day. This is an example of appropriate fall protection. Photo courtesy of The Event Safety Shop, LTD.

19.6.4 Employers must establish rescue plans to ensure safe rescue of any fallen person, considering all potential fall locations, time to safely accomplish a rescue, and emergency evacuation and medical treatment.

19.7 Protection from Falling Objects

19.7.1 While structures are being erected or disassembled, do not lift materials over the heads of people working or passing below. Such areas should be clear to prevent death or injury. Create secure "no go" areas below working areas to prevent harm to people. For example, it is recommended that the ground riggers be easily identifiable (i.e., wearing a high visibility vest) and responsible for controlling the ground area beneath overhead work. When it is necessary to lift lighting or sound equipment overhead, it is mandatory that all items in the rig are secured or tied off securely.

19.7.2 The employer has the responsibility to provide a workplace free from recognized hazards that may cause injury or fatal injury to employees, audience members and performers and comply with applicable standards, rules and regulations.

19.8 Use of Lifting Equipment

19.8.1 Any organization using lifting equipment such as forklifts and scissor lifts has a duty to provide qualified operators and physical evidence to health and safety inspectors to demonstrate that the last inspection has been carried out. People hiring lifting equipment should make sure that it is accompanied by the necessary documentation and is in good repair.

19.8.2 After positioning rigging and similar equipment that is integral to the erection or operation of the structure, the user should ensure that a competent person inspects the lifting equipment before it is put into use to make sure it is safe to operate. The user then has the duty to manage the subsequent lifting operations in a safe manner.

19.8.3 Everyone involved in erecting and dismantling temporary structures must be appropriately trained in the safe technique of high level rigging. Training is now commercially available in safe techniques for high level rigging, and those working at high level must have undergone training and assessment.

19.8.4 Everyone involved in erecting and dismantling temporary structures should be appropriately trained.

19.9 Essential Documentation

19.9.1 All proper designs should include calculations to determine design forces on the structure and to establish the structural integrity and stability of the structural design. The engineer should provide:
- A statement about the structural system concept and what the structure system is designed to do;
- A list of specific critical structural items or connections that require checking by calculation prior to each time the structure is erected;
- Details of the methods of stabilizing the structure to resist all horizontal forces, resulting from rigging forces, distribution of weights, etc.; and
- For outdoor structures, details of the methods of resisting all vertical and horizontal environmental forces (e.g. wind).

19.9.2 Construction drawings will normally be required for all but the simplest temporary structures. These should be accompanied by full calculations, design loads and any relevant test results. These documents should normally be sent to the local authority in advance of the event. Check with the authority during the planning phases of an event to determine their permitting/review timeline and requirements.

19.9.3 Risk Assessment

19.9.3.1 The structure provider should implement a safety plan and risk assessment that includes the design, erection, and suitability of the temporary structure for each configuration. For example, a festival application may require multiple assessments due to multiple performances utilizing the same structure. Note, even the simple addition of a banner or sign may change the risks and safety plan due to the change in wind loading.

19.9.3.2 These risk assessments should be submitted with the initial design plans and calculations to the local authority and should be specific to the structure intended to be used at the venue.

19.9.4 Supervising the Installation

19.9.4.1 Monitor all activities at the venue relating to the erection and construction of temporary structures to ensure that they are erected to the detailed specification and that the safety plan and safe working practices are followed.

19.9.4.2 Ensure that all structures are checked by a qualified person after they have been erected and before they are used, to ensure that they conform to the drawings and specified details. It is preferable to have the design engineer perform an evaluation of the structure prior to use at an event. A letter should be requested from the engineer that states the structure complies with the design intent of the engineer's calculations and drawings.

19.9.4.3 It is critical that the engineer of record determine that the as-built structure meets his or her design intent, and that the Authority Having Jurisdiction approves the structure as built. It would also be prudent for the owner/operator (whoever is responsible for the structure) to require a "special inspection" by someone with distinct expertise to ensure compliance with approved construction documents and referenced standards (as described in the International Building Code, Sections 1702 and 1704).

19.10 Managing the Completed Structure

19.10.1 Before Admitting the Audience

19.10.1.1 As noted above, temporary structures should comply fully with the design documentation before the audience is admitted to the site. If modifications to the structure are necessary, then changes must be approved by the qualified engineer, subject to approval by the authority having jurisdiction.

19.10.1.2 Loads on temporary structures can be applied in various ways but must not exceed the specified allowable loads. Therefore, adequate measures must be taken to prevent overload by:
- Unauthorized additions (e.g., banners, billboards, projection screens, scrims, scenic facades, lighting units, speakers, etc.) should not be added to temporary structures without the prior consent of the qualified engineer;
- Equipment loads (e.g., lighting, special effects, sound systems, video and TV screens) can be significant; therefore, it is important that the installer confirms the weights prior to use, and that accurate or conservative weights are included in the qualified engineer's calculations; and
- Check with the design engineer to determine if scrims, screens, or banners are allowed to be anchored/restrained at their bottom edge, or if the engineer analyzed and approved those elements to be free hanging.

19.10.2 Monitoring After Erection

19.10.2.1 A competent person should continuously monitor a structure that is susceptible to the effects of the weather and/or misuse (by overloading the roof structure, for instance). This person could be a representative of the supplier of the structure or a properly trained and qualified rigger.

19.10.2.2 For structures that may be unused between performances, the structure must be secured in a way that does not require constant onsite monitoring.

19.10.2.3 Before using a structure that has previously been secured, a thorough evaluation should be conducted to ensure the integrity of the structure and its suitability for continued use.

19.10.3 Managing the Loads

19.10.3.1 Loads on temporary structures can be applied in various ways but must not exceed the allowable loads considered by the engineer. Therefore, adequate measures must be taken to prevent overload by people and/or equipment due to overcrowding any part of a temporary structure. Be particularly cautious of overloading during load-in, erection, dismantling and load-out phases of an event.

19.11 Requirements for Outdoor Events Structures

19.11.1 The structure must be a purpose-built system used according to its intended purpose. All aspects of the structure must comply with the local building codes and standards including but not limited to the most current versions of ASCE/SEI-7 and ASCE/SEI-37. Building codes contain design requirements suitable for permanent buildings that are often onerous for temporary structures, and many building codes at present do not address temporary structures. As a result, the designer should consider utilizing the provisions of ANSI E1.21 or an equivalent standard as identified by a qualified person, subject to the approval of the Authority Having Jurisdiction.

19.11.2 The event producer must submit the required documentation including the structure event specific plan to the artist team no later than seven days before the planned commencement of the construction. Any changes, additions and/or modifications to the structure or the plan must be completed and submitted by three days before construction begins so that the changes can be reviewed by the engineer. Changes or modifications after this date should not be accepted.

19.11.3 If local jurisdiction permits/approvals are required, the above timing requirements must be submitted in compliance with the permit submittal schedule rather than the start of construction.

19.11.4 Compliance of the structure to applicable codes and standards, including ASCE/SEI-7, ASCE/SEI-37 and ANSI E1.21, or applicable code or authority having jurisdiction requirement, must be demonstrated by providing a Letter of Conformity/Compliance before the day of the event. This letter must include the following supporting documentation:

19.11.4.1 <u>Drawings</u>: Erection diagrams, plan and elevation views, identifying:

(a) Positions and connections for all structural elements including lifting devices (motors), lock-off devices, guy line connections, guy line anchorage/ballast requirements, all supported loads, etc.
(b) Areas and limits of allowable surfaces that receive wind loading (pressure and suction) including:
   i.   Stage roof covering;
   ii. Side/back walls;
   iii. Additional signage;
   iv. Suspended audio, lighting, video and scenic equipment.

19.11.4.2 <u>Calculations</u>: Allowable load documentation from licensed engineer indicating limits of the structure for:

(a) Allowable equipment loads for show equipment (live load) for audio, lighting, scenery, video, etc., including permissible connection locations to structure;
(b) Environmental limits including seismic and wind loading;
(c) Ground conditions, capacities and foundation requirements;
(d) Lateral stability requirements including anchorage details
    i. Differences between ground anchorage and ballast requirements must be clearly identified;
(e) Load assumptions and considerations;
(f) Operations Management Plan minimum requirements.

19.11.4.3 <u>Event Specific Compliance</u>: Summary of the specific event loads (rigging plot, audio loads, etc.) and conditions that demonstrates the structure is sufficient for the intended purpose including:
(a) Suitability of the site:
    i. Access for equipment and erection of structure;
    ii. Ground conditions;
    iii. Underground interferences (sprinkler systems, utilities, etc.);
    iv. Condition of elevation changes affecting structure (changes in ground height);
(b) Overall site layout diagram showing:
    i. Location of structure;
    ii. Access locations;
    iii. Audience location(s).
(c) Rigging plot overlaid on structure indicating compatibility;
(d) Copy of permit(s) from the local jurisdiction event is to be held if applicable;
(e) Inspection records of the structure and its components should be available upon request as needed;
(f) Upon the completion of the structure erection, a certificate of completion should be provided by the installer.

19.11.4.4 <u>Operations Management Plan</u>: An action plan indicating the various stages of operating conditions of the structure and the actions to be taken when those conditions are met. This is a plan describing the various conditions structures may encounter (e.g., regular and emergency maintenance, overloading, wind, rain, snow, ice, hail, lightning, flood, fire, etc.) and the actions to be taken should the structures be confronted with any of the described conditions.

(a) Wind:
    i. Monitoring
        1. A specific qualified individual responsible for monitoring on site conditions and forecasted conditions must be identified before the event;
        2. Wind must always be monitored on site with an anemometer or other appropriate device; and,
        3. A reliable means of communication must be established with appropriate weather resources to monitor impending conditions.

ii. Actions: Identify what specific actions will happen at specific reported or forecast wind speeds. But, removal of equipment must be done in safe wind conditions, which may be notably lower than the design threshold wind speed. These actions MUST consider the nature of the loads on the structure specific to this event. For example, if there is equipment on the structure that prevents it from being lowered to the ground, then lowering the structures to the ground is not an option to be considered in the action plan.

iii. Although wind is mentioned above, similar load thresholds must be established for all potential environmental hazards such as snow, ice, hail, rain, lightning, etc., along with appropriate actions to be taken when these thresholds are met. The International Building Code (IBC) and ASCE/SEI-7 should be referenced for the relevant applicable design loads.

iv. Training/Communication: Before the event, it must be clear to all relevant parties what actions are to be taken and at which identified thresholds.

(b) Responsibility:
   i. Individuals responsible for various tasks must be identified before event. This must include, but not be limited to:
      1. Weather monitor;
      2. Stage Manager;
      3. Security Personnel;
      4. Artists Representative;
      5. Promoter Representative;
      6. Temporary structure vendor crew lead;

   ii. Additionally, if a specific chain of command to any of the key responsible positions is in place, then all individuals in the chain must be aware of their immediate supervisor;

   iii. The contact information for key responsibility positions including names, phone numbers and work locations must be provided before the event.

(c) Suspension/Cancelation: Any entity with definitive responsibility for a portion of the event—including the temporary structure vendor, production manager, promoter or state/local authority—can make the non-negotiable decision to suspend or cancel an event if public safety is jeopardized. The method of initiating the event suspension or cancelation must be outlined explicitly before the event so responsibilities and liabilities are identified and the decision, when made, can take effect immediately.

19.11.4.4.1 A thorough operations management plan should encompass as many response time related details as possible. Knowing these details will enable the command team to set decision deadlines to ensure timely responses. For example, ask and answer the following questions:

- How long will it take to fully evacuate a maximum capacity audience?
- How long will it take to bring equipment to a less threatening state (e.g., land any "flown rigging" elements as determined by the engineers, such as the PA)?

See Appendix C for *Requirements for Outdoor Event Structures, Preparation Checklist*; and, Appendix D for *Requirements for Outdoor Event Structures, Key Personnel.*

20. Pyrotechnics, Fireworks and Flame Effects

20.0.1 The application of this subject matter is so potentially hazardous and highly regulated that the first and most important recommendation must be that anyone interested in providing pyrotechnics, fireworks, and/or flame effects at an event should engage a qualified vendor…

- Who has been fully vetted and found to be qualified, competent, licensed, safe, and reputable;
- Who has provided references with which you have followed up, including vetting them through the state regulatory authority on fireworks (often the State Fire Marshal);
- With relevant and well documented experience in exactly what is being asked of them;
- Who has an excellent and well documented safety record related to pyrotechnics, fireworks, and/or flame effects;
- Who can show documented training in contemporary methods and procedures related to pyrotechnics, fireworks, and/or flame effects;
- Who possesses commercial general liability insurance with an appropriate endorsement for pyrotechnics (a limit of no less than $5 million is recommended) and workers' compensation insurance covering all the vendor's employees;
- Who employs for the event only individuals licensed by the Authority Having Jurisdiction (AHJ) who are 21 years of age or older;
- Who provides résumés for every employee working on the event; and
- Who is, ideally, known to the local fire official.

20.0.2 Nothing, including this document, can replace the use of a reputable, qualified, licensed and competent pyrotechnics vendor. In fact, compliance with the technical and regulatory requirements related to pyrotechnics, fireworks, and/or flame effects at an event are so important to the safety of guests and staff at the event that only those with significant experience in their safe use should be dealing with them at all. This chapter is not meant to replace the identification and use of a qualified and competent vendor of pyrotechnics, fireworks, and/or flame effects at an event. However, this chapter will provide some background, general guidance, checklists, and resources that event organizers might find helpful when familiarizing themselves with the laws, regulations, and safety issues related to the use of pyrotechnics. It remains the responsibility of the event organizer to identify and engage an acceptable vendor.

20.0.3 Chapter outline. This chapter is organized into five general parts: Definitions, Getting Started, Guidelines, Checklists, and Safety Resources. (The section number is indicated in parentheses.)

- The "Definitions" section (20.1) describes the relevant terms used in this chapter, as well as citing some of the standards and rules that apply to the subject matter.

- The "Getting Started" section (20.2) describes some important initial steps one should take to incorporate pyrotechnics, fireworks and/or flame effects into a show. It also lists the codes and standards relevant to the subject matter.
- The "Guidelines" section provides standard-based guidance on the use of flame effects (20.3), flame performers (20.4), and pyrotechnics (20.5). Each of these sections include:
  a. A description of the contents of what should be in the required design plan for each of these activity types;
  b. The required written procedures and documents for each of the activity types;
  c. A description of the required qualifications and experience documentation that must be collected from and submitted for each operator; and
  d. Other guidance related to the activity type.
- The "Checklist" section includes standard-based checklists for using gas flames (20.6), flame performers (20.7), proximate pyrotechnics (20.8), and fireworks (20.9) at an event. This section also includes the OSHA Display Fireworks Sites checklist (20.10) and two examples of question-based checklists the State Fire Marshal in Ohio uses for outdoor exhibits of fireworks (20.11) and indoor fireworks or flame effects exhibits (20.12).
- The last section in this chapter is "Safety Resources," which includes a list of Internet-based resources that can be accessed for additional information on pyrotechnics, fireworks and flame effects safety.

20.1 Definitions

20.1.1 An "explosive" is any article that is designed to function by explosion (i.e., an extremely rapid release of gas and heat) or which, by chemical reaction within itself, is able to function in a similar manner even if not designed to function by explosion. The U.S Government defines three classes of explosive materials: high explosives, low explosives, and blasting agents (27 CFR 555, Subpart K). Fireworks are pyrotechnic articles that, for transportation purposes, usually fall into the "low explosives" class. No person may offer a new fireworks explosive for transportation in the U.S. unless the device has been classed and approved by the Pipeline and Hazardous Materials Safety Administration's (PHMSA) Associate Administrator for Hazardous Materials Safety. PHMSA is the agency within the U.S. Department of Transportation (USDOT) responsible for issuing the hazardous materials regulations (HMR; 49 CFR, Parts 171-180). Under the USDOT classification system, explosives are divided into six divisions: 1.1, 1.2, 1.3, 1.4, 1.5, and 1.6 (division 1.1 is the most powerful). Fireworks most often fall into two of these divisions (1.3 [commercial, display fireworks] and 1.4 [consumer fireworks]).

20.1.2 NFPA 1126, *Standard for the Use of Pyrotechnics Before a Proximate Audience*, defines "Pyrotechnics" as, "controlled exothermic chemical reactions that are timed to create the effects of heat, gas, sound, dispersion of aerosols, emission of visible electromagnetic radiation, or a combination of these effects to provide the maximum effect from the least volume" (NFPA 1126, 2011, p. 1126-7).

- The National Fire Protection Association (NFPA; www.nfpa.org) is an international, nonprofit organization that provides and advocates codes and standards using a full consensus-based process. The name of each code or standard is referenced by a number and year of publication. For example, NFPA 1126 was most recently published in 2011 and would be cited simply as "NFPA 1126, 2011."

20.1.3 "Fireworks" are a class of explosive pyrotechnic articles used for aesthetic and entertainment purposes. They can be either consumer or display (professional) grade and are generally classed as to their chemical compounds and how and where they perform, i.e., ground or aerial. According to the NFPA, "fireworks" are, "any composition or device for the purpose of producing a visible or an audible effect for entertainment purposes by combustion (burning), deflagration (rapid and intense burning), or detonation (exploding)" (parentheticals added; NFPA 1123, 2014, p. 1123-7).

20.1.3.1 Consumer fireworks (a.k.a., 1.4G fireworks) are the small fireworks usually sold at stands around the Fourth of July holiday. These include some small devices designed to produce audible effects, ground devices containing 50 mg or less of flash powder, and aerial devices containing 130 mg or less of flash powder.
- The U.S. Department of Transportation (USDOT) classifies consumer fireworks as UN0336, UN0337.
- The U.S. Department of Justice, Bureau of Alcohol, Tobacco, Firearms, and Explosives (ATF), does not regulate the importation, distribution, or storage of completed consumer fireworks. However, any person manufacturing consumer fireworks for commercial use must obtain a Federal explosives manufacturer's license, and compliance with other federal, State, and local agency regulations is required.

20.1.3.2 Display fireworks (a.k.a., 1.3G fireworks) are the large fireworks used in fireworks display shows, generally under the supervision of a trained and licensed pyrotechnician. These fireworks are designed primarily to produce visible or audible effects by combustion, deflagration, or detonation. They include, but are not limited to, salutes containing more than 2 grains (130 mg) of flash powder, aerial shells containing more than 40 grams of pyrotechnic compositions (including any break charge and visible/audible effect composition but exclusive of lift charge), and other display pieces which exceed the limits of explosive materials for classification as "consumer fireworks." They also include fused set pieces containing components which together exceed 50 mg of flash powder.
- Display fireworks are classified as fireworks UN0333, UN0334 or UN0335 by the USDOT.
- The USDOT now classifies aerial shells that are 8 inches (0.2032 m) or larger as 1.1G and therefore they have different (stricter) transportation and storage requirements than 1.3G.

20.1.3.3 A pyrotechnic special effect is used for theatrical productions, concerts, or other special events and labeled UN0337 by the USDOT. The NFPA defines a "special effect" as, "a visual or audible effect used for entertainment purposes, often produced to create an illusion" (NFPA 1126, 2011, p. 1126-7). Some pyrotechnic special effects fireworks can be labeled 1.4G and classified as "Article, Pyrotechnic" (labeled UN0431 by USDOT) but, like pyrotechnic special effects, are not intended to be sold to the public. Both consumer and commercial display fireworks are intended to be used outdoors, whereas pyrotechnic special effects may be used at either indoor or outdoor displays. Pyrotechnic devices intended for indoor use must be marked as such by the manufacturer (NFPA 1126, 2011, 7.2.1[4]).

20.1.4 A "Flame effect" is, "the combustion of solids, liquids, or gases to produce thermal, physical, visual, or audible phenomena before an audience" (NFPA 160, 2011, p. 160-5). A flame effect can be automatic (fired by an automatic control system), manual (operated without an automated control system), or portable (designed to be moved in the course of operation).

20.2 Getting Started

20.2.1 As early as possible in the planning phase of the event, alert the local authority having jurisdiction (AHJ) of your intentions. This is usually done through the fire department. The AHJ will provide guidance and may be able to provide input or feedback regarding a vendor. Once vetted and approved, the vendor should take over communicating with the AHJ and dealing with any permitting and regulatory issues.

20.2.2 Generating a site map and stage plot that clearly shows the locations of all fireworks, flame effects, and any resources required to manage them, is an important early step. With most of these effects, there will be a restricted set-up or "holding" area (or a "pyro room," if indoors) cordoned off where the equipment is arranged. There may also be a "fall-out zone" where debris descends to the ground after a pyrotechnic or flame effect is triggered. All of these areas should be noted on all maps, diagrams and stage plots.

20.2.2.1 With display fireworks, one important and often overlooked map marking is the prevailing wind direction. Although there may be variations in the wind direction from the time a map is made until the date and time of the event, marking the prevailing (expected) wind direction on the map gives both event organizers and the AHJ a starting point from which to plan. Wind direction and velocity must be monitored before and during an event, and adjustments to the setup and plan, such as expanding the fallout zone, may need to be carried out based on this information.

20.2.2.2 Also note on the map the location of security posts to help keep the areas clear of non-working personnel during the set-up and the triggering of the display.

20.2.3 The installation and operation of all flame special effects and pyrotechnics, and operations involving all flame performers, should all be conducted in accordance with the following codes and standards, as applicable:
- NFPA 30, *Flammable and Combustible Liquids Code.*
- NFPA 54, *National Fuel Gas Code.*
- NFPA 55, *Compressed Gases and Cryogenic Fluids Code*; formerly "Standard for the Storage, Use and Handling of Compressed and Liquefied Gases in Portable Cylinders."
- NFPA 58, *Liquefied Petroleum Gas Code.*
- NFPA 59A, *Standard for the Production, Storage, and Handling of Liquefied Natural Gas (LNG).*
- NFPA 101, *Life Safety Code*®.
- NFPA 140, *Standard on Motion Picture and Television Production Studio Soundstages, Approved Production Facilities, and Production Locations.*
- NFPA 160, *Standard for the Use of Flame Effects Before an Audience.*
- NFPA 430, *Code for the Storage of Liquid and Solid Oxidizers.*

- NFPA 701, *Standard Methods of Fire Tests for Flame Propagation of Textiles and Films*.
- NFPA 1123, *Code for Fireworks Display*.
- NFPA 1124, *Code for the Manufacture, Transportation, Storage, and Retail Sales of Fireworks and Pyrotechnic Articles*.
- NFPA 1126, *Standard for the Use of Pyrotechnics Before a Proximate Audience*.
- Title 27, Code of Federal Regulations, Part 18, Bureau of Alcohol, Tobacco and Firearms, Part 181, Commerce in Explosives.
- Title 49, Code of Federal Regulations, Parts 171-177, U.S. Department of Transportation
- All pressure vessels must comply with the ASME *International Boiler and Pressure Vessel Code*, Section VIII.
- State and Local laws, ordinances, rules and regulations (AHJ).

20.2.4 This chapter describes selected best practices for events that include pyrotechnic performances before a live audience. This chapter is not intended to be used where an audience is not present, as would be the case with certain television and motion picture performances.

- NFPA 1126, *Standard for the Use of Pyrotechnics Before a Proximate Audience*, requires that when no audience is present, NFPA 140, *Standard on Motion Picture and Television Production Studio Soundstages, Approved Production Facilities, and Production Locations*, must be used (NFPA 1126, 2011, 1.3.5.2).
- NFPA 1126 applies at any outdoor use of pyrotechnics at distances less than those required by NFPA 1123, *Code for Fireworks Display* (NFPA 1126, 2011, 1.3.4.1).

20.3 Guidelines – Flame Effect

20.3.1 When flame special effects will be used, a design plan must be incorporated into the event operations plan and submitted in writing. A flame effect design plan includes at least the following:

20.3.1.1 Certain weather conditions can result in the termination of flame special effects, so weather must be monitored (see Chapter 7, *Weather Preparedness*, for more information on monitoring weather). In the plan, define the weather conditions that may require specific actions, up to and including termination of the effects. For an outdoor event, include a weather forecast report in the design plan. For an indoor performance, identify ventilation options and plan for HVAC/detector adjustment and fire-watch procedures, as needed.

20.3.1.2 A site plan (map of the site drawn to scale) must be developed that indicates at least the following:
- Location of all effects;
- Operator control areas;
- Restricted zones;
- Area affected by the flame special effect—identify materials for flame retardant treatment and type;
- Proximity sensors as applicable or means of monitoring intrusion (security);
- Audience proximity;
- Animal proximity;

- Performer and cast member proximity to each effect;
- Storage location of flammable and combustible liquids or gases;
- Location and construction specifications of flammable materials, piping, and accumulators;
- If outdoors, prevailing wind direction;
- Means of egress; and
- Location of firefighting equipment, quantity and type.

20.3.1.3 A list and description of each effect including size, location, fuel load, and maximum height, width and BTU of all flame effects.
- Identify the flame effect group as outlined in NFPA 160, *Standard for the Use of Flame Effects Before an Audience.*
- Identify whether the installation will be permanent or temporary and install everything accordingly as required by NFPA 160, considering amount of use and timeframe.

20.3.1.4 A description of the safety control mechanisms, emergency control devices, and the means of activation or firing of the system.

20.3.1.5 Material Safety Data Sheets (MSDS), a.k.a., Safety Data Sheets (SDS), for all products being used and corresponding effect.

20.3.1.6 A description of any device pre-engineered from a manufacturer including name, address, and phone number of the manufacture.

20.3.1.7 Documentation of inspection by a professional engineer for all flame special effects manufactured by the operator or user. This should include review of all fuel system components, materials, and design for proper use and application per NFPA standards and/or local fire codes. All control systems and fuel components should be listed (i.e., acceptable to the AHJ, tested and found suitable for a specific purpose) and acceptable for intended uses.

20.3.2 When flame special effects will be used, the following written procedures and documents must be developed and maintained:
- Pre-show and start-up safety reviews and equipment checks along with documentation/forms to support them.
- A show operations manual that includes applicable documentation for show performance and process verification, and that identifies the sequence of effects and application within the performance.
- Emergency procedures including a fire hazards evaluation for permanent installations to determine level of fire protection required and in place.
- Normal and emergency shut-down procedures for the effects.
- Maintenance schedules and procedures for all equipment involved.
- A copy of all applicable documentation, including records of any inspections and operational tests, should be included in the written materials for verification of operability of protective control systems such as fire protection systems.

- Schematic and detailed drawing of all special effects including electronic controls, valves, hoses, piping, fuel control mechanisms, accumulators, pressure gauges and proximity devices.

20.3.3 When a flame effect will be involved, an operator must submit detailed documentation of his or her qualifications and experience, including the same for all operators, crew members, designers, and fabricators. These documents must include at least the following:

- Name, address, and contact information of company.
- Commercial general liability insurance with an appropriate endorsement for pyrotechnics (a limit of no less than $5 million is recommended) and workers' compensation insurance covering all the vendor's employees. Confirm that the coverage meets all the requirements of the AHJ for permitting and operations.
- History of past shows so they can be thoroughly vetted. Make sure to identify the types of flame effects and systems used and the description of any incidents involving the use of flame special effects, especially where someone was injured or killed.
- Licenses or certifications held in any state and/or country.
- Collect and review a résumé for each person working for the contracted company that will work at your show or manage any portion thereof. Make sure the résumé includes all relevant experience and that all flame effect operators are at least 21 years of age.
- Event management should review all information and documentation prior to its submission to the AHJ to meet permitting requirements. A copy of all approved permits must be maintained on site and made available to inspectors and regulators upon request.

20.3.4 Performers, support personnel and non-participants exposed to the associated hazards of the flame special effects must be informed by the operator of the hazards and trained in emergency procedures.

- Also identify the effect on animal areas and procedures for proper animal handling.

20.3.5 Identify the personal protective equipment (PPE) to be used that is applicable to the hazards associated with the exposure to, and handling and use of, the flame effects utilized for the performance.

20.3.6 All flammable and combustible liquids not connected for use must be stored in holding areas designed and maintained in accordance with applicable NFPA and local fire codes. They must also be properly labeled and stored in areas that are locked and secured.

20.3.7 Documentation must be developed and maintained that shows that all affected combustible set, scenery and rigging materials, are treated with flame retardant materials. All flame special effects devices must also be properly secured to noncombustible materials.

20.3.8 The flame special effects operator must inspect and document all areas of the site where flame special effects and pyrotechnics effects and devices are ignited before start-up and after shut-down.

20.3.9 Written documentation should be maintained confirming that all flame special effects operators and crew members have successfully passed a drug test in accordance with an accepted drug and alcohol policy. Although this is not currently required in all jurisdictions, it is a recommended best practice.

20.3.10 All effects must have an emergency stop system to cause shut-down of the flame effects system. This system must bring the device to a zero energy state causing all ignition sources to be unpowered and detached from their respective fuel sources. This includes:
- Removal of power to all gas valves;
- Removal of power to ignition devices; and
- Operator action to acknowledge the cause and reset the system.

20.3.11 A number of detailed technical system requirements are included in the standards related to the use of flame effects. Some examples are provided here but operators must refer to the NFPA standards and local AHJ for specific guidance:
- All control mechanisms including valves, sensors, and switches must be designed and located to prevent tampering and accidental firing or release of fuel.
- Each special effect must be provided a means of primary lock-out that interrupts the supply of fuel and control energy.
- All burners must be provided with a listed automatic ignition device or flame safeguard that supervises ignition.
- Flame failure response time should not exceed 4 seconds.
- Ignition of the device or pilot must be proven prior to attempting to light a burner.
- Gaseous flame effects must be capable of maintaining system pressure at 1-1/2 times the maximum operating pressure for a period of 24 hours.
- With the complete system in place, all piping, accumulator valves, hoses, etc. must pass a pressure test of 1-1/2 times design pressure with an inert gas.
- The system pressure must be recorded with atmospheric temperature and pressure and there must be no observable pressure loss within 24 hours.

20.3.12 For temporary installations, at least two pressurized water 2-A fire extinguishers and two 10-BC fire extinguishers must be provided in the venue in addition to those required by NFPA 10, *Standard for Portable Fire Extinguishers*.

20.4 Guidelines – Flame Performer

20.4.1 When a flame performer will be used, a design plan must be incorporated into the event operations plan and submitted in writing. A flame performance design plan includes at least the following:

20.4.1.1 Certain weather conditions can result in the termination of the flame performance, so weather must be monitored (see Chapter 7, *Weather Preparedness*, for more information on monitoring weather). In the plan, define the weather conditions that may require specific actions, up to and including termination of the performance. For an outdoor event, include a weather forecast report in the design plan. For an indoor performance, identify ventilation options and plan for HVAC/detector adjustment and fire-watch procedures, as needed.

20.4.1.2 A site plan (map of the site drawn to scale) must be developed that indicates at least the following:
- Location and size of flame performance area;
- Identify location for preparation of equipment involved with flame performance;
- Means of ignition and extinguishing of flame equipment;
- Restricted zones;
- Area affected by the flame effect and performance—identify materials for flame retardant treatment and type;
- Proximity sensors as applicable or means of monitoring intrusion (security);
- Audience proximity to performance and preparation areas;
- Animal proximity;
- Performer and cast member proximity to each effect;
- Storage location of flammable and combustible liquids or gases;
- Location and construction specifications of flammable props and materials;
- If outdoors, prevailing wind direction;
- Means of egress; and
- Location of firefighting equipment, quantity and type.

20.4.1.3 A list and description of each effect including size, location, fuel load, and maximum height, width and BTU of all flame effects.
- Identify the flame effect group as outlined in NFPA 160, *Standard for the Use of Flame Effects Before an Audience.*

20.4.1.4 A description of safety control mechanisms, as applicable, as well as the means of activation or ignition of props for the flame performance and the process for extinguishing flame equipment within the flame performance.

20.4.1.5 Material Safety Data Sheets (MSDS), a.k.a., Safety Data Sheets (SDS), for all products being used and corresponding effect.

20.4.1.6 A description of any device pre-engineered from a manufacturer including name, address, and phone number of the manufacture.

20.4.1.7 Documentation verifying that all props and equipment manufactured by the operator meet the requirements for proper use and application per NFPA standards and/or local fire codes. All materials should be listed (i.e., acceptable to the AHJ, tested and found suitable for a specific purpose) and acceptable for intended use.

20.4.2 When a flame performer is involved, the following written procedures and documents must be developed and maintained:
- Pre-show and start-up safety reviews and equipment checks along with documentation/forms to support them.
- A show operations manual that includes applicable documentation for show performance and process verification, and that identifies the sequence of effects and application within the performance.

- Emergency procedures including a fire hazards evaluation for permanent installations to determine level of fire protection required and in place.
  - In all cases, at least two pressurized water 2-A extinguishers and two 10-BC extinguishers must be provided, in addition to those required by NFPA 10, *Standard for Portable Fire Extinguishers*, for the venue. The extinguishers must be placed so that at least one each is located on opposing sides of the performance where flame effects are used.
- Normal and emergency shut-down procedures for the effects.
- Maintenance schedules and procedures for all equipment involved.
- A copy of all applicable documentation, including records of any inspections and operational tests, should be included in the written materials for verification of operability of protective control systems such as fire protection systems.
- Schematic and detailed drawing of all flame props and special effects.

20.4.3 When a flame performer is involved, an operator must submit detailed documentation of his or her qualifications and experience, including the same for all operators, crew members, designers, and fabricators. These documents must include at least the following:
- Name, address, and contact information of company.
- Commercial general liability insurance with an appropriate endorsement for pyrotechnics (a limit of no less than $5 million is recommended) and workers' compensation insurance covering all the vendor's employees. Confirm that the coverage meets all the requirements of the AHJ for permitting and operations.
- History of past shows so they can be thoroughly vetted. Make sure to identify the types of flame effects and systems used and the description of any incidents involving the use of flame special effects, especially where someone was injured or killed.
- Licenses or certifications held in any state and/or country.
- Collect and review a résumé for each person working for the contracted company that will work at your show or manage any portion thereof. Make sure the résumé includes all relevant experience and that all flame operators are at least 21 years of age.
- Event and venue management should review all information and documentation prior to its submission to the AHJ to meet permitting requirements. A copy of all approved permits must be maintained on site and made available to inspectors and regulators upon request.
- If a permit is not required, confirm this fact, document it, and maintain a record of it.

20.4.4 Performers, support personnel and non-participants exposed to the associated hazards of the flame special effects must be informed by the operator of the hazards and trained in emergency procedures.
- Identify the effect on animal areas as well as procedures for proper animal handling.

20.4.5 Identify the personal protective equipment (PPE) to be used that is applicable to the hazards associated with the exposure to, and handling and use of, the flame effects utilized for the performance.

20.4.6 All flammable and combustible liquids not vital to and approved for the performance must be stored in holding areas designed and maintained in accordance with applicable NFPA and local fire codes. They must also be properly labeled and stored in areas that are locked and secured.

20.4.7 Documentation must be maintained confirming that all affected combustible set, scenery and rigging materials, have been treated with flame retardant materials.

20.4.8 The flame effect operator must inspect and document all areas of the site where flame props and effect devices will be ignited before start-up and after shut-down.

20.4.9 Written documentation should be maintained confirming that all flame special effects operators, crew members and flame performers have successfully passed a drug test in accordance with an acceptable drug and alcohol policy. Although this is not currently required in all jurisdictions, it is a recommended best practice.

20.4.10 For flame effects requiring pressurized expulsion of fuels for effect, see the requirements under "Guidelines - Flame Effect" (section 20.3) above.

20.5 Guidelines – Pyrotechnics

20.5.1 When pyrotechnics will be used, a design plan must be incorporated into the event operations plan and submitted in writing. A pyrotechnics design plan includes at least the following:

20.5.1.1 Certain weather conditions can result in modification of the execution of the performance and termination of the pyrotechnic effects, so weather must be monitored (see Chapter 7, *Weather Preparedness*, for more information on monitoring weather). In the plan, define the weather conditions that may require specific modifications to the execution of the performance, up to and including termination. For an outdoor event, include a weather forecast report in the design plan. For an indoor performance, identify ventilation options and plan for HVAC/detector adjustment and fire-watch procedures, as needed.

20.5.1.2 A site plan (map of the site drawn to scale) must be developed that indicates at least the following:
- Location of all pyrotechnic materials including classification and trajectory;
- Identify the location for the preparation of any product involved with the pyrotechnic display;
- Operator control areas;
- Restricted zones;
- Area(s) affected by pyrotechnic materials—identify materials for flame retardant treatment and type;
- Proximity sensors, as applicable, or means of monitoring and securing discharge and fallout locations;
- Audience proximity to performance and preparation areas;
- Performer and cast member proximity to each effect;

- Storage location of pyrotechnic materials;
- If outdoors, prevailing wind direction;
- Means of egress; and
- Location of firefighting equipment, quantity and type.

20.5.1.3 A list of all pyrotechnic materials including:
- Size;
- Classification;
- Quantities;
- Maximum height and duration of the product performance specification;
- Types of debris produced; and
- Manufacturer with contact information.

20.5.1.4 A description of the safety control mechanisms and emergency control devices.

20.5.1.5 A description of the means of ignition or electronic firing of the system. Manual ignition (hand-lighting) of pyrotechnic materials for display before an audience is not permitted.

20.5.1.6 Material Safety Data Sheets (MSDS), a.k.a., Safety Data Sheets (SDS), for all products being used and corresponding effect.

20.5.1.7 Documentation of inspection by a professional engineer for all fire control systems manufactured by the operator or user. This should include review of all safety features, materials, and design for proper use and application per NFPA standards and/or local fire codes. All control systems should be listed (i.e., acceptable to the AHJ, tested and found suitable for a specific purpose) and acceptable for intended uses.

20.5.2 When pyrotechnics are involved, the following written procedures and documents must be developed and maintained.
- Pre-show and start-up safety reviews and equipment checks along with documentation/forms to support them;
- A show operations manual that includes applicable documentation for show performance and process verification, and that identifies the sequence of effects and application within the performance;
- Emergency procedures including a fire hazards evaluation for permanent installations to determine level of fire protection required and in place;
- Normal and emergency shut-down procedures for the effects;
- Maintenance schedules and procedures for all equipment involved;
- A copy of all applicable documentation, including records of any inspections and operational tests, should be included in the written materials for verification of operability of protective control systems such as fire protection systems;
- Description and schematic (detailed scale drawing) of all effect holders and means of securing them to assure proper function. This must include at least the following:
  - Identify positioning and trajectories for each pyrotechnic device;

- o Pyrotechnic materials must only be fired from equipment specifically constructed for the purpose of firing pyrotechnic materials, and mounted in a secure manner to maintain their proper positions and orientations when fired according to the submitted plan;
- o Mortars in racks must be of sufficient design to prevent repositioning by a shell exploding in a mortar, which causes the mortar to burst;
- o Where there is doubt concerning the strength of racks holding chain-fused mortars, the separation distances from those racks to spectators must be twice that required for the size of shell used. In addition, a screen barrier must be utilized that is of sufficient strength for containment. In this situation, a schematic on the rack configuration and containment must be provided for approval prior to use;
- o Base side supports for rack formations must extend at least 3/4 the distance from the height of the tallest mortar, with a minimum of three securing points for each support to the rack frame;
- o Account for recoil forces generated from pyrotechnic ignition; and
- o Where a pyrotechnic special effect is placed near on or in contact with a performer's body, a means of shielding or containment adequate to prevent any injury to the performer must be provided.

20.5.3 When pyrotechnics are involved, an operator must submit detailed documentation of his or her qualifications and experience, including the same for all operators, crew members, designers, and fabricators. These documents must include at least the following:

- Name, address, and contact information of company;
- Commercial general liability insurance with an appropriate endorsement for pyrotechnics (a limit of no less than $5 million is recommended) and workers' compensation insurance covering all the vendor's employees. Confirm that the coverage meets all the requirements of the AHJ for permitting and operations;
- History of past shows so they can be thoroughly vetted. Make sure to identify the type of materials, classification and performance size as well as a description of any incidents involving the use of pyrotechnics, especially where someone was injured or killed;
- Licenses or certifications held in any state and/or country;
- Collect and review a résumé for each person working for the contracted company that will work at your show or manage any portion thereof. Make sure the résumé includes all relevant experience and that all pyrotechnic operators are at least 21 years of age; and
- Event and venue management should review all information and documentation prior to its submission to the AHJ to meet permitting requirements. A copy of all approved permits must be maintained on site and made available to inspectors and regulators upon request.

20.5.4 Performers, support personnel and non-participants exposed to the associated hazards of the flame special effects must be informed by the operator of the hazards and trained in emergency procedures.

- Identify the effect on animal areas as well as procedures for proper animal handling.

20.5.5 Identify the personal protective equipment (PPE) to be used that is applicable to the hazards associated with the exposure to, and handling and use of, the pyrotechnic materials used in the performance.

20.5.6 For multiple performances, all pyrotechnic materials for the day's scheduled performances must be stored in holding areas designed and maintained in accordance with applicable NFPA and local fire codes. They must also be properly labeled and stored in areas that are locked and secured.

20.5.7 Documentation must be maintained confirming that all affected combustible set, scenery and rigging materials, have been treated with flame retardant materials. All pyrotechnic materials must be properly secured to noncombustible materials.
- All materials worn by performers in the fallout area during use of pyrotechnic effects must be inherently flame-retardant or have been treated to be flame-retardant.

20.5.8 Pyrotechnic operator must inspect and document all areas of the site where pyrotechnics effects are ignited before start-up and after shut-down.

20.5.9 Written documentation should be maintained confirming that all pyrotechnic operators and crew members have successfully passed a drug test in accordance with an accepted drug and alcohol policy. Although this is not currently required in all jurisdictions, it is a recommended best practice.

20.5.10 All pyrotechnic operators and approved spotters must have a clear view of the effects, and all spotters must be in communications with the operators to verify proper clearances and safe operations of the effect performance.

20.5.11 All firing control systems, mechanisms and switches should be designed and located to prevent tampering and accidental firing of pyrotechnic materials.

20.5.12 All firing control systems must have an emergency stop system to cause shut-down of the pyrotechnic effects system. This system must bring the device to a zero energy state causing all ignition sources to be unpowered.
- Control firing system should have capability to omit firing of identified pyrotechnic materials within the operation of the performance sequence.

20.5.13 Each control system for ignition of pyrotechnic material must be provided with a means of primary lock-out that interrupts the supply of control energy.

20.6 Checklist – Gas Flame

20.6.1 When using gas flame effects, ensure that the following tasks are completed:

20.6.1.1 Required permitting by the AHJ has been confirmed and documented.

20.6.1.2 Confirm that the characteristics of the following items are sufficient to support the fire performance:

- Ceiling height;
- Building construction materials;
- Décor;
- Fire suppression;
- Exits; and
- Flooring material.

20.6.1.3 Confirm that the site plan matches the visual inspection of the identified location.

20.6.1.4 Where flammable liquids and gases will be used, confirm the following:

- That all flammable liquids are properly stored and labeled in approved containers;
- That general housekeeping is according to proper standards;
- That there is convenient access to critical areas;
- That any items not directly related to the storage or management of the fuel supply have been removed;
- That propane and other pressure vessels are secured and stored outside with "No Smoking" signs and that they are protected from vehicle traffic; and
- That all flammable liquids and gasses are kept at least 25 feet (7.62 m) away from heat sources.

20.6.1.5 Visually inspect the performance area(s), checking for the following:

- Confirm that general housekeeping is according to proper standards;
- Confirm that there is convenient access to critical areas;
- Confirm that there is no accumulation of incidental flammable materials;
- Confirm the appropriate condition of sets and props; and
- Confirm that intrusion prevention (security) is in place.

20.6.1.6 Visually inspect the fire effects equipment and systems, checking for the following:

- Assure proper clearances (to combustibles) as outlined in the plot plan;
- For indoor effects, inspect the ventilation system and equipment;
- Confirm that all hoses and valves for pressurized gasses are approved;
- Confirm that flame effects are properly mounted and secured;
- Confirm that the fuel supply is properly secured;
- Confirm that control systems and safety sensors are fully operational;
- Confirm that the emergency shutdown for the effect, including fuel management and fire suppression measures, are present and/or operating normally; and
- Assure that all flame effect devices have been tested to verify they operate in accordance with their designs.

20.6.1.7 Review and evaluate personnel, checking for the following:

- Identify and confirm that any performers in close proximity to flame effects are protected by protective clothing and/or means suitable for their exposure; and

- Confirm that the pyrotechnician (operator) has demonstrated competency by training or experience.

20.6.1.8 Review the emergency operations (safety plan) and check for the following:
- Confirm that all necessary first aid personnel and equipment are available;
- Assure that the emergency exit plan is in place and that the retreat path for the operational crew is clear;
- Confirm that all fire suppression personnel and equipment are available;
- Review the description and location of all safety and suppression equipment;
- If weather conditions can impact the event, confirm that weather monitoring services and related communication channels are in place;
    o Review with everyone involved the conditions that would prompt the termination of the use of the effects;
    o Establish exactly who will make weather related termination decisions and communicate it to everyone involved;
- Review the emergency response and emergency communications plan;
    o Review the need and process for interrupting any automated life safety systems (i.e., disabling smoke detectors, etc.) and take this action only if and when necessary;
    o Confirm that all response personnel have a means of communication and raising the alarm; and
- Identify and confirm the position(s) and suitability of all fire safety personnel, including flame effect assistants.

20.6.1.9 After the performance, confirm that all post-show tasks have been completed, including the following:
- The fire performer must adequately clean-up the hazard area (the area where the flame effect is ignited, including an appropriate safety perimeter) and the holding area (a staging area where flame effect materials, tools and/or fuels are held and fueled prior to use before an audience) and perform a final walk-through confirming that the areas are clean; and
- All interrupted life safety systems (i.e., smoke detectors, fire protection systems, etc.) must be restored to their pre-show condition.

20.7 Checklist – Flame Performer

20.7.1 When using a flame performer, ensure that the following tasks are completed:

20.7.1.1 Required permitting by the AHJ has been confirmed and documented.

20.7.1.2 Confirm that the holding area (a staging area where flame effect materials, tools and/or fuels are held and fueled prior to use before an audience) has the following characteristics:
- It is well lit;
- It is isolated from any audience traffic;
- There are a sufficient number of "No Smoking" Signs posted;

- All flame effect materials and devices that are not in use are stored so as to prevent accidental ignition and properly labeled in approved container;
- The spin-out area is an appropriate size with sufficient ventilation and security;
- There are at least two fire extinguishers fully charged and operational for quick access standing by; and
- There are no additional hazards are located within the holding area.

20.7.1.3 Confirm that the hazard area (the area where the flame effect is ignited, including an appropriate safety perimeter), has the following characteristics:
- There are no excess fuels within the performance area;
- The hazard area is in a neat and orderly condition;
- There are no additional hazards located within the hazard area; and
- The following items are sufficient to support the fire performance:
  - Stage;
  - Safety perimeter;
  - Stage entrances and exits;
  - Tool/prop extinguishment location(s);
  - Ventilation.

20.7.1.4 Visually inspect the fire effects equipment, checking for the following:
- All flame effect devices are in good working condition—For example, looking for any visible signs of wick decomposition, loose fasteners or frayed grips;
- All flame effect devices have been tested to verify they operate in accordance with their designs;
- A walk-through or representative demonstration of the flame effect devices has been provided;
- At least four fire extinguishers (at least two pressurized water 2-A and two 10-BC) are fully charged, operational and ready for quick access placed with at least one on each end of the performance area—these are in addition to those required by NFPA 10, *Standard for Portable Fire Extinguishers*, for the venue; and
- Suitable methods for extinguishing flame effect devices are readily accessible for the performance, including, but not limited to, fire-retardant duvetyn (a soft, short-napped fabric with a twill weave) and/or damp towels.

20.7.1.5 Review and evaluate performers, checking for the following:
- The fire performer and flame effect assistants are protected by clothing or means suitable for their exposure to flame effects (bare skin is permissible for performer);
- The fire performer demonstrated competency by training or experience; and
- Persons that are under the influence of intoxicating beverages, narcotics, prescription drugs and/or non-prescription drugs that could affect judgment, mobility, or stability are prohibited from participating or being in any area associated with the performance.

20.7.1.6 Review operational procedures and check for the following:
- The Fire Performer provided a detailed verbal description of the show;

- The fire performer has advised all performers, flame effect assistants and crew members that they are exposed to a potentially hazardous situation during the execution of the flame effects;
- The security of the holding area and hazard area are sufficient for flame effect preparation and performance;
- If weather conditions can impact the event, confirm that weather monitoring services and related communication channels are in place;
  - o Review with everyone involved the conditions that would prompt the termination of the use of the flame performance;
  - o Establish exactly who will make weather related termination decisions and communicate it to everyone involved;
- There are an appropriate number of trained response personnel (safety officers or fire watch) for the performance; and
- Response personnel have a means of communicating and raising the alarm.

20.7.1.7 Review the emergency operations (safety plan) and check for the following:

- Confirm that all necessary first aid personnel and equipment are available;
- Assure that the emergency exit plan is in place and that the retreat path for the operational crew is clear;
- Confirm that all fire suppression personnel and equipment are available;
- Review the description and location of all safety and suppression equipment;
- Review the emergency response and emergency communications plan;
  - o Review the need and process for interrupting any automated life safety systems (i.e., disabling smoke detectors, etc.) and take this action only if and when necessary;
  - o Confirm that all response personnel have a means of communication and raising the alarm; and
- Identify and confirm the position(s) and suitability of all fire safety personnel, including flame effect assistants.

20.7.1.8 After the performance, confirm that all post-show tasks have been completed, including the following:

- The fire performer must adequately clean-up the hazard area (the area where the flame effect is ignited, including an appropriate safety perimeter) and the holding area (a staging area where flame effect materials, tools and/or fuels are held and fueled prior to use before an audience) and perform a final walk-through confirming that the areas are clean; and
- All interrupted life safety systems (i.e., smoke detectors, fire protection systems, etc.) must be restored to their pre-show condition.

20.8 Checklist – Proximate Pyrotechnics

20.8.1 When using proximate pyrotechnics, ensure that the following tasks are completed:

20.8.1.1 During a walkthrough of the site, make sure that any required permitting by the AHJ has been confirmed and documented, and that the site plan matches the visual inspection of the site and the actual site dimensions.

20.8.1.2 Confirm that any pyrotechnic devices to be used are intended for indoor/proximate use and verify that the performance specifications match the intended use, including quantities, locations and trajectories. Also, make sure what is physically at the site matches what is on the site/plot plan.

20.8.1.3 Verify that appropriate security and methods for restricting unauthorized persons from entering are in place for the site.

20.8.1.4 Verify that all pyrotechnic devices are securely fixed at the proper distance (separation) from the audience:
- The distance from the audience must be a minimum of 15 feet (4.572 m) or 2 times the fallout radius.
- Concussion mortars must be a minimum of 25 feet (7.62 m) from the audience and in a secured area.
  - Warning signal lights must be used to indicate the impending firing of a concussion special effect. The warning signal lights must be located at least 25 feet (7.62 m) from the concussion effect but within a distance to warn working personnel and other individuals of the impending concussion effect firing.
- The trajectory of the effect must match the plot plan.
  - "Comets/mines" must not be fired over an audience.
- Any "waterfall" effect area(s) must be kept free of all flammable materials.
- Any "grid rocket" effects must be properly secured and terminated.
- Where permitted, any "airbursts" over an audience must be a minimum height of 3 times the diameter of the effect and no sparks must fall within 15 feet (4.572 m) of the floor.

20.8.1.5 Review the emergency operations (safety plan) and check for the following:
- Confirm that all necessary first aid personnel and equipment are available;
- Assure that the emergency exit plan is in place and that the retreat path for the operational crew is clear;
- Confirm that all required fire suppression personnel and equipment are available, including at least two pressurized water 2-A extinguishers and any other firefighting equipment required by the AHJ and/or outlined in the operational procedures;
- Review the description and location of all safety and suppression equipment;
- Review the emergency response and emergency communications plan;
  - Review the need and process for interrupting any automated life safety systems (i.e., disabling smoke detectors, etc.) and take this action only if and when necessary;
  - Confirm that all response personnel have a means of communication and raising the alarm;
  - Identify and confirm the position(s) and suitability of all fire safety personnel, including flame effect assistants;
- If weather conditions can impact the event, confirm that weather monitoring services and related communication channels are in place;

     o Review with everyone involved the conditions that would prompt the termination of the use of the pyrotechnics; and

     o Establish exactly who will make weather related termination decisions and communicate it to everyone involved.

20.8.1.6 Identify the personal protective equipment (PPE) to be used that is applicable to the hazards associated with the exposure to, and handling, preparation, and loading of, the pyrotechnic materials used in the performance.

20.8.1.7 Verify that the operational and safety features and functions of the pyrotechnic firing systems are working correctly.

20.8.1.8 Confirm that all pyrotechnic operators and approved spotters have a clear view of the effects, and that all spotters are in communications with the operators to verify proper clearances and safe operations of the performance.

20.8.1.9 All pyrotechnic materials for the day's scheduled performances must be stored in holding areas and containers designed and maintained in accordance with applicable NFPA and local fire codes.

- Any time pyrotechnic material (1.4G only) is stored within a facility, it must be stored in a type 3 storage "day box" (type 3 explosive magazine for the temporary storage of explosives while attended) approved by the AHJ;
- Required separation of any pyrotechnic material from heat, flame and sparks must be maintained;
- Confirm that general housekeeping is according to proper standards; and
- All pyrotechnic materials, including the storage day box in which it is stored, must be properly labeled indicating its classification (i.e., 1.4G) and posted with a sign that indicates that "No Smoking" is permitted within 25 feet (7.62 m).

20.8.1.10 Assure that the pyrotechnic operator has demonstrated competency by training or experience, and prohibit persons in any area associated with the performance who are under the influence of intoxicating beverages, narcotics, prescription drugs and/or non-prescription drugs that could affect judgment, mobility, or stability.

20.8.1.11 Confirm that the following tasks have been completed as part of the rehearsal and show checklists:

- Hold a safety meeting with all participants;
- Use the scheduled pyrotechnics at rehearsal, if required, so that the performers can acclimate to them within the performance and confirm the sequence of events;
- Before the performance, determine who will give, and how the final alert will be given, to all performers, crew and support personnel;
- Conduct a final inspection immediately before the performance and complete the following tasks:
    - o Confirm that the pyrotechnic wiring connections and firing system are configured correctly and working;

o Confirm that the pyrotechnic devices are placed, mounted and oriented correctly, and that everything connected to it is ready;

o Confirm that all necessary firefighting equipment is in position and ready to be deployed;

o Confirm that when all operators and approved spotters are in place, they will have an unobstructed, clear view of the effects; and

o Confirm that all approved spotters will be in communications with the operators during the performance to verify proper clearances and safe operations.

- During the execution of the show, assure that the following tasks are completed:

  o Confirm and maintain the required audience separation;

  o Confirm that all performers and support personnel are ready and positioned correctly;

  o Confirm that all approved spotters are in communications with the operators to verify proper clearances and safe operations of the effect performance.

  o Confirm that all operators and approved spotters have an unobstructed, clear view of the effects; and

  o Confirm that someone (safety officer?) is observing and monitoring the effect performance for safe operation.

20.8.1.12 After the performance, confirm that all post-show tasks have been completed, including the following:

- The fire performer must adequately clean-up the hazard area (the area where the flame effect is ignited, including an appropriate safety perimeter) and the holding area (a staging area where flame effect materials, tools and/or fuels are held and fueled prior to use before an audience) and perform a final walk-through confirming that the areas are clean; and

- All interrupted life safety systems (i.e., smoke detectors, fire protection systems, etc.) must be restored to their pre-show condition.

20.9 Checklist – Fireworks

20.9.1 When an event includes fireworks, ensure that the following tasks are completed:

20.9.1.1 During a walkthrough of the display area, make sure that any required permitting by the AHJ has been confirmed and documented, and that the site plan (map of the site drawn to scale) matches the visual inspection of the display area and the actual site dimensions.

20.9.1.2 Confirm that a site plan has been developed that indicates at least the following:

- The locations of nearby buildings, roads, and overhead obstructions; and

- The separation distances between the audience and the fireworks display area, which must be a minimum of 70 feet (21.34 m) radius from the display for every inch of internal mortar diameter (unless increased per operational procedures).

20.9.1.3 Verify that the types of devices to be used and their classifications are appropriate.

- Confirm that the performance specifications, quantities, locations and trajectories of the fireworks to be used match the planned and intended use.

20.9.1.4 Verify that appropriate security and methods for restricting unauthorized persons from entering are in place for the display site.

20.9.1.5 For all transportation and safety equipment, do the following:
- Confirm that all required vehicle placards are in place;
- Confirm that all fireworks are stored in an approved manner and location; and
- Confirm that any storage containers used for fireworks are locked and secured.

20.9.1.6 For all fire protection and safety equipment, do the following:
- Confirm that all required fire extinguishers are available and accessible;
- Confirm that there is a shovel and container available for retrieving duds; and
- Confirm that all appropriate personal protective equipment (PPE) is in use (e.g., hard hat, long sleeved shirt, etc.).

20.9.1.7 Conduct an inspection of every mortar prior to the display that includes at least the following:
- Confirm that all mortars are approved for use with type of shells to be fired;
- Confirm that all mortars are inspected prior to use for defects, debris, etc.; and
- Confirm that all mortars are the proper length for the size of shells to be fired.

20.9.1.8 Conduct an inspection of every rack prior to the display that includes at least the following:
- Confirm that all racks have a base and side supports that extend 3/4 the length of tallest mortar;
- Confirm that all racks are secured to prevent them from falling over; and
- Confirm that racks are in good condition.

20.9.1.9 Conduct an inspection of every aerial shell prior to the display that includes at least the following:
- Confirm that the number of shells on site matches the quantity listed on the permit; and
- Confirm that the operator examines all of the shells for damage.

20.9.1.10 Conduct an inspection of every ground device prior to the display that includes at least the following:
- Confirm that all ground devices match the items listed on the permit;
- Confirm that the location and spacing of devices is the same as is on the approved site plan; and
- Confirm that all necessary equipment is available to secure and brace ground devices.

20.9.1.11 Conduct an inspection of the discharge and fallout area prior to the display that includes at least the following:
- Confirm that the location is as per the approved operational plan;
- Confirm that at least the minimum required fallout area is provided; and
- Confirm that proper security has been established to keep the area clear.

20.9.1.12 Conduct an inspection of the firing control system prior to the display that includes at least the following:
- Confirm that the firing system is protected against accidental firing;
- Confirm that a method for testing the circuit is available; and
- Confirm that the firing panel is located a minimum of 75 feet (22.86 m) from any devices or as identified within the approved operational plan.
    - For less than minimum required distance, an appropriate safety barrier must be properly secured in place.

20.9.1.13 During the operation of the display, confirm the following:
- That the discharge, display and fallout areas are cleared of all unauthorized personnel and remain effectively secured;
- That the discharge area is cleared of all persons prior to electrical circuit testing and when power has been applied to the discharge area; and
- That during discharging of shells, the operator closely observes all activities including monitoring the wind and the fireworks' trajectories and making adjustments, if necessary.

20.9.1.14 After the performance, confirm that all post-display tasks have been completed, including the following:
- No one is permitted to enter the display, discharge or fallout areas until the pyrotechnic operator has scanned these areas for duds and determined the area to be safe ("all clear");
- Once an "all clear" has been established for the fallout area, it may be open to the general public only after a minimum of 20 minutes past the end of the display or after the "all clear" has been called;
- Unfired shells removed and returned directly to supplier or stored in manner approved within operational plan; and
- Discharge site has been cleared and safe to open for general access.

20.9.1.15 Any situation during the display that causes an injury may require the premature termination of the show and the notification of the AHJ so the situation can be investigated. Follow the instructions of the AHJ for these situations and include these required procedures in the operational plan.

20.10 Checklist – Display Fireworks Sites (OSHA)

20.10.1 The following fireworks safety checklists is provided by OSHA and designed to advise display fireworks operators and other affected employers of some procedures that may be followed to help ensure that display fireworks are used safely (*OSHA Safety Guidelines for Display Sites*, 2004). It is important to understand that due to their sensitivity, display fireworks can present hazards when improperly handled or used. Employers are encouraged to follow these or other more protective safety guidelines when using display fireworks. The following safety guidelines do not supersede any regulatory requirements adopted at the Federal, State, or local levels. This information is also available online on OSHA's web site (www.osha.gov).

20.10.1.1 Pre-Display Approval Checklist.
- Obtain required Bureau of Alcohol, Tobacco, Firearms and Explosives (ATF) licenses and permits.

- Obtain U.S. Coast Guard approval for displays fired from harbors or navigable waterways.
- Obtain Federal Aviation Administration (FAA) approval if close to an airport or heliport.
- Ensure pilots are warned through issuance of a Notice to Airmen (NOTAM).
- Submit required applications to the State and/or local Authority Having Jurisdiction (AHJ) and obtain necessary approval, licenses and permits. Minimum items to address include:
  - Qualified operator in charge
  - Properly trained assistants
  - Site layout with proper separation distances
  - Event description
  - Firing procedures
  - Termination procedures
  - Emergency procedures
- Arrange for inspections required by State/local AHJ or Federal authorities.
- Obtain approval from appropriate authorities to close roads or restrict access.
- Arrange for fire service and EMS to be available for the display.
- Obtain the required or appropriate insurance.

20.10.1.2 Pre -Display Site Checklist.
- Establish site security prior to arrival of pyrotechnic materials.
- Protect all fireworks, pyrotechnic materials, and launching equipment from inclement weather and keep them dry at all times.
- Prohibit smoking material, matches, lighters or open flames within 50 feet (15.24 m) of fireworks or pyrotechnic material.
- Only necessary personnel required to perform the display set up and show must be allowed at the display site.
- Prohibit persons in the display site who are under the influence of alcohol, narcotics, or medication that could adversely affect judgment, mobility, or stability.
- No cell phones or radio frequency (RF) generating devices are permitted within the immediate discharge area while electrically ignited fireworks or pyrotechnic devices are prepared, loaded, or set up.
- Wear all personal protective equipment appropriate for setup duties.
- Verify that all mortars and racks are made of approved materials, and are of sufficient strength, length and durability to allow shells to be propelled to safe deflagration heights.
- Make sure all mortars, mortar racks, bundles, pre-loaded box items, cakes, candles, and ground displays have been thoroughly inspected and deemed inherently stable.
- Avoid placing any portion of your body over mortars during loading, wiring, or igniting, and immediately after the display has been fired.
- Use safe handling and loading procedures for all pyrotechnic devices.
- Pre-load larger shells as required.
- Check proper fit of shells in mortars.
- Designate spotter(s).

20.10.1.3 Display Checklist.

- Verify fire service and emergency medical service (EMS) units are available and ready to respond.
- Establish good communications between crew, event sponsor, AHJ, and fire service/EMS units.
- Maintain crowd control, utilizing monitors and/or barriers.
- Use all required personal protective equipment especially protection for head, eye, hearing, and foot.
- Wear long-sleeved and long-legged clothing made of cotton, wool or similar flame resistant cloth.
- Avoid placing any portion of your body over mortars when manually igniting them.
- Monitor weather and crowd conditions to maintain safety.
- Comply with directions given by the AHJ, spotter(s), or fire/EMS units
- Use only flashlights or other nonincendive (circuits which may spark under normal operating conditions, but which may not release enough energy to cause ignition) lighting in firing and ready box areas (A "ready box" is a storage container for aerial devices for use during set-up and display).

20.10.1.4 <u>Post - Display Checklist</u>.
- Wear personal protective equipment appropriate for cleanup duties.
- Disable any electric firing switches and disconnect all electric cables.
- After at least 15 minutes, conduct search of the display and fallout areas.
- Follow proper marking and warning procedures for unexploded shells.
- Ensure that all unused live product and duds are accounted for, properly handled, repackaged and secured according to Federal, State and local regulations.
- Conduct a second site search at first light.

20.11 Checklist – Outdoor Exhibition of Fireworks (example)

20.11.1 To assist with compliance with their fireworks laws and regulations, one state (Ohio) provides a checklist of questions that, when answered in the affirmative, help confirm compliance by both the fire official and the exhibitor with the relevant requirements of NFPA 1123, *Code for Fireworks Display*, NFPA 1126, *Standard for the Use of Pyrotechnics Before a Proximate Audience*, and State fireworks laws and regulations. The full form can be found online as part of the document at: www.com.ohio.gov/documents/2013fireworksredbook.pdf. This same checklist also includes "additional requirements" (see below) that help focus the exhibitor on the essential elements of a safe event. Although the specific requirements for fireworks displays vary by state and local jurisdiction, this question checklist approach has been successful in facilitating communications between the invested parties and effectively documenting compliance with the requirements. This checklist of questions is intended to, first, have each question answered and initialed by the fireworks exhibitor, then verified and initialed by the fire official (AHJ). This is only an example. Please verify all specific requirements with the AHJ.
- Was insurance bond of $1,000,000 or more verified?
- Was product purchase verified to be from a licensed Ohio Wholesaler, Ohio Manufacturer, or Out-of-State Shipper with a shipping permit?

- Has exhibitor identified names of all assistants who will be present? Has exhibitor provided copies of all licenses and permits?
- Was the site inspection conducted prior to issuing permit?
- Have locations, distances, and details of the site plan been verified for accuracy?
- Does the proposed location for temporary storage (up to 14 days) of fireworks comply with this rule?
- Is adequate fire protection available?
- Does distance to spectator area comply with a minimum of 70 feet (21.34 m) per inch (0.0254 m) of largest shell, and/or close proximity separation distances for audiences in accordance with NFPA 1126?
- Is the display site selection in accordance with the relevant sections of NFPA 1123 and/or NFPA 1126?
- When the fireworks arrived at the site, were they in a properly placarded vehicle/trailer?
- If the show requires reloading of fireworks, are the extra fireworks properly secured in a ready box, 25 feet (7.62 m) upwind of the mortars?
- Was the condition of mortars checked for damage, dents, broken plugs, etc.?
- Are all shells greater than or equal to 8 inches in diameter provided with electronic ignition?
- Are the mortars made of approved material, of sufficient strength, length and durability to cause shells to be propelled to safe altitudes?
- Are buried mortars installed to comply with NFPA 1123?
- Are mortar racks and/or bundles that are not inherently stable, secured or braced to stabilize them?
- Are the racks properly positioned to prevent them from firing towards the spectators?
- Are the racks braced in a manner that secures the balance of the mortars should one fail?
- Are all security persons and monitors in place and positioned to prevent unauthorized persons in discharge site?
- Is smoking or open flame prohibited where fireworks are present?
- Have all mortars, mortar racks, bundles, box items, ground displays, cakes, and candle placements been examined, and any items not inherently stable secured properly or braced for stability?
- Are only the licensed exhibitors and designated registered assistants within the discharge perimeter?

20.11.1.1 Additional requirements (on the state checklist):
- Both the Fire Official and licensed exhibitor should mutually agree in advance on a method used to communicate during the exhibition. Agreement should also be made in advance pertaining to the exact location the Fire Official will be stationed to maintain safety for all involved.
- If a condition arises requiring the entry of fire protection or other emergency response personnel into the fallout area security perimeter, the display must be halted until the situation is resolved.
- If a significant hazard exists due to weather, lack of crowd control, or other condition, the exhibition must be halted until resolved.
- The security of the display site must be maintained until released by the exhibitor.

- Check with the safety person and monitors for any signs of problems.
- Allow registered assistants to enter the area and attend to extinguishing fires, smoldering embers, and debris in the firing area and fallout area.
- Before entering the area, wait a minimum period of time, that the exhibitor deems necessary, to include letting the area cool with resulting inspection by fireworks crew.
- Confer with the licensed exhibitor and, if mutually agreed, release fire crew and equipment from the scene. Do not release security or monitors. Maintain barricades for area until spectators have left.
- Be sure all live product and duds are properly repackaged and secured into vehicle. Replace placards on vehicle.
- Where fireworks are displayed at night, appropriate morning re-inspection of the site should be mutually agreed to.

20.12 Checklist – Indoor Fireworks or Flame Effects Exhibition (example)

20.12.1 To assist with compliance with their fireworks/flame effects laws and regulations, one state (Ohio) provides a checklist of questions that, when answered in the affirmative, help confirm compliance by both the fire official and the exhibitor with the relevant requirements of NFPA 160, *Standard for the Use of Flame Effects Before an Audience*, NFPA 1126, *Standard for the Use of Pyrotechnics Before a Proximate Audience*, and State fireworks and flame effect laws and regulations. The full form and checklist can be found online as part of the document at: www.com.ohio.gov/documents/2013fireworksredbook.pdf. This same checklist also includes "additional requirements" (see 20.11.1.1, above) that help focus the exhibitor on the essential elements of a safe event. Although the specific requirements for fireworks displays vary by state and local jurisdiction, this question checklist approach has been successful in facilitating communications between the invested parties and effectively documenting compliance with the requirements. This checklist of questions is intended to, first, have each question answered and initialed by the fireworks exhibitor, then verified and initialed by the fire official (AHJ). This is only an example. Please verify all specific requirements with the AHJ.

- Was insurance bond of $1,000,000 or more verified?
- Was product purchase verified to be from a licensed Ohio Wholesaler, Ohio Manufacturer, or Out-of-State Shipper with a shipping permit?
- Has exhibitor provided the number, names and ages of all assistants who will be present?
- Was the site inspection conducted prior to issuing permit?
- Has all planning and use of pyrotechnics and flame effects been coordinated with the venue owner, manager, or producer?
- Was a written plan submitted to AHJ identifying firing and fallout area for each device, location of the audience, number & types of devices to be fired, and a description of each effect?
- Have locations, distances, and details of the site plan been verified for accuracy?
- Was certification provided that the set, scenery, rigging materials, and all materials worn by performers in the fallout area are inherently flame retardant or have been treated to be so?
- Was a walk through and demonstration provided and approved?

- Was a MSDS sheet provided for all pyrotechnic materials and fuels to be used in the effects?
- Do all electrical firing systems have written instructions and/or a description of performance specification of flame effect created?
- Does the flame effect system have a control system for emergency stop and complete shutdown?
- Are all devices mounted so no fallout or flame damages property, causes personal injury, or death?
- Is the fire protection system going to be interrupted during the performance? If yes, the owner must be notified and an approved fire watch present during that time.
- Is all required portable firefighting equipment present and ready for use?
- Are pyrotechnic product, binary systems, and mixing containers identified or marked by manufacturer?
- Has type of communication been determined that will be used during the exhibition?
- Are all pyrotechnic materials/devices and fuel stored in a secured, inaccessible, or supervised area?
- Are all firing devices constructed and secured to remain in a fixed position during firing?
- Is smoking or open flame prohibited where pyrotechnic materials or fuel is present?
- Are all mortars and flash pots constructed so that they do not fragment or become distorted in shape when pyrotechnic material is fired?
- Have measures been established to provide crowd management, security, fire protection, and emergency services?

20.13 Safety Resources

20.13.1 OSHA offers a number of relevant resources that might be useful to event organizers who will be dealing with pyrotechnics, fireworks, and/or flame effects. All of them can be accessed through their web site (www.osha.gov/SLTC/pyrotechnic/display/display.html) and they include the following:

- General industry OSHA standards (29 CFR 1910);
- Maritime activities, such as the launching aerial displays from barges, must comply with OSHA maritime standards (29 CFR 1918);
- Construction activities, such as the building and removal of the display structures, must comply with OSHA construction standards (29 CFR 1926);
- Fireworks safety display poster (in conjunction with American Pyrotechnics Association);
- OSHA Compliance Policy for Manufacture, Storage, Sale, Handling, Use and Display of Pyrotechnics;
- Standard interpretations, for example, "1910.119 does not apply to public displays of flame effects" (1997, May 16);
- Example enforcement actions; and
- OSHA Safety Guidelines for Display Fireworks Sites (June 2004).

20.13.2 All documents included in the U.S. Code of Federal Regulations (e.g., 29 CFR 1910) are available through the Electronic Code of Federal Regulations (e-CFR) at www.ecfr.gov.

20.13.3 The U.S. Department of Justice (DOJ), Bureau of Alcohol, Tobacco, Firearms, and Explosives (ATF) is authorized to inspect the site of any accident or fire where there is reason to believe that explosive materials were involved. Other Federal agencies, or state and local agencies, might also investigate such incidents, depending on the circumstances (www.atf.gov/).

- ATF occasionally gets requests to temporarily store fireworks in a vehicle prior to the staging of a large event such as the Fourth of July. Variances are approved and this newsletter lists the minimum conditions which must be observed: (www.atf.gov/explosives/how-to/request-variance-exemption-or-determination.html).
- ATF has prepared a selection of those questions that it receives most often about commerce in explosives (29 CFR 55 and 18 USC 40). These questions and answers can be found at: www.atf.gov/files/publications/newsletters/fel/fel-newsletter-2006-09.pdf#page=3
- ATF rulings, general Information for the fireworks industry, explosive dealer's and user's basic guide to federal explosives regulation and additional information can be found at: (www.atf.gov/files/publications/download/p/atf-p-5400-7.pdf).
- All storage of all pyrotechnics and fireworks must be stored in compliance with the ATF regulations at 27 CFR, Part 555, Subpart K—Storage, as well as all relevant State and local regulations.

20.13.4 Consumer fireworks (1.4G) and any professional (commercial), display fireworks device that contains a specific defect that presents a risk to consumers, such as those observing fireworks displays, may be subject to the provisions of Section 15 of the Consumer Product Safety Act, 15 U.S.C. Part 2064 (www.law.cornell.edu/uscode/text/15/2064).

- The Consumer Product Safety Commission (CPSC) fireworks business guidance regulations for the general public can be found at: (www.cpsc.gov/en/Business--Manufacturing/Business-Education/Business-Guidance/Fireworks/).

20.13.5 U.S. Department of Transportation (USDOT), Office of Hazardous Materials Safety, Pipeline and Hazardous Materials Safety Administration (PHMSA) has a web page dedicated to information regarding common mistakes made on Federal fireworks applications, links to the approvals database search, procedures for obtaining EX numbers, guidance and criteria for fireworks novelties, and instructions/required information for the designated agent. It can be found at: (www.phmsa.dot.gov/hazmat/regs/sp-a/approvals/fireworks).

- Some of 49 CFR Subtitle B, *Other Regulations Relating to Transportation*, is also relevant to the transportation of fireworks.

20.13.6 The American Pyrotechnic Association (APA, www.american pyro.com) is a fireworks industry group that promotes safety in the design and use of all types of legal fireworks. Its members are committed to safety and regulatory compliance and are a good source of industry experts.

- On its web site, the APA includes a directory of state laws related to pyrotechnics and fireworks. It can be found at: (http://www.americanpyro.com/state-law-directory).

20.13.7 The Ohio State Fire Marshal's Office, like similar offices in other states, publishes basic safety and regulatory information for fireworks exhibitors and event organizers on their web site. In Ohio, a very helpful document is published each year that includes the latest laws and industry guidance for producing a safe and enjoyable show. All event organizers may benefit from reviewing this document and seeking out its equivalent for the state in which the event will be held. This document can be found at: www.com.ohio.gov/documents/2013fireworksredbook.pdf. California's fireworks guidance can be found online at the Office of the State Fire Marshal: http://osfm.fire.ca.gov/strucfireengineer/strucfireengineer_fireworks.php and similar information for Texas can be found here: (http://www.tdi.texas.gov/fire/fmlifirework.html).

20.13.8 Natural Resource Canada (www.nrcan.gc.ca/explosives/fireworks/9883) has an informative web site that provides information on fireworks and explosives. For organizers of events in Canada, this is essential reading. Although Canada uses some of the same standards as those used in the United States, Canadian fireworks regulations differ from U.S. regulations, sometimes significantly. Make sure to become familiar with, and comply with, the laws and rules that apply to the location in which your event will take place.

- Natural Resources Canada publishes the Display Fireworks Manual (2010), which applies in Canada to commercial fireworks displays but not to pyrotechnic special effects. It can be found online at: (www.nrcan.gc.ca/explosives/fireworks/9903).

20.13.9 The Journal of Pyrotechnics (www.jpyro.com) is an organization "dedicated to the advancement of pyrotechnics through the sharing of information" that publishes a technical journal on pyrotechnics, including fireworks, pyrotechnic special effects, propellants and rocketry, and civilian pyrotechnics. Articles encompass reports on research, reviews, and tutorials. The Journal is published twice a year.

21. Special Effects and Lasers, Other than Pyrotechnics

21.1 Special Effects

21.1.1 Special effects include, but are not limited to the following:
- Mechanical effects such as hydraulics, flying or suspended props, scenery or performers;
- Mock fires and/or explosions;
- Atmospheric effects such as wind, rain, fog, snow, and clouds;
- Puppets and prosthetics;
- Some wardrobe, hair and prosthetic make-up enhancements;
- Computer generated imagery (CGI);
- Other optical effects also known as "photographic effects;" and
- A magic show's physical elements.

21.1.2 A person seeking to hire a special effects company should educate him- or herself on the desired effect and the companies that specialize in providing that effect.

21.1.3 Because some special effects can be expensive, producers and event organizers may experience "sticker shock" when they first see cost estimates for the desired special effects. Although price is an important consideration, producers should base their selection on a variety of factors, including the qualifications of the vendor and quality of materials used. Producers should avoid making product and service selections based solely on price or cost.

21.1.4 Special effects are often customized for a particular purpose and created from scratch. There is often a significant amount of Research and Development (R&D) required to create the desired effect.

21.1.5 Generally, producers should employ a "buyer beware" perspective when seeking a vendor. Be clear in the request for proposal and diligent when comparing quotations between potential vendors.

21.1.6 As with all contracts, once a decision is made on the vendor and you are ready to move forward with a contract, take extra care to clearly define the scope of work, the indemnity and liability clauses and the remedies in the event of a breach of contract.

21.2 Lasers

21.2.1 The term "LASER" is an acronym for "Light Amplification by Stimulated Emission of Radiation."

21.2.2 Laser Hazard Classifications (OSHA Technical Manual, 1999)

21.2.2.1 The intent of a laser hazard classification is to provide warning to users by identifying the hazards associated with the corresponding levels of accessible laser radiation through the use of labels and instruction. It also serves as a basis for defining control measures and medical surveillance.

21.2.2.2 Lasers and laser systems are required by Federal law, 21 CFR Part 1000, to be classified and appropriately labeled by the manufacturer. It should be stressed, however, that the classification may change whenever the laser or laser system is modified to accomplish a given task. Virtually all of the U.S. domestic, as well as all international standards, divide lasers into four major hazard categories called the laser hazard classifications. The classes are based upon a scheme of graded risk, which is essentially the ability of a beam to cause damage to the eye or skin.

- Class I: cannot emit laser radiation at known hazard levels (typically continuous wave [cw]: 0.4 µW at visible wavelengths). Users of Class I laser products are generally exempt from radiation hazard controls during operation and maintenance (but not necessarily during service). Since lasers are not classified on beam access during service, many Class I industrial lasers will consist of a higher class (high power) laser enclosed in a properly interlocked and labeled protective enclosure. In some cases, the enclosure may be a room (walk-in protective housing) which requires a means to prevent operation when operators are inside the room.
- Class I.A: a special designation that is based upon a 1000-second exposure and applies only to lasers that are "not intended for viewing" such as a supermarket laser scanner. The upper power limit of Class I.A. is 4.0 mW. The emission from a Class I.A laser is defined such that the emission does not exceed the Class I limit for emission duration of 1000 seconds.
- Class II: low-power visible lasers that emit above Class I levels but at a radiant power not above 1 mW. The concept is that the human aversion reaction to bright light will protect a person. Only limited controls are specified.
- Class IIIA: intermediate power lasers (cw: 1-5 mW). Only hazardous for intrabeam viewing where the eye is exposed to all or part of a direct laser beam or a specular reflection. Some limited controls are usually recommended.
- Class IIIB: moderate power lasers (cw: 5-500 mW, pulsed: 10 J/cm2 or the diffuse reflection limit, whichever is lower). In general Class IIIB lasers will not be a fire hazard, nor are they generally capable of producing a hazardous diffuse reflection. Specific controls are recommended.
- Class IV: High power lasers (cw: 500 mW, pulsed: 10 J/cm2 or the diffuse reflection limit) are hazardous to view under any condition (directly or diffusely scattered) and are a potential fire hazard and a skin hazard. Significant controls are required of Class IV laser facilities.

21.2.3 Event Related Laser Safety Regulations and Considerations

21.2.3.1 In the United States, several organizations concern themselves with laser safety. These organizations include

- The American National Standards Institute (ANSI)(www.ansi.org);
- The Center for Devices and Radiological Health (CDRH) of the U.S. Food and Drug Administration (FDA)(www.fda.gov/MedicalDevices/);
- The U.S. Department of Labor's Occupational Safety and Health Administration (OSHA)(www.osha.gov);
- The Council of Radiation Control Program Directors (CRCPD)(www.crcpd.org); and
- Several state governments and the CRCPD have developed a model state standard for laser safety.

21.2.3.2 As described below, both the laser presentation and the projector must comply with generally recognized laser safety requirements and laws. In the U.S., this means compliance with federal laws and having a valid variance (formal permission to deviate from a requirement of the regulations). Some states and localities may also have requirements. If lasers are used outdoors, the appropriate aviation authority may have additional requirements.

21.2.3.2.1 For laser light shows and devices, a variance permits use of laser radiation levels that exceed the limits (Class IIIA) for demonstration laser products as specified in 21 CFR 1040.11(c). A variance for laser light shows and devices is generally granted based on a determination that the product is required to perform a function which cannot be performed with equipment in compliance with the standard and that suitable means of radiation safety and protection will be provided. These suitable means are specified in the conditions of the variance and constitute, together with the balance of the laser product performance standard, an individual performance standard for a specific manufacturer of those specific laser products that may be certified by the manufacturer under the variance. Refer to 21 CFR 1010.4, *Variances*, for more information on laser variances. Review of the *Compliance Guide for Laser Products* (June 1992, U.S. Department of Health and Human Services, Public Health Service, FDA, CDRH) may also be helpful and can be found on the FDA web site (www.fda.gov/).

21.2.3.3 The laser show company is responsible for, and should handle, all reporting requirements. These requirements should not be assigned to the producer or venue.

21.2.3.4 The following documents are required and event planners in the U.S. should review them prior to the event:
- The laser company variance issued by Food and Drug Administration (FDA) / Center for Devices and Radiological Health (CDRH);
- Show reports or notifications, if applicable;
- Company's proof of applicable insurance coverage, e.g. General Liability and Workers Compensation;
- Proof of other necessary coverage based on the event location or event; and
- If the event is outdoors, an approval letter from the FAA may be required, especially when the lasers are not able to be terminated into a non-reflective structure.

21.2.3.5 In the U.S., a given laser's beam is typically required to be 10 feet (3.048 m) above the floor surface where an audience will stand, and 8 feet (2.44 m) laterally from where the audience can reach sideways. In practical terms, this means the venue ceiling must be a minimum of 12 feet (3.66 m) high, giving a foot (0.3048 m) or so for the beam effects. It also means that if a projection is coming from behind the audience, towards a screen, the beam must always be at least 10 feet (3.048 m) above the floor where the audience is located.

21.2.3.6 If the laser projection could bounce off reflective surfaces, such as mirror strips and chandeliers into audience areas, the beam must be masked to protect the audience from any stray beams. Other reflective surfaces may be present, including mirrors on intelligent lights, silver trusses, Mylar balloons, and reflective posters.

21.2.4 General Laser Use Considerations

21.2.4.1 Adequate time for aiming and fine-tuning laser projections is required and should always be built into the production schedule. There should be no one in the laser areas, except the laser company's technicians during these tests. A good time for setup is when other crews are taking a meal break, or after they have finished their calls. Generally, the laser company will alert the producer prior to laser setup and will verify all personnel have left the event area, including catwalks, backstage areas, etc.

21.2.4.2 During the pre-event briefing, the laser show company should be included in the relevant event management, house management and security briefings to address safety of the staff prior to public entering the area. Signs will be posted on areas where laser beams might be accessed, such as backstage and catwalk areas. A security person should be posted by any area that cannot be secured.

21.2.4.3 Ideally, each laser operator and technician will be in direct communications with event staff and management. If this is not possible, then the operator at the main laser controls must be on headset with the producer, his agent or the lighting director while the other laser technicians can be in communication with the laser operator. If the event is a high-noise environment, suitable earmuff style headsets will be required to enable the technicians to effectively communicate.

21.2.4.4 There are many creative ways lasers can be used at an event. Most of these uses fall into two broad categories: "seeing beams in mid-air" and "seeing graphics on a surface."

21.2.4.5 There are three general types of laser show options: stock, semi-custom, and custom. When using lasers at an event, planners will need to consider:
- Laser Color: Green is the most visible color of light. You can use a less powerful (and thus easier-to-use and less expensive) laser if green is an acceptable color for your event.
- Ambient Light: Keep your event as dark as possible during the laser show so that the laser won't need extra wattage to "punch through" ambient light. This can be challenging under the best of circumstances. Even though all event elements are supposed to be on the site plans and you built your production per those plans, surprises still occur. Beware of "pop up" concession or merchandise locations that occasionally are set up between the laser set up time and the event time. These locations may unwittingly set up in an area where the laser beams will be scanning, and they can emit additional light reducing the effectiveness of the laser show.
- Laser Beam Divergence: Some types of lasers have tighter beams than others. Low-divergence beams look brighter since the light is concentrated in a smaller area.

- Amount of Fog and Smoke: Fog improves the visibility of laser beams. If large quantities of fog are problematic, the event is outside, or the event is in an indoor venue where atmospheric haze is not permitted, then a more powerful laser may be needed for the presentation to be effective. Physical conditions should be discussed with the laser vendor in order to balance the applicable environmental conditions with the appropriately powered laser.
- Area Projected: If the audience is spread out over a large area, then the laser's power will also be spread over a large area and a higher power laser will be required to achieve the best laser presentation.
- Audience Safety: In many parts of the world, the audience is scanned with laser beams. The beam power and divergence must be sufficient to produce effective beams and effects without producing an eye hazard. One solution is to have the beam at full power when above the audience, but at a lower power when scanning the audience. Consult your laser vendor for the best options for the location of your event.

21.2.4.6 Before 2000, most large laser shows used bulky argon or krypton ion gas lasers that required 220 to 440 volts of electricity, and around two gallons of water per minute for cooling. Fortunately, in recent years, new solid-state lasers such as "DPSS" (diode-pumped solid-state) and "YAG" (yttrium aluminum garnet) types have become widespread. The lasers themselves are small enough to be easily carried by one person and some have a form factor and features similar to conventional lighting instruments. These new lasers have revolutionized laser presentations. They make all logistical aspects of show production easier, from freight to location flexibility. Because these new lasers are similar to conventional lighting instruments, handlers and operators must be extra aware and extremely cautious while handling them, because laser light, unlike other lighting, can cause serious injury.

21.2.4.7 There are numerous factors to be considered when using lasers, and each should be very familiar to a company producing laser shows. They include, but are not limited to the following:
- Beam Direction: For beams, the laser equipment is usually positioned in front of the audience. Beams will be aimed over their heads.
- Beam Brightness: Laser beams appear brightest when parallel to the viewer's field of vision, i.e., when they come straight toward you. They appear second-brightest when they come from straight behind you; and they appear least bright when they are crossing your field of vision. The terms for these effects are "on axis" when the beams are parallel with the viewer's field of vision, and "off axis" when crossing.
- Graphics Screen: For graphics, rear-projection is generally preferred over front-projection. The graphics projector should be located no closer than the largest dimension of the screen area. For example, if projecting onto a 20 feet x 30 feet (6.096 m x 9.144 m) screen, the laser should be no closer than 30 feet (9.144 m) to the screen. The farthest distance you would locate the graphics projector is roughly 100 feet (30.48 m) from the screen. These dimensions can vary if lenses are used for wide-angle or beam sharpening. They can also vary depending on how close the audience is to the screen.
- Signage: Your laser provider must post signs to alert people and deter them from entering the projection "cone," i.e., the area from the point of light emission to the outer-most edges of the projected image on the screen.
- Audience Scanning: This is a new form of delivery for laser media.

- During the event, an agent from the laser company must be on hand to limit access in the area from projector to screen and always be present when the laser is on.

21.2.4.8 Some equipment positioning considerations include the following:
- Direct-feed Projectors: These devices join the laser and scanners as a single "laser projector" unit. These may be compact enough that the entire projector can be mounted on a stand or flown in the rigging.
- Fiber-fed Projectors: With these projectors, a fiber-optic cable brings laser light to a remote scan head and thus they have even more flexibility in positioning. They can be put on a stand or flown. With these units, 100 feet (30.48 m) is a typical maximum distance for a cable run. Because the fiber-optic cable is delicate, precautions must be taken to protect the cable from being run over, severely bent, and/or kinked.
- Control Positioning and Setup Time: Lasers require roughly the same setup time and control console space requirements as required by the lighting or audio departments. The laser control console location should allow the operator to see the audience. If this is not possible, a laser safety observer must be positioned in an appropriate location to allow a complete field of vision of the laser presentation. This safety observer must be provided with a headset, walkie-talkie or other device that will allow for immediate communication with the laser operator. Always consult with the laser operator prior to finalizing the control location so they can review with you the pros and cons of the options.

21.2.4.9 For more information relating to lasers and laser safety, please refer to the following:
- For additional information and/or questions about using lasers for an event, contact the International Laser Display Association (www.laserist.org/).
- The U.S. Department of Health & Human Services, Food and Drug Administration, Center for Devices and Radiological Health (CDRH), Radiation-Emitting Products, Laser Light Shows (www.fda.gov/radiationemittingproducts/radiationemitting
- productsandprocedures/homebusinessandentertainment/ucm118907.htm).
- American National Standards Institute, ANSI Z136.1, *Safe Use of Lasers*, available through Laser Institute of America (www.lia.org). It is in the best interest of employees and organizations to follow the standards set forth in the American National Standards Institute (ANSI) Z136 series of laser safety standards.
- Manufacturers of electronic radiation emitting products sold in the United States are responsible for compliance with the Federal Food, Drug and Cosmetic Act (FFDCA), Chapter V, Subchapter C - Electronic Product Radiation Control, which can be found on the Government Printing Office web site (www.gpo.gov).

22. Sound: Noise and Vibration

22.0.1 High sound levels present a risk to hearing, both for those working at an event and for the audience. High levels of vibration can affect the integrity of temporary and permanent structures if not properly constructed and assembled. Both sound and vibration can lead to noise nuisance outside the venue. Therefore, proper control and management of sound and vibration levels is needed both in rehearsal/sound check and during the event.

22.0.2 The risk to hearing from loud sounds is directly related to the dose of sound energy a person is exposed to. The risk of hearing damage increases the louder the sound and the longer a person is exposed to it. At high sound levels the risk of damage to hearing occurs at much shorter exposure times than at lower levels; at extreme high or impulsive levels the risk of injury to the ear is almost immediate.

22.1 Hearing Damage

22.1.1 Deafness is caused by damage to the structures within the cochlea. This damage results in loss of both frequency sensitivity and increase in hearing threshold, i.e., noises need to be louder to be able to hear them.

22.1.2 Sometimes after being subjected to loud noises people experience deafness that goes away after a while. This is called temporary threshold shift. But after sudden, extremely loud explosive noises, or more usually prolonged lower level exposures to noise over a number of years, permanent hearing loss can occur. It may be that the damage caused is only noticeable when it becomes severe enough to interfere with daily life. This incurable hearing loss may mean that the individual's family complains about the television being too loud, the individual cannot keep up with conversations in a group, or they have trouble using the telephone. Eventually everything becomes muffled and people find it difficult to catch sounds like "t," "d," and "s," so they confuse similar words. Social situations can become difficult.

22.1.3 Age and general fitness are no protection from hearing loss - young people can be damaged as easily as the old. Someone in their mid-twenties can have the hearing that would be expected in a 65-year old. Once ears have been damaged by noise there is no cure.

22.1.4 Hearing loss is not the only problem. Tinnitus or ringing in the ears may be caused as well. Most people suffer temporary tinnitus from time to time, often after a spell in a noisy place, but with noise-damaged ears it can become permanent. Some people find it more distressing than the hearing loss.

22.1.5 Most members of the audience will not attend events regularly enough to suffer serious hearing damage solely as a result of going to music events. However, the louder events can contribute to the overall sound exposure that members of the audience receive throughout their life, including noise from other leisure activities, at work and at home, therefore increasing the risk of damage to their hearing.

22.1.6 The OSHA Occupational Noise Exposure Standard (29 CFR 1910.95) establishes uniform requirements to make sure that the noise hazards associated with all U.S. workplaces are evaluated, and that the hazards associated with high sound/noise levels are communicated to all affected workers so that effective protective measures can be taken.

22.1.7 For the community impact of noise from events, many local authorities already have environmental music noise control protocols which they apply to venues in their district. Refer to this source for guidance for the control of environmental music noise and its impact on communities neighboring outdoor music events.

22.1.8 In terms of vibration impact, the effects off site will generally be much less significant than on site, with the nuisance aspect of vibration being most significant.

22.2 Workers

22.2.1 OSHA Occupational Noise Exposure Standard 29 CFR 1910.95 establishes employer requirements to prevent damage to the hearing of workers from excessive noise at work. The regulation sets out actions which must be taken when stated levels of noise exposure are reached. It should be noted that a main objective of a music event is to amplify sound and distribute it. As such, administrative procedures and personal protective equipment such as earplugs or earmuffs become an integral measure to mitigate overexposure. Engineering methods can also be used regarding the type of sound systems used and placement of those systems.

22.2.2 When employees are subjected to sound levels exceeding those listed in Table 22-1, OSHA 1910.95(b)(1) states that, "…feasible administrative or engineering controls shall be utilized." If such controls fail to reduce sound levels within the levels of Table 22-1, hearing protection must be provided and used to reduce sound levels to within the levels of the table. Earplugs will have a Noise Reduction Rating (NRR) such as NRR-33 (i.e., NRR-33 is a noise reduction of 33 decibels). To compensate for known differences between laboratory-derived attenuation values and the protection obtained by a worker in the real world, the labeled noise reduction ratings shall be derated as follows (*Criteria for a Recommended Standard: Occupational Noise Exposure*, Revised Criteria, June 1998, National Institute for Occupational Safety and Health [NIOSH]):

1. Earmuffs: subtract 25% from the manufacturer's rating labeled NRR;
2. Slow-recovery formable earplugs: subtract 50%; and
3. All other earplugs: subtract 70%.

22.2.3 These derating values must be used until such time as manufacturers test and label their products in accordance with a subject-fit method such as Method B of ANSI S12.6-1997.

22.2.3.1 Other methods exist for compensating for known differences between laboratory-derived attenuation values and the protection obtained by a worker in the real world. However, the use of the highest quality ear protection available can make the figures almost irrelevant.

22.2.4 Consideration needs to be given to the nature of the noise, including its component frequencies. In some locations such as work within stage pit areas, the proportion of low frequency sound may be very high, and the hearing protection provided must be able to properly attenuate at such frequencies.

Table 22-1

Permissible Noise Exposure (1) (from OSHA 29 CFR 1910.95, Table G-16)

Duration per Day, Hours	Sound Level dBA Slow Response
8	90
6	92
4	95
3	97
2	100
1-1/2	102
1	105
1/2	110
1/4 or less	115

Footnote(1) When the daily noise exposure is composed of two or more periods of noise exposure of different levels, their combined effect should be considered, rather than the individual effect of each. If the sum of the following fractions: C(1)/T(1) + C(2)/T(2) C(n)/T(n) exceeds unity, then, the mixed exposure should be considered to exceed the limit value. Cn indicates the total time of exposure at a specified noise level, and Tn indicates the total time of exposure permitted at that level. Exposure to impulsive or impact noise should not exceed 140 dB peak sound pressure level.

22.2.5 If noise exposure is likely to reach the action levels listed in Table 22-1, employers must: provide workers with information and training; reduce exposure as far as is reasonably practical by reducing sound levels or the time exposed to the noise or both (without hearing protection); provide hearing protection to all workers and ensure that they are used correctly. The regulation also requires workers to comply with the employer's instructions regarding noise exposure, including wearing hearing protection or taking breaks when necessary.

22.2.6 For more information refer to 29 CFR 1910.95 which can be found online at http://www.osha.gov/.

22.3 Audience

22.3.1 There is no specific legislation setting noise limits for the audience exposure to noise. However, reasonable efforts should be made to maintain sound levels that are rational.

22.3.2 When portions of the event are likely to exceed 96 dB(A), consider advising the audience of the risk to their hearing in advance, e.g., either on tickets, advertising or notices at entry points.

22.3.3 Sources of noise other than music also need to be properly controlled. In particular, the noise from pyrotechnics should be restricted so that at head height in the audience area, noise from pyrotechnics does not exceed peak sound pressure level 140 dB or 200 Pa. Discuss this requirement with the specialist pyrotechnic technicians before the event, as charge density and altitude of deployment may need adjusting to meet this requirement.

22.3.4 Noise sources such as music associated with mobile amusement rides/attractions and sound systems brought on site by merchandising concessions can also add to the overall noise levels produced by the event as a whole. Also consider an assessment and control of these sources.

22.4 Noise Assessments

22.4.1 To enable effective management of sound and vibration levels, both in terms of hearing protection and external nuisance to the nearby community, a pre-event assessment of likely sound levels, coupled with monitoring and control of sound levels during the event will be necessary.

22.4.2 This assessment should include the following:

- Sound levels in the audience area: if the sound levels likely in the audience area are expected to exceed the values in table 22-1, then advance warnings for the audience should be considered.
- If any worker's noise exposure reaches or exceeds any of the action levels in table 22-1, then hearing protection must be made available to them.
- Arrangements for monitoring and control of sound levels during the event should be considered.
- The type, placement, composition and articulation of sound systems used can have significant benefits in controlling and managing noise levels and vibration levels at the venue and outside in the nearby community.
- For vibrations of structures caused by high, prolonged intensity of sound, refer to Chapter 19, *Structures*, for more information. Most temporary structures are built to withstand the typical vibrations caused by sound systems; if there is cause for concern, alternative placement and/or arrangement/articulation of sub enclosures may need to be considered.

22.5 Myths

22.5.1 Within the music business there is a radically different approach to noise management than in most other industries. In other workplaces time, effort and investment is made trying to minimize the creation and transmission of noise; in the event industry we are more commonly deliberately trying to create the hazard. However, there is little doubt loud music – even music you like – can cause long term harm.

22.5.2 There are a number of myths and misconceptions regarding music, noise and hearing damage, and these cultural factors are serious impediments to taking effective steps to protect workers and others.

Myth	Reality
"Music is not really noise. If I like what I hear, it won't harm me."	Loud music you like is certainly less irritating than the sound of loud machinery; but there is no evidence that the physiological damage caused is any different. On the contrary, there is plenty of evidence to show that musicians and others have been seriously harmed by the music they enjoy.
"It doesn't affect me; I'm half-deaf already."	A common response from crew members when asked to wear hearing protection (HP). If you stop for just a second to consider this, you'll see how crazy this myth is. Firstly, the fact that you are "half-deaf" indicates that noise could be playing a real part. More importantly, being "half-deaf" doesn't mean that your remaining hearing function has somehow been toughened up. It means it is even more at risk and you need to take steps to look after your remaining hearing. The one-eyed man needs to take extra care of his eyesight – you wouldn't say, "Oh I don't bother with goggles, you see I lost one eye already!"
"You can't communicate properly with earplugs in."	Wearing ear protection takes a bit of getting used to, but reducing the overall level of sound reaching the ear will make it easier to distinguish speech. Correct selection of HP type will also help.
"If I go deaf, I'll just wear a hearing aid."	Noise induced hearing loss (going deaf from too much sound) is only one possible outcome. There are equally if not more unpleasant side effects such as Tinnitus, Hyperacusis & Diplacusis - a condition where each ear hears a pure tone as a different note, causing distortion in even the most harmonic sounds.
"It takes ages to get damaged."	The reality depends on several factors, including actual noise level, duration of exposure and your own physiology. In truth some people do take years to show any signs of damage – but that is usually at relatively modest levels. In extreme environments where the average noise level can be over 100 dB, the damage can happen rapidly indeed. For very loud peak levels, the damage from acoustic trauma can be instantaneous.
"So audiences will have to wear ugly ear protection at concerts in future."	No. The regulations do not apply to members of the public. When attending concerts they are making an informed choice to do so. They attend relatively infrequently when compared to workers. However, members of the public can and do buy their own earplugs.
"The levels set are way too low, I can shout at 85 dB."	Aside from extremely loud instantaneous peaks, all noise levels given in Noise at Work Regulations are average levels evened out over a nominal 8-hour working day. You may be able to shout over 85 dB, but try keeping it up continually for 8 hours.

"All loud leisure noise is dangerous noise."	No. There is a tendency when talking about the risk, for the less well informed, to consider only the level of noise exposure and not the duration of exposure. There is also a tendency to sensationalize the risks of non-occupational exposure. For example, a story may warn that rock concerts are typically '130 dB SPL' (sound pressure level). This is one of the highest levels reported for rock concert noise. The mean of published sound levels from rock concerts is closer to 100 dB. There is also confusion over the annoyance and temporary effects of a loud exposure (e.g., TTS or temporary threshold shift) which are widespread, and the risk of permanent hearing damage, which is minimal. Studies show that most listeners sustain moderate TTS and recover within a few hours to a few days after exposure. The risk of sustaining permanent hearing loss from attending rock concerts is small, and limited to those who frequently attend such events.

22.6 Noise Control Measures

22.6.1 The first, simplest and most effective measure is to turn down the volume wherever practical. Unfortunately this is often overlooked and goes against the 'Rock and Roll' attitude. However, the simple step of keeping levels under control at every stage of the instrument/signal/amplification/reinforcement chain is fundamental.

22.6.2 Loud stage noise levels can compromise the quality of the performance and the sound that is delivered to the audience. It has been known for stage monitoring levels to be so loud that the front-of-house engineer in an arena has been unable to hear his own mix. This seriously compromises the possibility of creating a suitable sound for the audience.

22.6.3 On-stage control measures include the following:

22.6.3.1 Turning it down does not necessarily mean reducing the overall output of the main PA, but requires an analysis of why things are so noisy and then targeting measures to control the main "offenders." This is particularly true on stage where amplification of individual instruments (backline) often competes with on-stage monitoring (fold-back, side-fills) and the PA itself.

22.6.3.2 First consider substituting quieter instruments and amps. High-quality amplifiers and speakers that operate without distortion are far preferable to driving inferior systems at higher rates. Introducing distortion makes the output less intelligible and leads to increases in sound level in attempting to achieve clarity. The result is often a spiral of increasing volume without ever achieving clear monitoring.

22.6.3.3 Consider increasing distance, isolation or shielding of noisier instruments where possible. Drum kits can be positioned and shielded/enclosed to minimize noise levels for performers and workers situated close by. Ideally shielding should be acoustically absorbent rather than reflective material.

22.6.3.4 Position and angle guitar amplifier/speakers (guitar combos) for maximum ease of listening for the player. Additionally simply raising a guitar combo on a flight-case could significantly reduce exposure for other players, have a marked reduction in overall stage noise and improve clarity for the player. Guitar combos could be positioned and mic'd in a separate area from the performance area.

22.6.3.5 Consider using technology that eliminates the need for loud backline amplifiers on stage. This could range from simply plugging instruments into a mixing desk by means of Direct Injection (DI) boxes rather than mic'ing up an amplifier, through to using amplifier modeling software, foot pedals or other hardware. Whatever system is used, sound engineers can achieve greater control of on-stage levels through careful management of monitor levels rather than expecting musicians to fight it out in a battle of escalating stage volume.

22.6.3.6 Use risers to separate sections of the band, and to elevate particularly noisy instruments above the heads of other performers – or move to the front of the stage, particularly where very loud instruments may be used – brass instruments, amplified guitar and snare drums can produce extremely high sound levels.

22.6.3.7 A "shaker" or "thumper" is especially useful for reducing drum monitor levels. Shakers allow performers to use hearing protection and monitor their performance while still maintaining contact with their instruments.

22.6.3.8 Consider altering the drum kit set-up to ensure cymbals etc. are not at ear-height. Experiment with raising or lowering the cymbals as necessary to protect the hearing of everyone who is close by. Try hanging small strips of cloth from each cymbal's center nut.

22.6.3.9 Some drummers are happy with headphones/in-ear monitors and a shaker rather than a traditional drum fill. The headphones should be selected to provide hearing protection; the devices that reproduce sound inside the headphones should be limited. This alone may save several dB of overall stage level.

22.6.3.10 Consideration should also be given to limiting the time necessary to work in high noise areas. If properly considered and applied these techniques may mean that the noise exposure dose for staff can be kept below the threshold at which HP is required.

22.6.4 On-Stage Monitoring

22.6.4.1 The need for musicians to hear their own performance and that of other performers is fundamental, but this can lead to an excessively loud and confusing stage environment if not planned and managed correctly.

22.6.4.2 A well-balanced monitor system should allow all the players to hear what they need at a comfortable level while maintaining a reasonable work environment for everyone else on the stage. This needs time and planning, as well as a skillful monitor engineer who understands the needs of musicians.

22.6.4.3 An effective means of avoiding monitor spill is to use monitor headphones or in-ear monitors (IEMs). IEMs and monitor headphones allow a quiet stage environment with benefits for all workers. IEMs' benefits include clarity, controllability and comfort. Be aware that IEMs themselves can produce harmful levels to the user and so the use of limiters is strongly recommended.

22.7 Hearing Protection (HP)

22.7.1 Employers are required to provide employees with hearing protection if they ask for it and their noise exposure is expected to approach exposures depicted in Table 22-1.

22.7.2 Employers must provide employees with hearing protectors and make sure they use them properly when their noise exposure exceeds the upper exposure action values depicted in Table 22-1.

22.7.3 The main types of hearing protection are:
- Earmuffs, which completely cover the ear;
- Earplugs, which are inserted in the ear canal;
- Semi-inserts (also called "canal caps"), which cover the entrance to the ear canal.

22.7.4 You should use the results from your noise assessment and the information from hearing protection suppliers to make the best choice of hearing protection. Aim to get below 85 dB at the ear, and ensure it is suitable for the employees' working environment and compatible with other protective equipment used by the employee (e.g., hard hats, dust mask, eye protection).

22.7.5 We can all more or less tell when something is loud and could cause harm, but selecting the right kind of hearing protection to ensure people are protected takes a bit more effort. The kind of noise produced at events is variable. Depending on the phase of the event cycle, noise may be produced by construction work. During the public phase, concert PA system and generators are likely to be the sources. There will be huge variations in not just overall level, but in the component frequencies which make up the sound.

22.7.6 Effective HP needs to absorb enough of the incoming energy to protect the ear, but it needs to target the frequencies which are most prominent and which contribute most to the overall energy of the sound. Simply getting HP with a high overall attenuation rating (the SNR value) is not necessarily the best approach.

22.7.7 As an example, consider the pit area of a large festival that receives a high proportion of low frequency sound. Effective HP for pit crews in this environment needs to be effective at absorbing low frequencies, which all HP do not necessarily provide. The protection provided by a particular HP device must be matched to the sound energy the device is intended to protect against.

22.8 Managing Use of HP

22.8.1 The noise level may subject people to exposure above established safe levels in a short period of time (less than a minute).Therefore consistent and proper use of HP throughout the exposure period is vital.

22.8.2 To be accepted and implemented workers need to understand the reason behind using HP and the basics of how to fit and use it. Fit, feel and comfort are just as important as the attenuation values discussed above. If the protection provided is bulky or uncomfortable, or it gets in the way of people carrying out normal duties, it is likely to be ignored or removed on a regular basis.

22.8.3 Overprotection should also be avoided. Bar and concession staff need to communicate clearly with customers, and providing excessively attenuating HP may appear to "err on the safe side," but may mean staff remove the plugs every time they speak to a customer – resulting in over-exposure.

22.8.4 Workers should be offered a choice of HP, so they can decide which is the most comfortable and usable. They must also be shown how to wear or insert muffs and plugs – particularly expanding foam plugs – and to recognize when they are improperly fitted or should be replaced.

22.8.5 In complex environments where many subcontractors operate, it will normally fall to the event organizer to implement a system for the enforcement of HP use among contractors and suppliers – whether they be carrying out technical operations on the stage or working in retail units. Such a requirement does not override the primary duty of the employer to conduct a Risk Assessment and make suitable provision for their staff; however the organizer has a duty to ensure that all contractors and third parties are properly warned of the extent and nature of the noise hazard on site.

22.8.6 In some instances noise maps may be appropriate to give an indication of the areas of site where noise levels can be expected to be above acceptable levels as defined in OSHA 1910.95. Such a plan will clearly identify mandatory hearing protection zones.

22.9 Using Contracts to Help with Noise Control

22.9.1 Contracts can help the planning process by setting out the arrangements for noise control. They have been found particularly helpful where there are several contractors working together with a producer/venue provider(s). Contracts can be useful when dealing with the specific requirements of the Noise Regulations and can form part of the overall health and safety considerations for the event/production.

22.9.2 A contractual approach is often more readily understood by the parties concerned as so many matters are already covered in this way - from performers' riders to equipment specifications. The contractual approach can also act as a memo. Experience shows joint meetings can often slip by because of time constraints, whereas specified contractual obligations for consultation are usually taken on board.

22.9.3 Including things in a contract can help principal contractors/producers to pass on relevant information to subcontractors. For example, a contract stipulating a hearing protection zone could insist that subcontractors' crews wear earmuffs.

22.9.4 For smaller-scale events, contracts may be the most direct way of ensuring noise-control issues are considered. Key points can easily form part of standard contracts for musicians. These may be of most help to those with individual contractual arrangements, particularly for short hire periods. Similarly, venue operators can include some standard points relating to their requirements from performers - for example, which instruments and equipment will be brought to the performance by the performers and what, if any, control measures will be carried out by them.

22.9.5 In small venues a contract should help remove gray areas about who would do what and identifying what needs to be done by laying down responsibilities early on (apart from the non-transferable legal responsibilities).

22.10 Noise Measurement

22.10.1 Regular review of noise control is important to assure that all the controls are operating properly. To make sure noise control actions are still effective the following is required:
- Ensure that performers and crew understand why they need to follow instructions on control measures. Provide additional training or team talks if necessary.
- Ensure that hearing protection is used correctly and consistently.
- Review the results of hearing health checks to see how well the noise controls are working.
- Review the steps you have taken.
- Regular spot checks of noise levels can help to monitor how well controls are working.

22.11 Sound-Related Terms

22.11.1 Please see Chapter 38, *Glossary of Useful Terms*, for definitions of terms related to sound, noise and vibration.

23. Barriers

23.0.1 Barriers and fences at events serve several different purposes. They can:
- Provide physical security, as in the case of a high perimeter fence at an outdoor event;
- Shield people and vehicles from hazards;
- Stream people into queue lanes or similar;
- Provide a protective barricade at the front of a stage; and
- Relieve and prevent overcrowding and the build-up of audience pressure.

23.1 Planning

23.1.1 The proposed use of barriers and fencing is an integral part of pre-event planning. Depending on the complexity of the site and the nature of special risks such as high-density crowds or mass movements, a source of competent advice may be needed during planning.

Fig. 23-1 – Crowd at entrance. Note the low level queue barriers in use. Photo courtesy of The Event Safety Shop, LTD.

23.1.2 If barriers and fencing are used, a risk assessment should be done. Although not an exhaustive list, the following should be taken into consideration when determining the location and type of barrier or fence to use:
- Their planned use;
- Layout;
- Ground conditions and topography;
- Weather;
- Loads on the barrier – wind and/or crowd pressure;
- Audience size, nature and behavior; and
- Any relevant factors unique to the location.

23.1.3 It is crucial that the type of barrier and fence does not present greater risks than those they are intended to control. Barriers have failed due to improper selection.

23.1.4 Barriers and fencing should only be erected, maintained and taken down by competent persons who fully understand the construction requirements and limitations of the equipment involved.

23.1.5 A competent barrier/fencing contractor should be able to provide the event organizer with system calculations, drawings, risk and safe work method statements. They should be knowledgeable of public events and should be able to offer advice and site visits. In the absence of previous experience, they should be able to demonstrate an appropriate level of technical ability. The contractor should also be able to supply a range of appropriate fencing/ barrier options to build the required configuration safely.

23.2 Deciding Which Types of Barrier/Fencing to Use

23.2.0 Barriers, in the context of crowd management at events, fall into two main categories: differentiating space barriers and pressure barriers.

23.2.1 Differentiating Space

23.2.1.1 Differentiating space barriers include tensor barriers, ropes, and tape. They are typically deployed where there is no expectation of crowd surges (pressure), such as defining queuing areas and waiting lines. Take into account the size of the crowd and the way people will arrive before deploying these types of barriers.

23.2.2 Pressure Barriers

23.2.2.1 Pressure barriers are specifically designed to resist horizontal loads such as those exerted by a crowd at the front of a stage. Pressure barriers need to be deployed when:

- The location can experience large crowds who may push towards the barrier line;
- Areas of sustained crowd movements where a dynamic load may be anticipated;
- Locations where security need to check and manage access (such as into a viewing enclosure).

Fig. 23-2 - An example of a Primary (silver, A) and Secondary (black, B) barrier lines being installed with block-and-mesh (C) being used to delineate public and working areas. Photo courtesy of The Event Safety Shop, LTD.

23.2.2.2 To assess if a pressure barrier or differential space barrier/fence is required an event organizer and their contractor should consider the number of people present at any one time, the size of the space, the nature of the crowd, the actions in that space (for example a waiting/queuing line may have surges when gates open or when busses arrive) and the general conditions (weather protection or duration of waiting time).

23.2.3 Product Selection, Design and Build

23.2.3.1 Various different types of fences or barriers are available and used for events. Detailed technical requirements for the various types of barriers referred to in this chapter can be found in the publication *Temporary Demountable Structures: Guidance on Procurement, Design and Use* (2007, Third Edition, The Institution of Structural Engineers [UK]) (http://www.istructe.org/).

23.3 Barriers for Differentiating Space

23.3.1 Basic Differentiating Barriers

23.3.1.1 Rope and pin, post and rope, and retractable fabric barriers are the most basic forms of barrier used to delineate an area.

23.3.1.2 "Rope and pin" is often used in parking areas on greenfield sites to create vehicle lanes and parking areas.

23.3.1.3 "Post and rope" is commonly used outside indoor venues as a way of maintaining a clear space at the entrance to the venue.

23.3.1.4 Retractable fabric barriers are a lightweight, fast system for create queuing systems in areas where compliance from the queue is predicted and no pressure likely to build.

23.3.2 Low Level Barriers

23.3.2.1 There are many different steel-framed, low level barriers on the market suitable for events. These tend to come at a height of 1.5m and in varying lengths. They are free standing and come with a number of different supporting systems: angle, arched feet, or flat footed. The flat footed variety is particularly useful for queue lanes and locations where people have to walk close to the dividing barrier.

23.3.2.2 There is a more robust type of low level barrier often referred to as "Police Barrier." This has a hooped foot construction and is used extensively at street events, parades and marches.

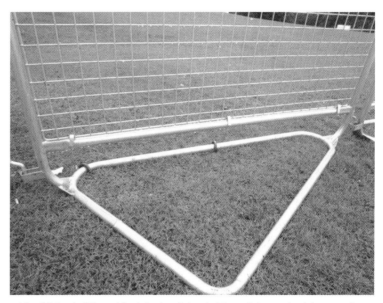

Fig. 23-3 - A different foot design, intended to resist a modest degree of pressure. Note the pinning of the foot to further stabilize the barrier. Photo courtesy of The Event Safety Shop, LTD.

23.3.2.3 It should be noted that these barriers have little structural strength to withstand crowd pressure. Their main uses are for restricting access, designating routes and queuing systems. They are normally delivered in stacks and can be deployed using a forklift truck and as a result reduce the amount of manual handling required.

23.3.3 Mesh Panel Fencing

23.3.3.1 This type of fencing is normally constructed of tubular steel frame with a steel wire mesh infill. Commonly it comes in panels 2m high and 3.5m long. It is supported by inserting the uprights into separate solid plastic or concrete block units and attaching them with an independent clip unit.

23.3.3.2 Mesh fencing is used extensively at event sites for creating perimeters that can be moved quickly and easily opened for access.

23.3.3.3 Mesh fencing has no structural resistance to crowd pressure. It is often commonly used to mount signs and branding panels or covered with plastic sheeting. In such instances it is critical to ensure sufficient braces and ballast are installed to prevent collapse in even modest winds.

23.3.3.4 When using branding or scrim on a roll, ensure there are regular breaks in the material should the barriers/fencing need to be broken for safety reasons to allow people through.

23.3.4 Temporary Fence Systems

23.3.4.1 Temporary fence systems are similar to mesh panel fencing systems. However the mesh is replaced with a corrugated thin solid steel infill. The panel size is usually reduced to 2 m x 2 m as the weight of the panel is considerably increased (Fig. 23-4).

23.3.4.2 This system should always be installed with the appropriate bracing at right angles to the fence panels and should either be staked directly into the ground or sufficiently weighted to give stability.

Fig. 23-4 - Solid temporary fence panels showing diagonal bracing. This is a pedestrian barrier used to channel queue lanes – perfect when no sideways pressure is anticipated. Photo courtesy of The Event Safety Shop, LTD.

23.3.4.3 These systems have limited resistance to lateral loads such as wind or crowd pressure. Manufacturer's instructions and guidelines regarding support systems should be directly followed when installing this type of fencing.

23.3.5 Steel Panel Fencing Systems

23.3.5.1 This is a solid panel system usually used for creating an enclosed perimeter. This system does offer a reasonably high degree of security e.g. to prevent people accessing and climbing onto a structure for a better view.

23.3.5.2 The fence panel size is normally 3m high x 2.4m wide and is formed of flat plastic coated steel over a fabricated steel frame. These overlapping frames are then bolted together, secured to the ground with pins and then supported by braces at a right angle. The braces should be positioned at every join of the panels to ensure stability. This system is designed to be load and wind bearing.

23.3.6 Roadway Panel Systems

23.4.6.1 This system is constructed from temporary roadway panels turned onto their side at right angles to the ground. This is a very efficient way of constructing a solid, secure perimeter (Fig. 23-5).

23.4.6.2 A roadway panel is of a similar size to that found in a steel panel system. Due to their weight, a panel needs to be supported by stronger braces and staked to the ground. If properly designed and installed this system should be load and wind bearing.

Fig. 23-5 - A custom-made fence designed to create an impregnable event perimeter. Block and mesh panels form an inner fence to create a moat in which security can operate.. Photo courtesy of The Event Safety Shop, LTD.

23.4 Pressure Barrier

23.4.1 A stage barrier (barrier at the front of a stage) is the principle type of pressure barrier (Fig. 23-6).

23.4.2 Because of the variety of uses, stage barriers have become an important piece of equipment for event organizers.

23.4.3 A stage barrier is designed around a basic "A Frame" to be load bearing and therefore normally used where there is a risk of crowd pressure.

23.4.4 Most stage barriers include the following design features:
- They are constructed of steel or aluminum, ideally fully welded.
- They should not be riveted in parts nor should soft materials such as wood used.
- Their individual sections are usually 1200 mm high and 1 meter wide.
- They have a footplate that the audience stands on to stabilize the system
- The top horizontal rail should be smooth and fall flush on the front vertical fascia (audience side).
- They should have a step on the rear (stage side) that working personnel can use.

23.4.5 To be fully effective, a stage barrier has to be built properly. It will only be a strong as it weakest link, so once a system has been built, the joining method must be in place. It is imperative that ALL joining mechanisms between barriers are properly engaged, whether these are bolts or locating tenon pins. Any joining bolts should be of suitable quality (high tensile steel) and not simply any bolt that fits.

23.4.6 In terms of appearance, the barrier should have:

- Smooth lines;
- No rust or disfigurement;
- All rivets should be in place and not rotating;
- Welds should be smooth and have no fractures;
- All bolts or fixings correctly installed.

23.4.7 Barrier sections with bent, corroded or missing connections should be rejected.

23.4.8 The barrier must be stable, and if used on uneven ground may require packing or infilling below the footplate to ensure it is properly seated.

Fig. 23-6 - A standard concert stage barrier, note the diagonal bracing stays which allow the barrier to resist horizontal load. Note also the extended rear plate allowing security personnel easy access and a safe platform from which to assist or rescue the public. Photo courtesy of The Event Safety Shop, LTD.

23.4.9 Voids beneath the audience footplate should be avoided; there is a possibility of audience members getting toes or even feet trapped beneath the footplate.

23.4.10 The barrier should not flex, and the junctions between sections should not open and close as the barrier is loaded. This could lead to pinching or even amputation of fingers etc. Joints between sections should be flush, or if this is not possible should be taped to prevent finger traps.

23.4.11 It is common for cables to be run to a control position within the audience, and a proper arrangement must be made where these cross the stage barrier. When ground conditions permit, these can be dug-in to transit the barrier. Be sure that any such trench does not undermine barrier stability. If a "cable gate" section is used (where cables can pass under the barrier because the gate has no footplate), then the gate must be securely attached on either side. It is normal for cables to run from the center point of the stage, and this is where one is likely to encounter peak density, so gates often come at locations where strength is a of critical importance.

23.4.12 The stage barrier should be inspected and signed-off by a competent person to ensure that it is both safe and secure prior to use.

23.4.13 Along with the overall strength and stability of the barrier, the organizer needs to consider its shape. Given that such barriers are designed to retain and resist audience pressure, it is critical to ensure that barrier location and shape does not lead to the creation of "pockets" in which people can become trapped or from which kinetic energy cannot safely be dissipated. A typical example would be preference for a convex and not a concave front face to a stage barrier, allowing crowd energy and movement to be transferred to the side of the barrier rather than "focusing" in the center. An alternative form of primary barricade at the stage involves creating a 4 to 6 foot (1.2 to 1.8 m) wide "thrust" into the audience area which serves to disrupt the direct crowd pressure at the front of the stage.

23.4.14 In many instances it is helpful to construct a "Secondary" barrier which replicates the convex shape of the primary barrier, but is some 50 meters or so further back into the crowd (Fig. 23-2). This barrier serves to minimize the risk of excessive crowding at the stage front or the transfer of potentially harmful kinetic energy in the form of crowd surges and shock-waves from the rear. A Secondary barrier will normally consist of two parallel runs of pressure barrier facing outwards, with a narrow "moat" between them where security personal, medics and others can monitor and access the audience.

23.4.15 Careful consideration needs to be given to the design and location of access gates, which may be required to enable workers to move from stage to auditorium or used to relieve pressure in high risk locations. In any such instance a properly experienced and competent person should advise on the location and design.

23.4.16 Where front stage barriers end, they will commonly join to another type of fence or barrier, continuing the secure area backstage. Such junctions need to be properly secured and be free of sharp or projecting edges. Consideration should be given to blanked-out or solid fence sections where sightlines to the stage are still good – this helps prevent a build-up of audience members where the pressure barrier ends and the less robust fencing type begins. If such a junction is required, ensure that the "continuation" barrier is mounted on the front face of the pressure barrier, allowing the load to be safely transferred to the more robust barricade.

24. Merchandising and Special Licensing

24.0.1 There are five aspects to merchandising that need to be planned and managed:
   (a) The merchandising facilities which include the structure of the stalls or stands;
   (b) The space requirements;
   (c) The setting up, dismantling and operation of the stall or stand;
   (d) The items for sale as merchandising; and,
   (e) The solid waste, sanitation, wash up and cleaning facilities. See Chapter 14, *Sanitary Facilities*.

24.1 Facilities

24.1.1 It is essential that merchandising stalls and stands are considered in the planning and management of the event.

24.1.2 Consider the following matters when planning the venue or site design:
   • The position, size and space requirements of the merchandising stalls or stands within the arena or venue to ensure that entrance and exit audience flows are not obstructed, or cause an audience build-up at any strategic points;
   • Whether stands and stalls are of a fixed or temporary nature;
   • Check that any structures will be erected properly and will satisfy any structural integrity requirements (see Chapter 19, *Structures*), as well as requirements in respect of fire safety (see Chapter 4, *Fire Safety*);
   • Power supplies, if required, need to be considered as part of the overall electrical supplies to the event (see Chapter 17, *Electrical Installations and Lighting*);
   • Any vehicle or vehicle movements associated with the stands or stalls; vendors generally need to unload their vehicle close to their vending space, then reposition the vehicle;
   • Allocation of parking spaces and camping accommodation for people working at the stalls or stands;
   • Waste accumulation and collection;
   • Security arrangements; and
   • Cash or scrip handling and pick up and accounting measures.

24.1.3 Ensure that people working on merchandising stalls and stands are informed of the site safety rules and local health department requirements particularly in relation to practices and the equipment that can or cannot be brought onto site or within the arena or venue. Also, make them aware of the space allocated to them on site and that they must not expand outside their allotted space.

24.1.4 It is recommended that exhibit spaces, vendor tents, stalls and kiosks that have anchoring stakes should have stakes and guy lines marked to reduce potential for injuries. Please refer to OSHA requirements for protection of works from exposed rebar (rods, posts), which can be found online at http://www.osha.gov/SLTC/etools/construction/falls/protruding_rebars.html.

24.2 Setting Up, Operation and Dismantling

24.2 1 Most workers employed by people running merchandising stalls and stands are temporary. Define the responsibilities for health and safety and agree on methods of communication with the merchandisers. Give a copy of the site safety rules to the merchandisers when they arrive on site and ensure that they and any subcontractors are informed of the site safety rules.

24.2.2 Checks should be made on any public and product's liability insurance certificates. Agree on the operation time of the merchandising stands with the operator and explain procedures to be taken in the event of a major incident or contingency. Any gas or electrical equipment brought onto site by merchandisers should be accompanied by relevant inspection certificates and have undergone the recommended testing. Other equipment should be examined to ensure that the relevant fire-fighting equipment is available in case of fire. Some health jurisdictions have special certifications for food trucks; check with your local health department for details.

24.2.3 In the case of permanent sites, information on health and safety policies within the premises will already be in place. Therefore, the procedures should be followed at all times by all concerned.

24.2.4 Stewards working on behalf of the merchandisers should be involved in the event briefing and lines of communication. Discuss the use of radio communication to avoid conflicting frequencies.

24.2.5 The storage of merchandising stock, particularly if flammable goods are for sale, should be discussed with the fire authority and local authority to ensure that the appropriate fire extinguishers are on hand on the stands or stalls. The control and movements of stock around the site should follow agreed procedures.

24.3 Items of Merchandising

24.3.1 The items for sale as merchandising should not breach any license requirements, trading standards, copyright, or trademark regulations.

24.3.2 Ensure that information concerning items that could cause injury or discomfort (e.g., glo-sticks) is given to the purchaser at the point of sale and the procedures for their correct use are prominently displayed.

24.3.3 The practice of tattooing, body piercing and massage may require a special license or permit from a local authority. Check that the necessary licenses and permits have been or will be issued by the local authority before allowing tattooing or body piercing to take place.

24.3.4 Offensive materials should be carefully considered and viewed in relation to the audience profile and perhaps not "actively displayed."

24.3.5 In the case of ticket scalpers and unwanted street traders, coordinate with the local law enforcement authority to determine the methods that can be legally used to deter such practices.

24.4 Emergency Communications to Vendors

24.4.1 It is essential that during an evacuation of the event site (bomb threat, fire, biological attack, etc.) that the vendors be told to cease operations and shutter their booths/kiosks. Guests who pause to peruse or purchase goods block the flow of egress for other patrons. Covering over the merchandise, food, drinks that would normally be for sale and making every appearance that the booths are closed for business is essential to maintaining a smooth and consistent flow of the crowd to a safer area. See Chapter 6, *Communications*, for more information.

25. Performers

25.0.1 The requirements and responsibilities of performers have to be considered in event planning. Contract negotiations provide an opportunity to raise concerns and resolve safety issues in advance. Performers have responsibilities related to the safety of the audience and site workers. Performers may be held directly responsible for injury that results from their behavior such as throwing things from the stage or not keeping to performance timings.

25.0.2 Determine whose representatives will create a full briefing document before the event, including: how to reach the site and a map of the site showing specific artists' entrance, stage, stage plan and accommodation plan; an itinerary of what is happening, site access times, sound check times, performance times; specific security arrangements as well as meet and greet, hospitality and press-related locations.

25.1 Performers' Areas and Accommodation

25.1.1 Arrival and Departure

25.1.1.1 Ensure that changing and "warm-up" facilities are weatherproof, well lit and secure. Provide toilet and dressing room facilities (if necessary) for male and female artists and consider separate toilet provision close to the stage.

25.1.1.2 Plan the arrival and departure times for performers. Their entry and exit points, if practical, should be different from those used by the audience. Where there is a risk of significant audience attention, try to keep their vehicles out of view or, at a minimum, separated by barriers or ribbon tape. Designate appropriate numbers of site staff and security to the area, if it is felt that performers will attract significant attention. Also consider the route to be taken to and from the venue. Some performers may arrive by helicopter so your risk assessment will need to cover the selection, marking and location of the landing zone.

25.1.2 Buses and Other Vehicles

25.1.2.1 Parking facilities for performers should, where possible, be separate from audience car parking and close to the stage. Where this is not possible, workers should be on hand, with appropriate transportation if necessary (e.g., a passenger van or golf cart to help move people and equipment).

25.1.2.2 The number of vehicles for performers should be kept to a minimum. Allocate a specific parking area for the vehicles, with the drivers available at all times in case they need to be moved.

25.1.2.3 Many vehicles carry on-board generators and it is undesirable to keep these powered by leaving engines running. The vehicle operator should carry cabling to connect to a site power supply where possible. Where practical, consider providing a site power supply commonly referred to as shore power for the vehicles to connect. (See Chapter 17, *Electrical Installations and Lighting*).

25.1.3 Workers and Guests

25.1.3.1 Control the number of workers and guests permitted into restricted areas to avoid overcrowding, especially on stage and performance areas. Try to keep workers associated with performers to a minimum and ensure that they have suitable security clearance, which should be graded with access to key areas such as dressing rooms.

25.2 Security of Performers

25.2.1 Ensure that performers are met and logged in on arrival at the venue, suitable security passes are issued and where any threat, such as mobbing by fans, seems likely, suitably trained security staff are employed. During the performance every effort should be made to secure the performance space. Artists and management should always be made aware of the part they play in this process.

25.2.2 Advise performers and their staff of evacuation procedures and the locations of medical facilities. If this is not practicable, advise a senior representative who can shadow performers while on site, keeping in mind security needs and escape routes.

25.3 Performers' Help in Emergency Planning

25.3.1 While also being aware of the site safety arrangements, performers or their representatives can participate in the emergency procedures planned, for example, by helping to calm a situation and asking the audience to stand back from a crowded stage barricade.

26. Camping

26.0.1 At many events, camping is an integral part of the event. The camping area should be provided within the defined event site and incorporated as part of the event planning. An adequate level of services and facilities must be planned for the duration of the camping event and not merely during the event's entertainment.

26.0.2 In isolated locations or where the music starts early or finishes late, contingency provision may have to be made for camping even when people were not intended to camp. Some consideration may also have to be given to crew camping and camping for booth vendors with their booths.

26.0.3 Services provided for people camping, including fire, security, guest services, medical facilities, and water supply, need to be available for the time that campers are allowed to remain on the site. Ensure your event publicity states the opening and closing times of the campsite. If large numbers of campers are likely to remain after the event, consider a gradual closing of the site to encourage those people to move, but without exposing them to risk.

26.1 Site Design

26.1.1 The camping area will need to be reasonably well drained and level with grass cut short to minimize the risk of fire spread. Camping should not be allowed on stubble. Break the camping areas up into discrete smaller areas to:
- Provide an identifiable camping area;
- Allow for the management of each area;
- Control the densities of each area; and
- Provide information and communications.

26.1.2 Music events involving camping are likely to attract a broad mix of people and it might be desirable to create a separate area for family camping. Separating areas can be carried out by using posts and tape while at larger events it may be necessary to provide some physical barrier to prevent camping such as metal trackways, road barriers, etc. Wherever possible the site layout should provide for an entertainment area in the middle of the site with camping on the periphery and parking beyond that. Crowd movements will therefore disperse away from the focus of the event. It is important that campsite layout plans are fully integrated between the various agencies involved, so that the site features and descriptions of locations will be identical for all the agencies.

26.1.3 Site arrangements and boundaries need to take account of natural hazards such as ponds, ditches, rivers, etc. Other hazards such as electricity pylons may need to be assessed to prevent access or risk of shock from activities such as kite flying and the use of tethered commercial balloons.

26.2 Site Densities

26.2.1 Many believe that a density of up to 430 tents per hectare (1 hectare = 10,000 square meters or 107,639 square feet or 2.471 acres) for rock/pop events is a realistic standard. At more family-orientated events, perhaps with larger tents with greater number of occupants, this density would need to be reduced possibly by half.

26.2.2 It is desirable to provide separation distances between individual tents to make the site safer from fire and trip hazards, etc. Provide people entering the site with information and maps showing the camping areas and ensure there are sufficient stewards to direct people to the appropriate areas as the campsite fills up.

26.3 Segregation of Vehicles/Live-In Vehicles

26.3.1 It is desirable to physically separate camping areas from vehicle parking areas. The reasons for this are to remove risks from cruising or joyriding, car fires or runaway vehicles.

26.3.2 Minimize the distance between parking areas and campsites. Consider providing internal transport—such as shuttles—for campers to and from the campsite. This is important for families with children who need to carry considerable amounts of equipment.

26.3.3 It may be justifiable to permit parking with camping in certain circumstances on a level site and where the audience is compliant (e.g., families). Where there is a desire to allow camping and car parking next to each other, the density will need to be substantially reduced to allow for increased roads and separation. The campsite should be designed in advance so that blockages of tents and cars cannot happen. It may be acceptable to allow vehicles and tents to mix in an area provided for campers with special needs.

26.3.4 If live-in vehicles (e.g., RVs, camper vans or camper-trailers or adapted vehicles) are to be allowed on site, set aside a special area for this purpose. Such vehicles should not be used for camping in a parking area.

26.4 Information, Organization and Supervision

26.4.1 Include right on the ticket information important on-site restrictions, such as no unauthorized PAs, campfires, etc. At strategic points on the site (including the campsites) provide information including a "you are here" map and key information to direct people to important facilities such as toilets, water, medical facilities, fire points, etc. Make information easily available, including site safety and restrictions. In a large event this could be by using a mobile patrol that would operate 24 hours a day. Ensure the mobile patrol has radio communication and can respond to information requests about emergency situations involving medical issues, fires, etc.

26.4.2 By breaking up the camping area into smaller discrete areas, people can be given an identifiable camping area to which they can more easily return. On complex sites involving many camping areas and a large entertainment area, provide all campers with maps on entry and/or preferably an information pack with safety advice.

26.4.3 Locate stewards within the camping areas before campers arrive to assist with the general build-up of the campsite, and to monitor key facilities such as toilets, fire provision, water supply, etc. These stewards will also have a role in helping to ensure that camping is dispersed in the best way over the designated camping areas.

26.5 Contingency Planning

26.5.1 Aspects of contingency planning that require particular attention where there is camping on site include:
* Adverse weather;
* Failure of water supply; and
* Other need to clear the area.

26.5.2 At certain types of events attracting young people, it is common for them to attend without tents. Similarly, people attending with tents may find that the tents are unusable so that they are without accommodation. Campers might also have their tents stolen. Contingency provisions should allow members of the audience to obtain shelter where they are unable to provide any themselves.

26.5.3 If temporary accommodation needs to be provided, existing canopies and tents may be suitable. In the case of adverse weather conditions, particularly wet weather combined with high winds, such structures may not be stable. A source of smaller tents may be advisable to provide emergency shelter.

26.5.4 At large events where people arrive in large numbers by public transportation it may be impossible to close the event and clear the camping area in an emergency. Facilities will have to be brought to the camping areas rather than the people removed to another place of safety.

26.6 Public Health

26.6.1 It is useful to provide advice to individuals on basic personal hygiene matters and the type of food that they should or should not bring with them. Given the undeveloped nature of a camping area, large numbers of people involved, basic sanitation and remoteness from care, it is essential to ensure that food outlets and personal hygiene are satisfactory. The consequences of an infectious disease outbreak would be significant in terms of both the numbers that could be involved and the likely amount of care that could be provided. Provide adequate catering facilities, some overnight, and outlets where campers can buy basic provisions such as food and beverages.

26.6.2 Sites that are grazed will naturally be contaminated with animal droppings and may expose campers to health risks such as Escherichia coli bacteria (a.k.a. E. coli) infection. Exclude animals from all areas other than parking lots for as long as possible before public access. E. coli can survive for long periods in most environments.

26.6.3 Dogs should not be permitted onsite and advance publicity should be given. Unnecessary health risks include fouling and dog bites, and stray dogs pose a nuisance. However, it is likely that people will bring dogs, in which case provision should be made to deal with strays.

26.7 Crime

26.7.1 Campers are vulnerable to having property stolen from tents but may be unable to carry around items that might be stolen if left unprotected in their tents. Consider providing secure accommodation on campsites where people can leave bulky or valuable items.

26.7.2 Campsites should be adequately lit and patrolled by stewards to deter both isolated and organized criminal activity. Patrols will also help to identify other matters such as fire outbreaks, unruly camp fires, etc.

26.8 Fire Safety

26.8.1 Campfires constitute a risk of burns, tent fires and can cause smoke pollution. They are undesirable and should be discouraged. At some types of events, however, it would be impossible to prohibit fires and for certain audience profiles more regulated (communal) fires are unlikely to be an attractive option. Where fires are allowed, consider providing chopped firewood to avoid destruction of trees and hedges and the potential for burning plastics and other material that could produce noxious fumes.

26.8.2 Consider the hazards and risks of camp fires in the event risk assessment to include the following:
- Suitably trained stewards or fire marshals;
- Fire points (locations where fire extinguishing materials are kept): as a minimum these should consist of a means of reporting a fire, such as a gong or triangle, and supplies of water and buckets, although these are probably of limited use in a tent fire;
- Watchtowers consisting of raised platforms staffed by trained personnel on fire watch with radios are a more effective means of observing for uncontrolled fires and suspicious behavior. They should be supplemented by the provision of fire extinguishers and, depending on the scale of the event, an on-site capability to attend to fires with specialized vehicles; and
- The fire points themselves becoming a hazard due to trash accumulation, etc.

26.9 Site Services

26.9.1 Ensure that facilities are maintained throughout the site 24 hours a day and services are provided for the duration that people are actually on site. All facilities must be lit at night.

26.10 First Aid

26.10.1 See Chapter 5, *Medical, Ambulance and First-aid Management*. At camping events that run through several days, it will not be sufficient to provide only a first-aid facility. Expect the demands that would be placed on a GP practice serving a community of similar size. Routine medical supplies, therapeutic drugs, etc., may need to be provided, including pharmacy facilities, dentistry and psychiatric facilities.

26.11 Welfare

26.11 1 Many children are likely to be on site and facilities will have to be provided, including potentially accommodating children overnight. Communications and availability of information on lost children, lost friends, etc., must be established.

26.12 Telephones

26.12.1 Ideally, provide land line telephones in suitable numbers, and ensure they are easily accessible and available 24 hours a day. However, with the increased use of cell phones, this may be less necessary in areas with good cellular service.

26.13 Sanitary Facilities

26.13.1 See Chapter 14, *Sanitary Facilities*. In the case of events with large camping areas, assess where and when facilities will be under pressure. There will inevitably be a peak morning demand.

26.13.2 It is suggested that a plan is established whereby sanitary accommodation, drinking water supplies, washing facilities and showers are all clustered together, creating an easily identifiable location for all facilities. Monitor the condition of sanitary accommodation to ensure they are regularly emptied and cleaned as required in addition to routine programmed servicing.

26.14 Trash

26.14.1 Provide trash receptacles along the walkways and access ways for vehicles and also at conspicuous points such as sanitary facilities, etc. Ensure that bins are emptied on a regular basis to encourage careful disposal and to avoid creating a fire hazard. On undeveloped sites with potentially difficult terrain this is likely to be achieved by tractors and trailers. Reductions in volume of trash are likely to be achieved by using recycling points to take separated waste.

26.15 Site Lighting

26.15.1 Provide adequate lighting to enable orientation at night, with higher levels of lighting at toilet areas, fire points, information and guard points, etc. Consider the nature of lighting. Lighting tower rigs are likely to be unsuitable for camping areas due to generator noise as well as providing an overly bright source of light. They may, however, be suitable for intersections, crossroads, facilities, etc. Festoon lighting can be tampered with so it won't work or becomes a safety hazard. Wherever possible, provide the camping areas with some illumination provided from "borrowed light" from other areas of higher lighting nearby, which can be supervised.

26.16 Access

26.16.1 Provide both vehicular and pedestrian tracks to and through camping areas to ensure ready access for emergency vehicles and also to provide safe routes for pedestrians free of trip hazards such as guy ropes, etc.

26.17 Noise

26.17.1 Plan for preventing or reducing the impact of potentially noisy activities within campsites or of dealing with any overnight activities that become problematic. Dependent upon the nature and proximity of residences to the site, restrictions may be needed in limiting the background music provided by concessionaires to avoid noise disturbances.

27. TV and Media

27.0.1 Entertainment events can attract a great deal of media interest from television crews, still photographers, print and radio journalists, as well as event patrons posting in real time on social media.

27.0.2 The management of media can be split into two areas: pre-event (just before the event) and during the event.

27.1 Pre-Event

27.1.1 As an aid to crowd management and public information, consider issuing a press release containing as much information as possible about the event: name, dates, times, location, line-up, ticket information, public transport information and contact name and telephone number.

27.1.2 Make sure that as well as national media outlets, all local media have been contacted with details of the event. If the event sells out or is cancelled or if a major incident occurs, good communications with local media will ensure that information is carried to the public quickly and efficiently.

27.1.3 Decide the amount of media that is manageable for the event. Setting an acceptable level of media attendance depends on available space and infrastructure, how many people are able to look after them and how long the event lasts.

27.1.4 All media can usefully provide advance advice to the public, such as conditions on site, travel arrangements, site facilities and restrictions. Ensure that each media representative who will attend your event, receives information on site safety arrangements.

27.2 During the Event

27.2.1 If producing medium- to large-scale events, consider setting up a press tent or press office within the VIP or guest hospitality area (if provided). Ideally this should be situated away from production or artist dressing room areas in a location with easy access to the front-of-house.

27.2.2 On occasion, to preserve the privacy of all artists, the dressing room compound is closed to press and media. Communicate this situation to the artists and the media before the event so all parties can plan accordingly.

27.2.3 Festival documentary crews are often allowed in the artist compound as part of the contract between the organizer and artist.

27.2.4 The press tent or office is where information about the event can be posted, interviews organized and a meeting place set up for photographers, film and radio crews before media activity. If possible, provide the press tent or office with Internet service and power points so that the media can recharge batteries, phones, etc. Water and beverages are always a hit with the press and some organizers even provide snacks or meals.

27.2.5 Photographers

27.2.5.1 Make sure photographers are escorted into and out of the pit area and display appropriate passes. Where possible, photographers should enter and exit the pit area from the same side to allow security and medical services total access from the opposite side. If for any reason the pit becomes crowded or the safety of the audience is compromised the photographers should be escorted out of the pit immediately. If there are a large number of photographers on site, it is recommended that they should be escorted to the pit area in smaller manageable groups to prevent overcrowding of the area.

27.2.5.2 If the photographer shoot location is in the pit, photographer platforms may be required to elevate them to a reasonable level to get flattering shots of the artists.

27.2.5.3 On occasion, a second media viewing/camera platform is required front-of-house near the mix area. This area should be equipped with an audio multi-box and power strips for those who need it.

27.2.5.4 When placing any camera platform, consider what the camera eye is going to capture around and behind the subject being photographed. The PR and Marketing department will always have an opinion on this location and it is good practice to accommodate their requests, if reasonable, whenever possible. The footage and stills captured by these photographers will be used to market the careers of the artists, the next event as well as any show-related commercial video or audio products resulting from the event.

27.2.5.5 The area square footage of all platforms including the space between the riser and any barriers used to protect them must be deducted from the usable public viewing area square footage.

27.2.6 Radio Broadcasters

27.2.6.1 Local radio stations often attend the site with a mobile or outside broadcast unit (OB unit) to feed live inserts or sound-bites back to the studios. The OB unit usually takes the form of a van or four-wheel drive vehicle with a large telescopic mast. Once the interviews are completed ensure the OB unit is moved off site or to allocated parking areas.

27.2.6.2 Sometimes a radio station is set up on site specifically to broadcast programs to the station's listening audience. This provides entertainment with interviews with performers and the audience and can be extremely useful in transmitting important safety information and messages for people. Plan how you will access the radio station with safety information you may wish to be transmitted.

27.2.7 Print Journalists

27.2.7.1 Print journalists normally require the least amount of attention as they attend to review the event as a whole rather than acquire individual interviews. Ensure any interviews with artists are pre-arranged before the event begins to limit the amount of on-site organization between the press office and artists.

27.2.8 TV Broadcasters

27.2.8.1 TV requires the most amount of attention and the type of TV crews and workers can be broken down into three main areas: event filming crews, TV news crews, and production companies.

27.2.8.1.1 Event Filming Units. A large event will generally be recorded by a dedicated production company or broadcaster for live or future broadcast. Plan these arrangements in advance as they require special facilities such as filming platforms, OB vehicles parked backstage, audio mixing trucks, video production trucks, front-of-house filming platforms, etc. These arrangements need to be considered in venue and site design.

27.2.8.1.2 TV News Crews. TV news crews will consist of local news crews, cable and satellite crews. These crews are normally small (two to four people). They will require only a short amount of time on site and therefore can be serviced relatively quickly. They should be supervised wherever possible and escorted quickly and efficiently to key locations (production offices, services offices, front-of-house, etc.). Local TV crews may also request to "go live" during newscasts. This coverage is generally desirable since it helps promote the event. These crews typically use a van or SUV with a telescoping mast with a microwave transmitting antenna on top. These transmitters require line-of-site to specific reception sites somewhere in the area (usually on top of tall hills or tall buildings). For large events, it is advisable to designate a media vehicle parking area that has the appropriate site line for their equipment. Coordination with a local television station's engineering department can help determine the best location for this area. TV stations prefer a location that meets the necessary technical requirements and provides a vantage point that allows the venue or event to serve as a backdrop for their reporter. After the location is established, the media coordinator should conduct a media walk-through prior to the event in plenty of time to address any concerns.

27.2.8.1.3 Production Companies. This is normally the largest part of the TV mix and is made up of television broadcast programs interested in covering the event. TV crews tend to need access to vehicles for equipment and storage, so consider space allocation close to the hospitality/VIP area as possible. Other non-essential vehicles can be allocated spaces in designated parking areas.

27.2.9 Foreign Media

27.2.9.1 Foreign media workers need to be clearly briefed in advance and given assistance to understand the safety requirements especially regarding the provision, compatibility and use of electrical equipment.

27.2.10 Student Media

27.2.10.1 Student media can be helpful to the event and useful for communicating with the audience at events attended by predominantly young people.

27.3 On-Site Structural Considerations

27.3.1 In addition to requirements already mentioned for facilities, vehicles and accommodation, the presence of media workers, and TV broadcasters in particular, will have to be considered in your venue and site design. Media provision such as camera cranes may restrict viewing areas for audiences and therefore cannot be counted in occupant capacity calculations.

27.3.2 Media may need to use platforms or other structures—as outlined in Chapter 19, *Structures*—such as scaffold towers and the use of barriers around media installations. Similarly, requirements for electricity supplies will need to conform to recommendations (e.g., burying cables).

27.4 On-Site Public Relations Staffing Requirements

27.4.1 The number of workers required to manage media will vary according to the size of the event, the number of days over which the event is held, the type of event, the capacity and the amount of media expected. As an example, for a large three-day music event with a 50,000 or more capacity, at least 10 people will be required to deal with the media. At smaller one-day events, four to six people should suffice.

27.4.2 Issue radios to all workers handling the press and media and assign them their own dedicated channel to avoid taking up unnecessary time on production channels dealing with media questions, guest lists, artists' whereabouts, etc. Ensure that all media workers are fully briefed and aware of any emergency procedures.

27.4.3 Media liaison workers need to have a base; normally this will be at the point where media representatives check-in or a press tent. These areas are best located close together to avoid large distances between the production areas, pit, press-tent and media check-in locations.

27.4.4 Over the course of an event, the event's media and press officers will get to know the individuals involved and this is extremely useful in the case of emergencies or important announcements. Ensure that the chief press officer is introduced to key security and stewarding staff, local authority officers, police spokespersons, welfare organizers, event film units, etc. This will enable a direct line of communication between the media and services that is controlled and efficient.

27.4.5 In the event of a large-scale incident, the above entities should convene together and determine what and how appropriate information should be communicated to the press and media. If a serious catastrophic incident has occurred, organizers and event related commercial PR agencies should anticipate the lead government agency on site will step in and control the dissemination of all information to the press and media.

28. Large Events

28.0.1 For the purposes of this chapter a "large event" normally has one or more of the following components:
- Multistage;
- Multi-performance;
- Multi-activity;
- Multiday; and/or
- Physical size of venue (generally outdoors).

28.0.1 The significant factor for this chapter is audience size, let's say greater than 15,000 patrons. It would be easy to regard a large event as being the same as any event but with more facilities, services and workers, etc. While reference should be made to the specialist chapters of this publication, there are a number of areas where the size of the event alone demands particular attention.

28.1 Planning and Management

28.1.1 The need for consultation and planning for a large event cannot be overemphasized. The formation of an event safety management team, comprising representatives of the emergency services and local authority, is a useful method of addressing the practicalities of event organization. Team meetings can be scheduled before, during and after the event and can run in parallel to any formal public permitting procedures.

28.1.2 See Chapter 2, *Planning and Management*, for more general discussion of this topic.

28.2 Crowd Management

28.2.1 While the proposed attendance figure is the key to the provision of services and facilities, account should be taken of the number of guests and staff. Dependent on the event, up to 10 percent of the capacity could be guests or staff at the event with the consequent additional load on site infrastructure.

28.2.2 Also consider easing local traffic congestion by opening the site early and restricting exits. Incremental occupation of the site should be accompanied by a similar incremental provision of services.

28.2.3 In some instances for nominally non-camping events, it may be useful to make contingency camping provision and low key entertainment on a normally silent night. There is, however, a danger of changing the nature of the event for subsequent years. Ticket pricing structures may control arrival, particularly for late Friday arrivals for a Saturday event.

28.2.4 Within the site there needs to be active and visible crowd management. The technical issues of stage layout, audience size and barriers are dealt with elsewhere. At a large event the layout should take account of audience movement across the site and minimize cross flow and points of congestion. A wheel layout, with entertainment at the hub and camping at the rim, could be combined with one or more of the following:

- Area or small team of event staff to maintain a controlled scale of audience movement;
- Dynamic entertainment management where the programs on separate stages are integrated into the audience management program;
- Ensuring that timing and running orders are followed in order to avoid conflicts at the end of performances;
- Gradual close down of main stages;
- Continuing (perhaps for 24 hours) low level entertainment such as cinema or markets; and
- No entertainment within the defined areas of campsites.

28.2.5 See Chapter 9, *Crowd Management*, and Chapter 23, *Barriers*, for a more general discussion on this topic.

28.3 Major Incident Planning

28.3.1 The size and complex infrastructure associated with a large event reinforces the need for a comprehensive major incident plan. The event safety management team in consultation with the local authority emergency planning officer, who would be familiar with local arrangements, should develop the plan. The following aspects should be considered:

- Is the evacuation of the entire site practical or would selective evacuation be preferable?
- Is the evacuation of the site desirable, given that under some circumstances food, water and sanitary facilities may still be operational on a scale unavailable elsewhere?
- What infrastructure is available elsewhere?
- What would be the impact of a mass exodus from one part of the site on other parts or on the locality?
- What implications are there for public address systems in various emergency situations?

28.3.2 See Chapter 3, *Major Incident Planning*, and Chapter 6, *Communications*, for more general details on this topic.

28.4 Transportation Management

28.4.1 If public transportation links are available they may be encouraged with integrated ticketing. Depending on the event and availability, many people may choose integrated bus/event travel. In rural locations, or where other transport is unavailable, much of the audience will, almost inevitably, travel by car and the logistics and impact on the locality should form an early item for consultation.

28.4.2 Traffic should be removed from the public road system onto site as quickly and efficiently as possible; the use of professional stewarding may be the best option. Within the site, parking areas should be divided into easily identifiable zones (perhaps associated with nearby camping) and traffic should be routed to avoid designated pedestrian routes/areas.

28.4.3 See Chapter 12, *Transportation Management*, for more general details on this topic.

28.5 Children

28.5.1 People may become more easily lost in a large event. This may be particularly true for children. Organizers of large events should consider ways to help reunite children with lost parents, and to accommodate the needs of children during the time until their parents return to them.

28.6 Information and Welfare

28.6.1 The provision of a comprehensive information and welfare service that can assimilate and coordinate information in an active way as the event progresses allows other agencies, such as emergency medical and police services, to undertake their specialist functions. Everything possible that an individual requires for the duration should be readily available on the site.

28.7 TV and Media

28.7.1 The presence of regional, national and international media may in itself influence the progress of an event. In particular, incorrect ticket availability broadcasts may cause problems. Ensure that channels of accurate information are available for coordinated release to the media.

28.7.2 See Chapter 27, *TV and Media*, for more information on this topic.

28.8 Venue and Site Design

28.8.1 The site design for a multiday event must recognize the need for 24-hour access to facilities for both the audience and for servicing the facilities.

28.8.2 See Chapter 8, *Venue and Site Design*, for more information on this topic.

28.9 Fire Safety

28.9.1 Discussion should take place, pre-event, on the areas of responsibility for fire safety teams. There needs to be a clear understanding of the circumstances under which the local fire department will respond and lines of communication must be established. There should be a policy and procedure in place for safely dealing with small arena fires.

28.9.2 See Chapter 4, *Fire Safety*, for more information on this topic.

28.10 Sanitation Facilities

28.10.1 Water availability is a limiting factor on the audience size at all events on undeveloped sites. In particular, the logistics of moving large quantities of liquid—whether water or effluent (sewage)—need to be addressed. While flush toilets are a preferred option, they are vulnerable to failure of water supply and can be difficult to bring back into use when the supply has been restored. The use of fewer toilet blocks with more units can, subject to proper access routes and efficient continuous servicing, mean that a greater number of toilets will remain in operation. For overnight or multiday events, there will inevitably be a peak morning demand, particularly if showers are provided in camping areas.

28.10.2 See Chapter 14, *Sanitary Facilities*, for more information on this topic.

28.11 Food and Drinking Water

28.11.1 Supplies of both food and drinking water must be adequate for the duration of the event; the facility for campers to buy basic commodities such as bread, milk, etc., needs to be available. To ensure sufficient supplies of water there will need to be a considerable amount of temporary pipework, which is susceptible to damage and vulnerable to contamination. Consideration should be given to splitting the water supply on the site into several independent supply zones. In this way the consequences of a serious incident affecting the water supply will not affect the whole site. It may be necessary to protect the quality of the supply by increasing chlorination above normal mains levels. The use of percussion taps will help reduce waste.

28.11.2 See Chapter 13, *Food, Drink and Water*, for more general details on this topic.

28.12 Health and Safety of Event Workers

28.12.1 Set up a proper management infrastructure with delegation of authority. The safety management team should include people with experience from previous or similar events. One of the issues that will be encountered with large events running over many days is one of fatigue among both management and contractors. All will be working long hours under stressful conditions and if this is not addressed, the quality of decisions, some of which may be critical, could be poor.

29. Unfenced or Unticketed Events, Including Radio Roadshows

29.0.1 Unfenced and/or unticketed events are popular at open-site venues such as local parks. Occasionally free events will be organized in existing arenas or stadiums. This chapter highlights specific health and safety issues at unfenced/unticketed events in open spaces. A few specific suggestions have been made for such events in arenas, stadiums and radio roadshows.

29.1 Planning and Management

29.1.1 The Planning and Management chapter provides information concerning the application of good health and safety management systems.

29.1.2 Risk Assessment

29.1.2.1 The entire event space should be inspected to determine if there are any particular hazards that present greater risks with a large number of people in attendance. Events taking place near a water feature such as a lake, river or pond will need a means of preventing people from falling or swimming in the water. Security and other event staff trained in life-saving skills may need to be employed and extra warning signs erected. In certain circumstances, it may be necessary to physically separate areas of the park or open space from the area chosen for the event.

29.1.2.2 If there are rivers, lakes, or ponds next to or near the event, additional provision may need to be made to prevent any runoff making its way into them. Examples might include oil and fuel from vehicles or overflow from fueling, sewage from any source, and any discarded waste.

29.1.3 Build-up/Breakdown

29.1.3.1 The fact that there is no perimeter fencing can cause added problems for contractors working on site. The public will often want to wander around the site to see what is happening. Vehicle movement should keep to dedicated paths, observe strict speed restrictions (5 mph), and use yellow "emergency-type" revolving lights (use of flashing hazard lights while driving is not recommended because it does not allow the driver to use turn indicators). If the park or open space is heavily used, it may be necessary to consider having a person on lookout while cautiously walking in front of the moving vehicle.

29.1.3.2 Areas where work is being undertaken can be temporarily cordoned off for security. Greater security will be needed especially at night to ensure that the temporary structures erected are not vandalized or tampered with. Consideration should be given to temporary barriers and the provision of specialist security guards.

29.1.3.3 When erecting temporary structures, follow the guidance contained in Chapter 19, *Structures*. Radio roadshows tend to use rapid deployment mobile stages, vehicles specially adapted for the purpose that contain an integral stage. These vehicles need to be situated on firm level ground that has adequate drainage. If the vehicle is to be placed on grass and there is the possibility of rain, temporary hard pads may need to be considered to avoid penetration into the surface beneath the stage. It is important that sufficient space be provided for the vehicle/stages in the venue design. Keep in mind the vehicle must be driven to the area and also consider the weight of the vehicle as well as the terrain and surfaces it will drive over to access the location.

29.1.4 Crowd Management

29.1.4.1 The benefit of free events held in parks or similar locations is that there is no enclosed arena, so there is no physical restraint to crowd dynamics. However, the numbers that are likely to turn up on the day are always difficult to predict and should be carefully considered. As always, planning for the safety and welfare of those in attendance is directly related to the size and nature of the audience attending the event. The number of stewards required is dependent upon the overall risk assessment. The fact that members of the audience are likely to be spread out over a greater area should be a constant consideration.

29.1.4.2 In these circumstances, you will need to estimate the expected audience levels. This estimate can vary considerably depending upon factors including the performer's popularity, the weather, other events at the same time in the local area and the amount of media attention. For health and safety purposes, it is better to overestimate the audience numbers rather than underestimate.

29.1.4.3 Free or un-ticketed events organized in existing fenced venues may cause problems when ensuring that the occupant capacity determined for the premises is not exceeded. It may be appropriate to issue free tickets to gain entry to the event or a system for counting audience members in and out of the venue.

29.2 A March Before the Event

29.2.1 A march sometimes precedes an event so that the majority of people arrive at the same time. Care must be taken to ensure that the site and services are ready and able to cope with the large number of people arriving within a limited time. The training of security and event personnel is essential at this type of event to ensure that the crowd is directed to where it is expected.

29.3 Crowd information

29.3.1 When an event has tickets, information on event times and transportation routes can be given on the reverse of the ticket. When this medium is unavailable, emphasis needs to be placed on providing information about the event on leaflets (flyers), local radio, newspapers, the event website, social media sites and smartphone apps. Information could also be made available throughout the event by electronic notice boards.

29.4 Major Incident Planning and Emergency Access Routes

29.4.1 Emergency planning and the design of dedicated emergency access routes can prove more difficult at unfenced events as the audience members are not contained in one area. At fenced events, there is relatively easy access around the site once members of the audience are in the arena watching the event. At unfenced events, members of the audience are able to move to all parts of the park or open space and this can hamper the movement of emergency vehicles. Consider providing cordons with appropriate security and stewarding to dedicated access routes. Adjustments may need to be made to the existing perimeter fencing of the park to allow for the safe evacuation of the audience from the park, other than through restricted park entrances and exits.

29.5 Communication

29.5.1 Good communication systems are vitally important to health and safety management. At unfenced events, planning for the location of security and event personnel around the site can be a problem as there are fewer easily defined positions and posts, e.g., entrances and exits to the fenced arena. Security and event personnel need to exhibit greater discipline to remain in the area that they have been stationed and not to wander around the site. Greater reliance on radio communication may be needed at a large site and event staff will need to have clearly gridded plans so that they can be more accurate in summoning assistance and identifying their own position.

29.6 Performers

29.6.1 It may be necessary to provide a secure backstage area for performers that is securely fenced to prevent members of the audience trying to get access to the performer. Planning for the arrival and departure of the performer may require cordoning separate areas and road closures.

29.7 Children

29.7.1 There may be a greater proportion of families with children attending this type of event compared to the traditional ticket/fenced concerts. There is also a higher probability that children and young adults will attend the event on their own. Provide "help points" and a lost children's facility.

29.8 Information and Assistance

29.8.1 Make sure to provide facilities for information and assistance. Establishing meeting points, information booths and first aid stations for the audience will be more important as there will not be the usual entrances and exit points for audience members to identify with.

29.9 Venue and Site Design

29.9.1 Venue design should consider overflow areas if the audience is larger than predicted. Overflow areas are required to prevent audience members from blocking roads or designated emergency escape routes.

29.9.2 The quantity of food and merchandising concessions, toilets, first-aid stations and other site facilities will depend upon the predicted size of the audience. Careful consideration should be given to the location of the food and merchandising concessionaires, first-aid stations, assistance and information points and toilets. It is likely that the audience will be spread over a greater area than is usually calculated for a fenced or enclosed arena. The location of these facilities should reflect this.

29.9.3 In an unfenced venue, way-finding and directional signage requirements will increase because the attendees will potentially be arriving from all directions instead of through a primary venue entrance.

29.10 Food and Drink

29.10.1 Glass bottles should not be sold on the site. Local public houses and food outlets should be contacted to request that during the event food and drink is not sold in glass containers.

29.11 Waste

29.11.1 At unfenced events, it will be impossible to prevent members of the audience taking glass bottles and cans on to the site. Consider providing as much pre-publicity about this aspect as possible. Special containers should be provided to encourage the audience members to dispose of their glass containers safely and if possible encourage people to decant the contents of glass containers into another container, e.g., recyclable cups.

30. Electronic Music and Other All Night Events

30.0.1 This chapter highlights issues to consider when organizing an all-night event. All-night events may take place in any of the following: open field sites, warehouses, leisure centers/facilities, exhibition halls, parking lots, race tracks, purpose-built stadiums and arenas, night-clubs and convention centers.

30.0.2 Trends evolve quickly with this demographic so the best training ground for organizers of these events is going to similar events to see what works, what doesn't work and under what conditions. Production vendors of these events tend to do many of them and they are a valuable resource when it comes to past experience.

30.1 Audience Profile

30.1.1 Although the specifics of any particular event can vary widely, which is why it is essential for organizers of all-night events to research previous similar events, a few generalizations may be useful. All-night events tend to attract a relatively younger crowd, from teenagers through young adult. They often skew more male than female. People coming to these events will dress in all varieties of clothing, depending on the type of event. Excessive use of controlled substances may be reasonably foreseeable.

30.2 Duration of the Event

30.2.1 All-night events vary in duration, but 10 hours is not unusual for indoor events, with over 16 hours for outdoor weekend events. There will also be a load-in period, followed by load-out period after the show which will have an impact on the local environment.

30.3 Management

30.3.1 The arrangements for these events mirror those referred to in other chapters of this publication. For multiple night events, a night working crew is likely to be necessary to affect 'running repairs' around the site. Due to the duration of these events, ensure that adequate rest periods are taken by workers and contractors.

30.4 Format

30.4.1 The format of these events is that different types of music will be played in different areas if the venue layout permits. The audience will move from one area to another throughout the event. This has crowd management implications due to the crowd dynamics of people trying to get into particular locations where the main DJs or live acts are taking place.

30.4.2 Ensure that the "running order" with the artist's name, the stage name and its location is openly advertised in advance on the event website, flyers, on signage at the venue and queuing lanes where people are waiting to enter the venue, fence panels and at information points. Programming should ensure that crowds are safely distributed around the site according to the capacity of the different areas, to avoid over-crowding, pressure on access points and mass movement around the site.

30.5 Medical

30.5.1 The medical provider must be familiar with symptoms and treatment of heat exhaustion, dehydration and drug and alcohol intoxication.

30.5.2 Local government provided medical services, e.g., paramedics and advanced life support (ALS) ambulance services will likely be required by the local authorities. If not, the organizer should seriously consider these services, regardless of the disposition of local authorities. In some communities the number of available resources in-market may be limited and local authorities may require the organizer to augment their resources with the additional services of a private emergency medical response company.

30.5.3 An on-site medical room or triage location, such as a large room or a tent should be considered. This location will need hot and cold running water and restrooms exclusively dedicated to this room immediately adjacent to the room with a member of the event staff to monitor access, restock of supplies and cleanliness of the units.

30.5.4 When resources are limited or the site is in a remote location, organizers should consider the services of experienced emergency room medical doctors and nurse staff to work in this room in addition to first aid and EMT staff, to assist in the determination of whether 'emergency transport' of patients to a hospital is required. The presence of this additional experienced emergency staff in the medical room can help the on-site medical supervisor better manage the resources available on site and in the environment.

30.6 Admission

30.6.1 To minimize the guest's time in the queue, consider a pre-check of ticket, bag and ID as the people enter the queue line chutes at the entry portal. At the head of the line where actual ID checks occur and tickets are torn, it is advisable to place a "secondary ID check" area to send people with ID or ticket problems who have arrived at the head of the line. This area should be staffed with people familiar with government issued identification.

30.7 Queue Line Layout and Management

30.7.1 It is good practice to use actual measurements from advance surveys or scouts when ordering fencing and bike rack barriers required at entrances. The entrance area often uses more equipment than anticipated so it is recommended you do not rely on a "scale diagram" as diagrams rarely indicate current site topography and landscaping.

30.7.2 An "ejection lane" should be designed into the entrance plan to expedite exiting from the area by persons who will not be allowed into the venue. This lane should be a discrete route so the person being ejected has no interaction with those in line behind them.

30.7.3 The following is an example of a successful sequence at the head of the line:
- ID CHECK (1 staff per lane)
- BAG CHECK (1 security guard minimum and a table per lane)
- PAT DOWN (2 security guards per lane)
- TICKET TEAR or SWAP (1 staff per lane)

30.7.4 A list of prohibited items, and acceptable forms of identification for entry and the size of bag they are allowed to bring into the venue must be publicized in advance.

30.7.5 VIP and other premium tickets often have their own dedicated entrances, away from the general public entrance. All infrastructure and checks required at the main entrances are also required for these entrances. Once inside the venue, VIPs are typically given unrestricted access to exclusive areas behind and/or beside the stages. These areas generally have their own sanitation facilities, bars and concessions.

30.8 Occupancy

30.8.1 Organizers need to agree on occupancy levels with authority having jurisdiction and resist the temptation to "oversell" the venue. For information on venue capacity, see Chapter 8, *Venue and Site Design*.

30.9 Venue and Site Design

30.9.1 For information on this topic, please refer to Chapter 8, *Venue and Site Design*. The general principals noted in this chapter should be followed for all-night events.

30.10 Tents

30.10.1 Outdoor events are usually held in dance music tents and/or the open air. An issue affecting the audience safety is the availability of tented cover for the occupant capacity of the event. An outdoor daytime concert could last up to 13 hours, with no cover provided against the weather for the audience. However, at night the air temperature can drop rapidly and with the possibility of limited public transport or remoteness of the site from other facilities, the risk assessment should consider the possibility of hypothermia.

30.10.2 A reasonable percentage of the audience should be able to find cover, particularly in bad weather. All tented accommodation must comply with the relevant structural and fire safety standards referred to in Chapter 19, *Structures*, and Chapter 4, *Fire Safety*.

30.11 "Chill-Out" Areas

30.11.1 Fast dancing can result in rising body temperatures in the participants and this can be exacerbated by the effects of some drugs. It is essential to provide a "chill-out" area (or possibly several). This will allow people to cool down in a more calming environment. There may be music, but it will be quieter and more relaxing. It can take a variety of forms such as a room, tent, roofed structure with seats, or space outdoors.

30.11.2 If an outdoor "chill-out" facility is provided in the winter, the air temperature may be so low that heating may be necessary in the area set aside for this use. Staff should maintain a presence in this area(s), and look out for individuals who may need medical attention or assistance. If youth/drug counselors are on site, they should also pay particular attention to these areas.

30.12 Ventilation at Indoor Venues

30.12.1 Due to the quantity of hot and humid air that needs to be moved, high velocity fans (forced or induced draft) will be needed to achieve sufficient air changes. If possible, a "balanced" system should be used where extracted air is replaced by fresh air drawn into the venue. The use of "smoke" machines or similar effects will need to be carefully assessed.

30.12.2 If the premises cannot be ventilated as previously described, temporary facilities can be brought in to give some relief to the audience, such as high velocity fans placed at approximately 5 feet (1.524 m) height, or about face level. They can be positioned next to areas where individuals will stand, e.g., bars.

30.12.3 In small arenas or rooms, portable air conditioning units can be used, and will bring about a reasonable reduction in temperature. Lightweight structures will quickly cool in the winter months and will become extremely warm during the summer.

30.13 Drinking Water

30.13.1 Drinking water fountains ensure that waste water can be retained and floors do not become slippery and dangerous. Drinking water taps should always be labeled as such. The pressure of the water supply should be adequate for the number of taps being used from it. Seek technical advice from the local water company.

30.13.2 Water is the most important aspect of maintaining personal safety at dance events. Individuals may perspire profusely and need rehydration. A good guideline of 16 ounces (556 ml) of fluid per hour can be used, although events where people are engaged in particularly vigorous activity for long periods of time may require more. This is a medical issue for which organizers of all-night events should consult a knowledgeable health care professional.

30.13.3 It is essential for the "core" of the human body to be kept cool; otherwise it could overheat (heat stroke). However, a cautionary note: too much water, consumed too quickly can also be hazardous and may cause medical problems with serious consequences.

30.13.4 The provision of a free drinking water supply, regardless of the venue type, is an absolute necessity. The staff and security staff should know the location of these facilities.

30.14 Alcohol and Sports Drinks

30.14.1 The consumption of alcohol at all-night events varies depending on the nature of the event. High sugar content drinks, such as sports drinks or fruit juices, help to replace body salts and minerals lost through dancing in a hot environment.

30.15 Relief Assistance

30.15.1 Some additional relief provisions may be needed. As the majority of people attending such events are young people it is essential to have trained youth or drug workers onsite so that they can identify people who may require support or assistance. These staff should be identifiable by the wearing of suitable external identification. They should have an accessible base on site so that they can be easily contacted when their services are needed. A clearly visible meeting point will be required for missing people. The local fire department, medical provider, department of health or peer group security firm may be good sources of this expertise.

30.16 Mass Transportation

30.16.1 Management of the audience arriving and leaving the arena should be discussed with the local public safety officials. Extra security may be needed at venue exits during egress to direct guests leaving the venue as to where to find taxis, shuttle buses, or public transportation. The majority of the audience members will leave the event at the end. It is important that there has been proper consultation with the public transport providers to ensure that sufficient public transport is available. See Chapter 12, *Transportation Management*, for more information.

30.17 Weather Forecasts and Warnings

30.17.1 Up-to-date weather information will assist both the production team and the audience. If night temperatures drop, lightly clothed participants may suffer hypothermia if they are ill prepared.

30.17.2 The use of a weather warning service that can send real time alerts is highly recommended. This type of service can keep the organizer informed of actionable drops in temperature in addition to providing general weather updates and information on important weather conditions (e.g., winds, precipitation, storms, etc.). See Chapter 7, *Weather Preparedness*, for more information.

30.18 Controlled Drugs

30.18.1 It may be prudent to arrange for an appropriate drug/alcohol counseling agency to be on site to help people are in need of advice or assistance. A number of staff should be retained after the event until the site is cleared or no problems are reported.

30.19 Amnesty Box

30.19.1 With the endorsement of law enforcement and local health authorities, some organizers have begun placing an "amnesty box" outside the event perimeter, near the entry portals. An amnesty box is a sealed one-way deposit container operated by a firm approved by the local authorities in which members of the audience, without consequence, may dispose of contraband objects brought to the venue. Certification and accreditation of disposal firms may vary; check with local authorities before hiring a disposal company. Law enforcement should be advised of the use of these boxes, and they should supervise the collection and disposal process and sign an affidavit of destruction.

31. Arena Events

31.0.1 This chapter aims to highlight factors to consider when organizing an event in an arena-type environment. This discussion addresses events held at purpose-built arenas specifically designed for hosting events as well as premises to which temporary structures are added in order to accommodate particular events.

31.1 Planning and Management

31.1.1 Arenas specifically built to host events may require a permit, which will be held by the arena operator. If you wish to stage your own event in an arena already holding a permit you will need to work directly with the arena operator. The most important planning aspect will be determining the responsibilities for health and safety between the respective parties and documenting the agreements.

31.1.2 Arena operators will likely already have a written safety policy, risk assessment and major incident and contingency planning documents for their own workers and events they promote themselves. If you are renting an arena (or part of it) and/or obtaining a permit in your own name you will also need to work with the arena operator so that information about the existing safety management policies and procedures can be exchanged.

31.1.3 Health and safety responsibilities need to be determined for the preparation of the risk assessment for every event. Agreements also need to be documented on the services supplied by the arena operator including labor sources and equipment. There may be a need to appoint a safety coordinator from one of the parties.

31.1.4 Arena operators are likely to have prepared their own in-house safety procedures which need to be communicated to any external contractors brought on site. In multi-occupied premises it is important that agreement is reached and health and safety responsibilities assigned between the parties, in relation to major incident planning. Ensure that the planning for the event is coordinated with the planning of the premises as a whole.

31.1.5 A system to ensure that health and safety information is communicated to other users of the building, especially if there is more than one event occurring at the same time, also needs to be agreed upon and documented.

31.1.6 The breakdown of the event may have to take place very quickly if the arena has been booked for other events. Working to tight deadlines needs careful planning to avoid a fatigued workforce more prone to physical and mental errors, which can lead to injuries.

31.1.7 All these matters should be discussed at the initial planning meetings with the arena operator.

31.2 Crowd Management

31.2.1 Peer Security

31.2.1.1 Employees of an arena's peer security staff (ticket takers, ushers, and security guards) should possess (where available) a security guard license issued by local or state authorities. Most security guard card applications generally require at least a background check and many jurisdictions require that newly licensed guards receive particular security training. Peer security should have a uniform that visibly sets their staff apart from members of the general public, and from other building staff such as concession and housekeeping workers. Security staff should be familiar with the arena's venue- and event-specific policies. Each staff member should have a clear understanding of the post orders for the position they're to which they are assigned.

31.2.1.2 In planning the deployment of peer security for an arena event, a "dot map" can be helpful to show security guard post positions. The dot map should ideally be agreed upon between the arena, security provider, and event organizer.

31.2.2 Public Safety

31.2.2.1 In addition to peer security staffing, it is recommended that local law enforcement officers be contracted for arena events. Anticipated attendance and the nature of the event should dictate the level of staffing for local police, sheriff's deputies, etc.

31.2.2.2 In planning for an arena event, the roles and responsibilities of local law enforcement should be agreed upon by the venue staff and the promoter.

31.2.2.3 In many U.S. cities, seating charts and egress plans have to be approved by the local fire department prior to the event going on sale to the public. The venue representatives and the local promoter should work together on the documents submitted to the fire department.

31.2.2.4 Fire department staffing is required for arena events in most U.S. cities. Firefighters assigned to arena events are responsible for ensuring that local fire codes are adhered to. Staffing levels of firefighters for arena events can vary depending on the anticipated attendance, and specific details of an event, such as the use of pyrotechnics or fireworks.

31.2.2.5 Most U.S. cities require emergency medical technicians (EMTs) be on-site as first responders for minor injuries and/or illnesses. Venue representatives should be familiar with minimum EMT staffing requirements for their arena. Recommended EMT staffing levels can vary depending on the anticipated attendance level, seating configuration (e.g., reserved, general admission, open floor, etc.) and the nature of the event.

31.2.2.6 Some arenas require their contracted security provider be used exclusively for all events, others do not. If outside security contractors are to work alongside the building's contracted security provider, clear lines of control and co-operation need to be established. All security staff should operate through a central control suitably trained and competent for their role. The roles and responsibilities of ticket takers, ushers, and security staff should also be clearly defined.

31.2.2.7 Some arenas may lack adequate queuing areas for early-arriving ticket holders. In these circumstances it will be necessary to deploy barriers to prevent audience members from blocking sidewalks and city streets. Communication with early arrivals is important to keep patrons informed.

31.2.2.8 Large amounts of litter can be generated in queues including glass bottles and cans. Trash and recycling receptacles should be deployed along the barrier line to keep the sidewalks clean, and to minimize cleanup after the doors open. You may need to provide portable toilet facilities outside the arena.

31.2.2.9 Management of the audience arriving and leaving the arena should be discussed with the local public safety officials. Extra security may be needed at venue exits during egress to direct guests leaving the venue as to where to find taxis, shuttle buses, or public transportation. The majority of the audience members will leave the event at the end. It is important that there has been proper consultation with the public transport providers to ensure that sufficient public transport is available (see Chapter 12, *Transportation Management*).

31.2.2.10 An agreement should be reached with the local law enforcement agency to control unruly behavior outside of the arena.

31.3 Venue design

31.3.1 When planning arena events, the event must conform to the physical limitations of the building, including the size of the premises, existing bathroom facilities and fixed entry and exit points. The occupant capacity will primarily depend upon the means of escape in case of fire, the limiting factor being the width and suitability of the exit doors for the different standing/seating configurations.

31.3.1 Arena operators need to agree with the fire department and the local jurisdiction the different standing/seating configurations that can be used within the arena. Event organizers can then be supplied with a copy of the various acceptable arrangements. Prior approval of specific arena layouts can be useful and may be required.

31.3.1 The positioning of all structures, no matter how small, should be discussed with the arena operator (e.g., food concessions and display stands) as these can have an effect on the safe evacuation of people in the event of fire or other emergency.

31.4 Structures

31.4.1 Arena operators may have their own facilities and staff to erect the necessary structures for a music event (e.g., stages and seating). However, the organizer may also need to install structures. Agreement should be clearly documented as to what structures and other equipment will be brought into the arena and who will be responsible for its correct positioning, safe erection and use. Incorrect positioning of stages can have a serious effect on viewing areas.

31.4.2 Health and safety management systems between an external workforce brought onto site and the existing internal workforce should be defined and documented. You will need to ensure the competence of external contractors brought onto site. The existing health and safety procedures must be brought to the attention of the contractors.

31.4.3 The organizer and arena operator should review the suitability of the arena for the event including but not limited to floor loading, roof capacity, electrical suitability, equipment receiving, and loading docks.

31.4.4 Engineering documentation should be provided by the arena operator validating the structural suitability. (For further information, see Chapter 18, *Rigging*.)

32. Conventions and Trade Shows

32.0.1 This chapter is aimed toward the producer of events at which products or services are showcased and marketed to potential customers. The presentation of one of these types of events combines the capabilities of a number of separate entities. Among them are the producer, venue or facility, exhibition's general contractor, exhibitors, exhibitor appointed contractors, audio-visual services, labor providers, contracted security services, official florists, transportation service providers, and many others.

32.0.2 The ultimate goal is to bring an audience together with an assembly of vendors who will want to maximize their selling opportunities. In addition to the complex operations that will need to be coordinated, the event producer will need to ensure the consumers and the exhibitors will have a safe and efficient environment in which to conduct their business.

32.0.3 Most trade shows and expositions boil down to a basic 10 feet x 10 feet (3.05 m x 3.05 m) or 8feet x 10 feet (2.4 m x 3.05 m) space which is sold to an exhibitor as a location at which they will interact and engage with consumers. The show producer will then do his or her best to attract a pre-qualified audience that wants to buy. The exposition can be as few as 10 or 12 of these basic units, or as much as 1,000,000 square feet (92,903 square meters) or more of exhibit space. The event's attendance can range from hundreds to tens of thousands of people per day. While the scale of the show will necessitate the addition or elimination of elements in the event safety plan, the factors below are common to any exposition.

32.1 Planning and Management

32.1.1 Once a group of vendors and a target audience have been identified, the event producer must establish how big the event space should be. A significant factor is whether the event is open to the public or only to a restricted group, e.g., a sponsoring organization. Attendee information from the sponsoring organization can often help in determining the size of the event.

32.1.2 To attract exhibitors, promotional literature is often produced. One of these pieces of literature is known as the Exhibitor's Manual. In this manual a potential exhibitor is informed of how the event will be managed. The manual will include the dates and times for the load-in, the actual exhibit, the load-out, and, importantly, it will include the event safety policy for the exhibitor. The rental agreement between the exhibitor and the producer—a separate document from the exhibitor's manual—should then require the exhibitor to comply with the safety policy described in the manual and posted on site.

32.1.3 The event producer will need to select the various contractors who will be servicing the event. Of the choices, the choice of a "general services contractor" may be the most important. During the vendor and contractor selection process, an important concern is the safety record of the contractor and the liability insurance they have in place. Consult your insurer to help determine the contractor coverage requirements and to verify that the correct coverage for the event is in place.

32.1.4 As with any public event, ensure that any contractors or subcontractors hired for the event are competent in managing their own health and safety on site. Require written safety policies and practices from these contractors and you will find many already have such documents in place.

32.1.5 A producer may require a liability insurance rider from contractors that covers the work performed by that contractor and their subcontractors.

32.1.6 Once the contractors are chosen, it is a good idea to hold a coordination meeting to establish communications among all parties. The key person for each contractor should be identified and contact lists distributed. For this contact list it is important to identify and differentiate the contractor's actual on-site operational personnel—the person(s) who should be contacted for on-site support—from the sales staff who have sold the services to the producer.

32.1.7 As with any event, all people working within the event should have a clear understanding of the event plan, what they are required to do as part of that plan, when to do it, and how to interface and communicate with other staff.

32.1.8 Most fixed facilities will have relationships in place with local public services. In some four-wall venue rental situations, you will need to make direct contact with the local authorities to provide them your event plan along with key contact information that can facilitate communications in the event of an incident. Regardless of who establishes and "owns" the relationship, it is important to obtain specific contact information for law enforcement, fire protection, emergency medical services and other potential emergency response organizations. This information can then be provided to key members of the event staff and posted in the show or event office.

32.2 Floor Plans

32.2.1 Once the size of the event is determined, the next major management task will be to contract a facility and develop a scaled floor plan for the exposition. Unlike many events, there is usually no central performance or stage area. The layout is broken into the unit spaces offered, e.g., 10 feet x 10 feet (3.05m x 3.05m).

32.2.2 The Authority Having Jurisdiction (AHJ) will usually require a review of the event site plans. In this review, the overall occupancy load and distribution and location of aisles and designated exits must meet local requirements and will be of primary concern. Clearly label your proposed occupant capacities and crowd densities on the plans for this review. For more information on capacities and crowd densities, see Chapter 8, *Venue and Site Design*.

32.2.3 The AHJ may approve or modify your proposed plan or set the overall maximum capacity inside the event space. The maximum capacity is computed by the building official working with the fire official and takes into account the type(s) of seating and configuration, travel distance to and capacities of aisles and exits. These local officials may also set a maximum number of people who can occupy any closed or confined space within an exhibit. Once determined and approved, these maximum capacities must be communicated to all entities and staff involved in the restricted area, and all reasonable means should be used to help enforce these limits during the event.

32.2.4 Often the exhibit services contractor will assist in developing the event plans. Consideration needs to be given not only to the entrance, movement and exit of staff and attendees, but also to the access to freight docks and the movement of material handling equipment necessary to move freight into and out of the exhibit space. The safe movement of what can be millions of pounds of freight needs to be a part of the plan.

32.2.5 Many exhibitors develop exhibits that are two stories in height. Most AHJs require the approval of plans for two-story exhibits and many of those will require a plan that bears the stamp of a professional engineer. The producer should acquire these engineered drawings in advance and to present them at the pre-event planning meeting with the AHJ.

32.2.6 For exhibits of two or more stories, the covered floors of the exhibit will likely require fire suppression equipment since overhead fire sprinklers will no longer reach those areas. The pre-event AHJ planning meeting is the time to review all such situations including the requirements necessary to allow time for the staff to remedy those situations.

32.2.7 Most AHJs will require a pre-opening inspection of the exhibit floor. All installations should be completed by this time and the floor should be "show ready." This inspection is to ensure that the approved floor plans have been followed and that exhibitors have not exceeded the space restrictions that may compromise the approved capacity, emergency exits or access to fire suppression equipment. During the inspection, the inspector may require last minute modifications, so producers should have their event staff ready and available to make the necessary changes.

32.3 Freight Operations

32.3.1 The movement of freight into and out of the exhibit space is often handled by the exhibit general contractor. However, in some cases the freight contract will have been given to a separate contractor. In addition, there are often smaller exhibitors who will want to deliver their own exhibit to their space. The event producer will need to ensure the exhibit general contractor or any other contractor handling freight for the event has the proper insurance in place at the time of the event.

32.3.1.1 When the expected number of trailer/truck traffic exceeds the number of available dock spaces at the venue, it is best to establish a truck marshaling area. This will prevent the overcrowding of the dock area with waiting trucks, which will in turn increase truck turnaround time at the docks. This marshaling area may be either adjacent to the venue or far off-site (less common). Time is often a critical issue so it is advisable for the dock master to have a plan and the tools on hand, such as communications and material handling equipment, and assistants to minimize the turnaround time between trucks at the docks.

32.3.2 The general freight contractor will need to have control over the load-in area (docks, streets, etc.) and "hand carry" exhibitors who do not need the assistance of a general freight contractor must also be accommodated. For those with smaller needs, it is usually best accommodate them in an area separate from the larger freight operations, especially if this equipment is being x-ray screened prior to installation.

32.3.2.1 When freight trailers are left at the dock without a tractor attached, the forward wheels need to be fully deployed. The landing gear (the retractable crank up legs under the forward part of the truck) must be deployed and, depending on the surface beneath where the trailer is positioned, the driver may need to position pads under the landing gear so as not to penetrate and sink into that surface.

32.3.2.2 Whenever a tractor is removed from a trailer that is left at the dock, the trailer should have its wheels chocked and a warning cone should be placed in front of the trailer to warn a driver that people may be working inside.

32.3.3 Inside the exhibit space, especially with large events, it is best to provide for clear freight aisles. These aisles are designed to allow the movement of freight in the exhibit area while allowing crates and other material necessary for large exhibits to be placed in an adjacent aisle while the exhibit is assembled. A good rule of thumb is to allow at least 20 feet (6.096 m) width for these freight aisles, so as to allow two forklifts moving in opposite directions to pass one another.

32.3.3.1 The provision of clear freight aisles also allows for access to all parts of the exhibit floor if it were necessary to respond to an accident with injuries. In the event of an emergency, a clogged floor can significantly delay the response time of first responders.

32.3.4 Forklifts are generally used to move the freight from the trucks to the exhibit floor. The operators of these lifts must be certified and carry the license while operating the equipment. A copy of all forklift and other equipment operators' licenses should be on file in the show office. Conditions on the exhibit or event floor are not like a warehouse or a construction site. Operators need to be constantly aware of and familiar with the pedestrian traffic that is a part of the exhibit space and be cognizant of the fact that pedestrians can appear from around blind corners at any time. In tight spaces or limited visibility situations, a "spotter" may be necessary to assist the forklift operator to safely navigate through the environment.

32.3.5 The fuel for most forklifts and personnel lifts is propane. Tanks containing spare fuel as well as empty or nearly empty tanks, need to be stored in an area separate from any combustible materials. It is important that fuel canisters, empty or full, be placed in a controlled space which can be locked, e.g., a cage, or strapped to a fence or wall. Verify with the AHJ the minimum distance required from the venue's perimeter wall to this propane storage location; it could be 100 feet (30.48 m) or more.

32.3.6 Once exhibits are set up, there will be a considerable amount of empty crates and transport equipment that will need to be stored while the show is open. An area for empty crate storage needs to be designated, which may have several names such as "dead storage" and "bone yard."

32.3.6.1 Empty crate storage is usually outside the facility in which the exhibits are displayed, thus they are subject to the weather. Care should be taken by the general contractor to secure them against possible wind (which can physically move an empty crate), water damage from rain or snow, and other environmental influences. Securing the empties may include ropes, nets or tarps tied securely to the ground. Empties may also be stowed in the trailers at the dock. Regardless of where the storage location is, consider the order you will want cases to come back into the venue as you execute the storage maneuver. The last cases placed in storage should be the first ones needed during the dismantling period.

32.3.6.2 Stacked equipment crates and cases can be unstable. Everyone should be prevented from climbing on any stack due to the significant risk of toppling the stack and causing injury to anyone in the area.

32.4 Exhibit Floor Set Up

32.4.1 Most exhibitions use pipe-and-drape to separate the individual exhibit booths from one another. This system is composed of upright poles of 8 to 10 feet (2.44 to 3.048 m) height mounted to a portable floor or "base" plate and a cross-bar on which the curtain or drape is attached, usually by threading the cross-bar through the drape. During set-up, the bases can become a trip hazard when left alone in the middle of a pedestrian zone floor. Care should be taken to always have a pole attached to the base. This can help prevent accidents caused by the base being hit by forklifts or tripping staff and exhibitors.

32.4.2 Carpets are often provided by the general contractor and installed before the exhibit staff arrives. All carpets need to be secured to the floor to prevent unwanted movement usually by using 2 to 4 inch (0.05 to 0.10 m) wide tape on the edge and occasionally double-stick tape underneath the carpet as well.

32.4.3 Exhibit carpets are often covered by plastic roll sheeting to prevent soiling the carpet. This material is often slippery and staff should be aware of this added risk.

32.4.4 Exhibit installations and dismantling will generate a considerable amount of trash. The general or cleaning contractor needs to be contracted and prepared to collect and remove this trash. With large scale exhibits additional trash containers and collection points may be necessary. Managing the volume of waste generated by many events is often an afterthought and underestimated; be certain to consult your contractor during the event planning phase.

32.5 Exhibit Materials

32.5.1 In any public building, materials used for decorations and exhibits must meet relevant fire code requirements and should be fire resistant. For textile materials, "fire resistant" (aka, flame resistant) usually means that the material is so impervious to fire that, for a specified temperature and time when exposed to flame, there will be no structural failure and the side away from the fire will not be hotter than a certain temperature. NFPA 701, *Standard Methods of Fire Tests for*

Flame Propagation of Textiles and Films, is often the standard used to address this need, but asking the local fire official is usually the easiest way to determine the details of any necessary requirements. A document certifying the required fire resistance of any material should be kept on file in the event office and shared with the both the general contractor and any individual exhibitors. It will likely be requested to be viewed by an inspector.

32.5.1.1 On-site (field) fire resistance tests can be performed for some textile materials, such as are described in NFPA 705, *Recommended Practice for a Field Flame Test for Textiles and Films*. But, in many jurisdictions this is not permitted. Make sure that both event management team and the individual exhibitors are prepared to present the proper documentation pertaining to fire resistance, if requested.

32.5.1.2 Some materials claim to be inherently fire resistant, such as Velon, a plastic sheeting material. These materials can be verified as fire resistant by the manufacturer but insist on written documentation and maintain the documents so that an inspector can view them, if requested. Although most inspectors are familiar with these materials, documentation should still be kept on file just in case it is requested.

32.5.2 Any electrical device used in an exhibit should bear an Underwriters Laboratory (UL) seal. Many products have a CE seal or both a CE and a UL seal. The CE is often taken to be an abbreviation of the French words "Conformité Européenne," which mean "European Conformity." A CE seal indicates compliance with European Union (EU) legislation of a product and is not a substitute for a "UL" seal. Electrical devices that can be plugged into an outlet should also be connected to a properly wired, grounded receptacle.

32.5.3 Some exhibitors will need to make use of a heat source to show their product. For example, in a cooking show, the food would need to be heated. Local authorities typically prefer electricity to be used for the heat source, but this should be verified in the pre-event planning meeting.

32.5.3.1 The use of propane as a heat source may be prohibited because it is heavier that air, can pool on the floor until it finds an ignition source, and cause a fire and/or explosion. Some local authorities will permit only gasses that are lighter than air, such as natural gas (NG, methane) or synthetic natural gas (SNG), and then in limited quantities.

32.5.3.2 The use of deep frying equipment can splash hot oil and cause injury. When a deep fryer is used, a splash shield between the fryer and the public must be in place.

32.5.4 Equipment which may throw sparks, splinters or flakes of metal, should be placed away from the area where attendees may be endangered by them.

32.5.5 The storage of approved combustible materials in the exhibit should be limited to no more than a one-day supply. Loose paper, empty cardboard boxes and other material must be disposed of as the need occurs. The waste disposal contractor should make regular collections to remove these materials.

32.5.6 Some exhibit materials, such as some oils and lubricants, can be flammable. The amount of such materials should be limited to only that which is necessary to exhibit the product and it must be done in compliance with local fire code requirements. Empty, sealed containers can be used to simulate a full display. Consult the type and quantity of flammable materials and verify with your AHJ any limitations and restrictions during the pre-event planning meeting.

32.5.7 Gas powered vehicles on display are usually permissible under certain conditions. These conditions may include the following: vehicles must have their fuel reduced to less than ¼ tank full, the battery must be disconnected and have a locked gas cap or have the gas spout taped to prevent accidental ignition sources from falling into the gas tank. Local AHJs may have alternative restrictions that must be followed.

32.5.8 Some exhibits involve the use of chemicals, most of which are classified by OSHA as hazardous and toxic substances (i.e., "…chemicals present in the workplace which are capable of causing harm" [29 CFR 1910, Subpart Z]). While the use of all chemicals in an exhibit should be performed in a safe manner, the movement and handling of chemicals from arrival to removal must also comply with all applicable laws, rules, codes and regulations. Verify these procedures in advance and influence all contractors to comply.

32.5.9 Packaging materials sent with exhibit materials are often not flame resistant. Materials such as these must be removed from the exhibit floor as soon as they are discarded because they can be a fire hazard.

32.6 Floor Coverings

32.6.1 Many shows provide carpeting for the aisles. This is often laid down the evening before the event opens. Carpeting is usually applied with double-stick tape. During the course of the event, the carpet can change position. Lumps or wrinkles can develop which may become a tripping hazard. It is important to require the general contractor maintain the carpet to prevent tripping.

32.6.2 Sometimes the surface on which an exhibit is located is already carpeted with permanent house carpet. When this is the case, an exhibitor may have a carpet installed over the house carpet. It is important to have this second layer also maintained to avoid trip hazards.

32.7 Electrical Services

32.7.1 In many cases the venue's existing electrical system was not designed with a trade show floor plan in mind. In these cases the general contractor, or an appointed contractor, will take the responsibility for electrical distribution. This contractor should have the necessary licenses for electrical service. The event producer should define the scope of this contractor specifically outlining the type of services provided and the responsibility they take for installation, operation and removal. For more information see the chapter on Electrical Installations and Lighting.

32.7.2 The most common form of electrical distribution is a "stringer" of outlets along a multi-cable which is connected to a power source. Exhibitors can plug into the outlets along the stringer for their individual needs. This electrical equipment must be maintained in good condition, as it is often placed along the back wall (either a hard wall or pipe and drape wall) of a line of exhibits and may not be easily accessed if there is a problem.

32.7.3 Some exhibits require electrical service which must be placed underneath the carpet or temporary flooring. This could provide the possibility of an electrical short, overheating situation or other malfunction, especially if improperly installed. The electrical contractor should do all such installations and be held responsible for any maintenance and repairs.

32.7.4 When it is necessary to make use of a transformer to provide larger amounts of power to an exhibitor, the high voltage supply side to the transformer will need to be routed in an area where the wiring is not exposed to either public or vehicular traffic.

32.7.5 When electrical or control cables, hoses etc. need to cross an aisle, a cable ramp or cover should be employed to mediate the trip hazard. The cable ramp or cover should also be a contrasting color to the surrounding floor surface to increase visibility to pedestrians and equipment operators. When covered by carpet, a contrasting colored tape can be applied to the carpet to show where the ramp is located.

32.8 Overhead Signage

32.8.1 Most large exhibitions will make use of overhead signage. At a minimum, it is necessary to provide directions to the attendees. Often exhibitors, advertisers or sponsors will also want to utilize the added exposure and visibility a sign provides.

32.8.2 Materials for the signs must be fire resistant (see 32.5.1, above) and structurally sound. Any equipment used for attachment to the overhead structure of the facility must have sufficient load carrying capacity for the weight of the sign as determined by a qualified engineer. Usually the attachments are made by the general contractor. The staff performing the work must be qualified and competent to do so.

32.8.2.1 Many facilities will have restrictions concerning the use of some types of equipment for overhead attachment. One of the most common is that the use of "bailing wire" or "stovepipe wire" is not permitted because this type of wire breaks too easily. Obtaining a list of permissible hanging devices is essential. The general contractor must be required to comply with the policies regarding the attachment of any elements to the venue's overhead structure.

32.8.3 To access the overhead structure it is usually necessary to use a high-lift or scissor lift. The driver of that type of equipment must be certified in its use and should carry the license at all times while operating the equipment. Operators should always have their training completion certificates in their possession. Any lift traveling the exhibit floor must have a ground person accompanying the lift to ensure safe operation during use. In addition, operators and workers in lifts should employ appropriate PPE and fall protection devices for the equipment they are operating.

32.8.4 Increasingly, exhibits are making use of rapid deployment materials and equipment such as chain hoist motors and trussing, which are commonly seen at concerts and other entertainment presentations. Chain hoist motors and rigging are used to lift and lower equipment overhead. Safe rigging practices must be followed in the installation of this equipment. See Chapter 18, *Rigging*, for more information.

32.9 Crowd Management

32.9.1 The crowd management requirements of public and closed or association exhibition events, while similar, do differ in a few respects. In a gated and ticketed public event, attendees buy a ticket in order to attend. This means that any person who gets into the exhibition without a ticket represents a loss of income. In a closed or association event, the total attendance to the exhibition translates into how much exposure and access an exhibitor will have to potential buyers and their return on investment.

32.9.2 The planning considerations outlined in Chapter 9, *Crowd Management*, should be applied with these differences in mind. Generally this will mean the public show will have a limited number of access points (turnstiles, ticket takers, etc.) while the association show will have many points of entry with badge checking as the key control of access.

32.9.3 In an association exhibition, attendees will obtain an ID badge, which will allow access to the exhibits and other portions of the event. The use of turnstiles and other one-at-a-time entry devices are not generally needed except in potential over-capacity situations; simple badge checking will permit access.

32.9.4 Exhibit aisles are often planned with the same 10 feet (3.048 m) width used for the individual exhibit booths. Care should be taken to ensure that there are clear pathways to the emergency exits. The visibility of signage at the exits also needs to be considered.

32.9.5 Extra aisle space may need to be allowed at individual exhibits that draw large groups of observers and some booths will contain an audience viewing area. Overflow of observers into adjacent aisles from these attractions can cause a blockage of those aisles and a hazard. Crowd management personnel need to be sensitive to these potential build ups and prepared to manage them before they become a safety hazard.

32.9.6 When the space is occupied, emergency exits must never be locked to exiting persons. Exit doors are generally opened from the inside and if not managed can allow unauthorized persons entry into the event. Plans should be made for crowd management personnel to keep these exits secure from improper entry.

32.9.6.1 Recent developments in electronic controls allow emergency doors to be held open or closed by electromagnetic means so individual doors can be locked or unlocked as needed. In the event of an evacuation of the facility, the magnets de-energize and the doors may be opened. In addition, electronic releases can be placed on doors to hold them open until an emergency requires them to be closed. This eliminates the need to block the doors open in possible violation of fire regulations.

32.10 Transportation

32.10.1 Provisions may need to be made for the transportation and/or parking needs of attendees and staff to an exhibition. See Chapter 12, *Transportation Management*, for additional information.

32.10.2 For a public event, sufficient parking should be available to accommodate the maximum occupancy. This may require a combination of parking structures and street parking.

32.10.2.1 Additional signage may need to be placed outside the venue to direct both vehicular and pedestrian traffic to the venue. This signage may need to be approved by the local authorities before placement in public areas.

32.10.3 In a closed or association event, the attendees are often housed in nearby hotels. In this situation, the need for vehicle parking will be reduced, but the need for alternate forms of transportation to the hotels will likely increase. Shuttle busses with regular routes and schedules to outlying areas may need to be planned. A bus passenger pick-up and drop-off zone will need to be created and should be established in an area away from other crossing pedestrian traffic. It should also be provided with sufficient staff to supervise the bus activity and assist with customer service while the area is active.

32.10.4 The quantity of taxi traffic the event may generate must also be considered. If necessary, taxi pick-up and drop-off zones may need to be established. The two zones should be separate to allow taxis dropping off to quickly do so without having to share the same curb space with standing taxis that are waiting to pick up passengers. In some facilities, a designated taxi zone exists and should be integrated into the overall transportation management plan.

32.11 Exhibit Removal

32.11.1 At the close of the event, attendees should be informed to vacate the exhibit area and associated passageways. An announcement via a public address system can be a simple way to accomplish this. Care should be taken to limit load-out activity while the general public is remain in the exhibit area. Access to the exhibit area may need monitored to prevent re-entry.

32.11.2 Exhibitors will often begin packing their exhibits before the official end of the exhibit, but placing any exhibit materials in the aisles—blocking pedestrian traffic egress—must be discouraged, monitored and/or prevented.

32.11.3 When aisle carpeting has been installed, the first thing to be removed is that carpet. The double-stick tape that was used to hold it to the floor must also be removed to prevent potential tripping hazards.

32.11.4 Even though the attendees have left the exhibit floor, there will be a considerable number of personnel still in the exhibit area. During the pre-load out safety meeting, forklift operators should be reminded to watch for pedestrians who may not be aware of them.

32.11.5 Returning of empty crates to the respective exhibits will begin as soon as the carpet is removed from the aisles. Clear freight aisles need to again be established to prevent obstruction of the routes for delivery of these crates. Outlining the routes in contrasting colored tape can help emphasize their existence to workers.

32.11.6 The tables provided to the exhibitors are often covered with a plastic sheet. It is common to remove this covering from the tables before storage. These coverings can become slippery and create a slip/fall hazard. Care should be taken to have these coverings collected and removed from the workspaces as needed to prevent risk of injury to workers and also expedite the movement of equipment in the environment.

32.11.7 During the removal of exhibits there may be many liquid spills that create slip hazards. Immediate containment and mediation of these spills needs to be included in the cleaning contractor's scope of work.

32.11.8 At the close of the exhibits, there may also be a wide variety of liquids in need of disposal. All potential hazardous and toxic substances need to be properly collected and removed by the cleaning contractor or other service contractor specializing in this type of waste. The event producer should communicate procedures for the proper disposal of waste material to the exhibitors. The cleaning and/or waste disposal contractors need to be aware of this and make appropriate provisions. Consider including language regarding the proper disposal of hazardous materials in all contracts.

32.11.9 As with the load-in, the bases for "pipe and drape" need to be monitored. If left without a pipe in place, a base becomes an immediate trip hazard. In addition, forklift operators may not see them and accidentally tip their loads.

32.11.10 The preparation of outgoing freight for transport will not necessarily occur in an orderly fashion. Because of this it will be necessary to again have a marshaling yard or area where trucks can wait until their loads are ready. Control of access to the dock area should be through the marshaling yard and monitored by the dock master. The event producer should make certain the general freight contractor is prepared to control this access.

32.12 Final Clean-up

32.12.1 Exhibitions generate a considerable amount of trash. The dismantling of an exhibition generates far more waste than any other show related activity. In addition to the usual trash, scrap lumber, nails and screws, much unwanted exhibit material, and the like will be left behind by exhibitors. The cleaning contractor will need to be prepared to deal with removing significant waste during this phase.

32.12.2 Once the venue is clear of all exhibits, debris and equipment, the final walkthrough and inspection can occur. Consider taking "before and after" photographs of the entire venue you occupied for archival reference once you have completed the event. These may be useful should issues arise later after other tenants have moved into the space.

33. Small Events

33.0.1 This chapter contains advice for the small-event organizer. The important factor to consider is not whether an event can be defined as "small" or "large" but the proportionate level and extent of facilities and safety management systems required to ensure the health, safety and welfare of patrons, performers, and event staff.

33.1 Planning and Management

33.1.1 It is suggested that small-event organizers use the chapter headings in this publication as a framework or checklist for event planning. All event organizers must be clearly aware of their responsibilities for the audience and other participants at their event, including performers, merchandisers, etc.

33.1.2 Although "small" events may have relatively few patrons, the magnitude of any individual health or safety issue is the same as with larger events, and requires the same attention to planning and event management. As always, the activity itself and the audience type will significantly influence the safety needs for any given event. It is just as important for a small-event organizer to carry out a risk assessment for the event, to identify which hazards are of greatest significance and therefore which parts of this publication are most relevant.

33.1.3 A risk assessment and safety policy need not be long or complicated, but should clearly demonstrate the approach taken to ensure the safety of all those involved in the event. Assistance in drawing up a risk assessment and safety policy can be found in Chapter 2, *Planning and Management.*

33.1.4 The safety management responsibilities could be handled by the existing management team for the event if they can effectively put the actions outlined in the safety policy into practice, otherwise a separate team should be established. All workers and/or volunteers need to be aware of safety procedures. Ensure that any contractors or subcontractors hired to build the stages, erect tents, booths or stalls, are competent in managing their own health and safety on site. If feasible, ask for copies of the contractors' safety policies, risk assessments for their work and safety method statements.

33.2 Staffing

33.2.1 Small events may operate with small budgets and rely on enthusiastic volunteers rather than paid employees or contracted service companies. The crucial aspect is good coordination by the event management team and close supervision, support and monitoring of volunteers. The organizing group can sometimes provide many services at small events such as catering and security, rather than buying them in from commercial companies.

33.2.2 Management of staff, including paid staff, vendor workers and volunteers requires clear job functions and responsibilities to be identified. It is important for inexperienced staff to receive proper briefing and supervision.

33.2.3 Everyone working or providing services at the event should be clear about what they are required to do, how to do it and when it needs to be done. This can be achieved by preparing a schedule when work is required to be carried out and by whom, and informing everyone involved.

33.3 Levels of Provision of Site Services and Facilities

33.3.1 While some of the recommended levels of provision in this publication may be reduced for small events, there are areas where a minimum provision will be required. For example, the number of toilets obviously cannot be below two. Realistically, the number of medical staff, security, etc., should never be less than two, to allow for contingencies.

33.4 Local Authority Liaison

33.4.1 Small-event organizers should consult with the relevant local authorities and emergency services representatives with responsibility for the event. These officers will be prepared to offer advice and assistance including whether any permits are required.

33.4.2 Provide the local authorities with sufficient written information to enable officers to understand the nature of the event. This documentation will in any case already have been prepared as part of your event planning and should include:
- A description of the event, including key event timing, e.g., load-in, show and load-out times, audience size, type of activities, etc.;
- A site plan showing relevant features and the event's relationship with any adjacent neighborhoods;
- A list of key members of the organizing team and their responsibilities;
- The risk management strategy, including a copy of the risk assessment, safety policy and site-safety rules.

33.4.3 Further documentation should be available on site during the event, including:
- The safety policies, risk assessments and safety method statements for any contractors or subcontractors hired to erect stages, tents, roofed structures, booths, stalls, etc.;
- Risk assessments and safety documentation of any activities associated with the event such as inflatable structures, trampolines, etc.;
- Appropriate certificates for any work equipment brought onto site, such as electrical equipment, generators, lifting equipment;
- Copies of flame certificates for treated materials.

34. Classical Music Events

34.0.1 A classical music event for the purposes of this chapter is defined as an outdoor performance on a greenfield site—typically parkland or an open air venue—with an audience who bring with them their own chairs, food and drink and sit where they want within a designated area.

34.0.2 As with any event, the initial planning meetings with the local authorities and emergency services are critical. It is common for a load-out immediately on concluding an evening performance.

34.1 Crowd Management

34.1.1 Patrons attending a classical music concert tend to be older and less active than at a typical rock concert. This is not always true, however, so the event organizer is advised to do the same type of crowd behavior research as for any other type of concert.

34.1.2 Local voluntary organizations may serve as event staff and will require training. An experienced event staff leader should be appointed who has been trained to deal with areas of potential conflict (e.g., extinguishing grills, moving audience to avoid over-crowding). A ratio of one non-law enforcement, event staff to 250 audience members has been found to be effective.

34.2 Transportation Management

34.2.1 It is unusual for a classical concert on a greenfield site to be located close to a major transportation route. Much of the audience will travel by car and vehicular access through minor roads and gated entrances, all of which may be limiting factors on audience capacity. Contingency planning is needed for bad weather including availability of tow vehicles and possible rerouting of traffic.

34.2.2 It is common for voluntary stewards to direct traffic on site and organize parking. A more experienced group may be required where the audience exceeds 4,000-5,000. Voluntary stewards must not direct traffic on or from a public road unless the police have specifically requested it.

34.3 Performers

34.3.1 A classical orchestra may comprise 75 or more musicians. The addition of choirs can greatly increase this number. It is important that they are provided with dedicated parking and welfare facilities.

34.4 Venue and Site Design

34.4.1 When using a field site as a music venue, problems arise which do not occur in a purpose-built arena. An open field may quickly show signs of strain in handling a large number of people in one evening. Narrow gateways and steps may become a hazard when used by thousands of people. Suitable access routes for site infrastructure (e.g., staging, portable toilets) need to be identified.

34.4.2 Evacuation of the site in an emergency, and dispersal of the audience, is not normally a problem in open parkland but a venue in a formal garden, with access through gateways, requires careful consideration.

34.4.3 Livestock and pets may be frightened by loud music or fireworks that may accompany such events. Arrangements should be made to move potentially affected animals before the event.

34.5 Sanitation facilities

34.5.1 The water supply to the site may be a limiting factor on audience size unless re-circulating or non-flush units are employed. A high expectation of the facilities should be anticipated and therefore all units should be serviced throughout the concert.

34.5.2 The number of toilets should be based on the recommended standard in Chapter 14, *Sanitary Facilities*. Toilets are used more efficiently if they are sited in the same location and easily accessible from the audience area. An exception to this would be wheelchair accessible facilities, which on larger sites could be located on either side of the audience to reduce travel distance.

34.6 Food

34.6.1 At a typical classical event, the audience will picnic. Personal barbecues are not normally permitted, so the catering facilities will need to serve food and drinks, which the audience cannot provide for themselves.

34.7 Waste

34.7.1 Greenfield sites are often the home of animals such as deer and cattle that may suffer considerable harm if the waste is not cleared efficiently. On sites with grazing animals it is important that as much waste as possible is collected on the night of the concert with a sweep the following morning to pick up loose material such as firework debris, nails, bolts and plastic fittings, etc.

34.7.2 Additionally, many venues may be open to the public on the day following the concert and it is important that the site is left in the same condition that it was found. Audiences will normally respect the venue and if issued with a trash bag (white for visibility after dark) will either take trash home or deposit it for collection.

35. Amusements, Attractions and Promotional Displays

35.0.1 Additional information on this topic can be found on the web site of the International Association of Amusement Parks and Attractions (IAAPA)(http://www.iaapa.org/).

35.1 Safety Standards, Regulations, and Policies

35.1.1 Standards on Amusement Rides and Devices are available through ASTM International and its F-24 Committee (http://www.astm.org/COMMITTEE/F24.htm). This ASTM Committee is composed of consumer advocates, government officials, amusement park operators, ride manufacturers and industry suppliers. The Committee helps to establish standards on design and manufacture, testing, operation, maintenance, inspection, quality assurance and more. These standards undergo periodic review and revision to keep up with new technologies, and have been adopted by many governmental jurisdictions.

35.1.2 Amusement parks are subject to varying combinations of state and local governmental codes, requirements, and safety reviews by government inspectors and/or insurance risk managers.

35.1.3 Safety inspections may be conducted by amusement park staff on a daily, weekly, monthly and yearly basis. They should follow detailed manufacturer guidelines for inspection and safety, and many parks use outside specialty companies to periodically re-inspect rides. ASTM International standards require fixed-site amusement industry operators and manufacturers report both incidents and ride-related defects, including notification of facilities when a ride develops a manufacturer-related safety issue.

35.1.4 The United States government does not regulate amusement parks. As of this writing, 44 states regulate amusement parks. The six who currently do not are Alabama, Mississippi, Nevada, South Dakota, Wyoming and Utah.

35.1.5 Generally, amusement parks follow the manufacturer's guidelines for inspection and safety and many parks contract third party companies to independently inspect their facilities. Amusement parks report incidents to state and local governments.

35.1.6 The International Association of Amusement Parks and Attractions (IAAPA; www.iaapa.org/) created a list of amusement ride safety tips for guest use.
- Obey listed age, height, weight, and health restrictions.
- Observe all posted ride safety rules.
- Keep hands, arms, legs and feet inside the ride always.
- Remain seated in the ride until it comes to a complete stop and you are instructed to exit.

- Follow all verbal instructions given by ride operators or provided by recorded announcements.
- Always use safety equipment provided and never attempt to wriggle free of or loosen restraints or other safety devices.
- Parents with young children should make sure that their children can understand safe and appropriate ride behavior.
- Never force anyone, especially children, to ride attractions they don't want to ride.
- If you see any unsafe behavior or condition on a ride, report it to a supervisor or manager immediately.

35.1.7 Guidance is already available about attractions, rides, amusement devices and stalls found at fairgrounds and amusement parks (see the ASTM International guidelines or the IAAPA website for more information on guidelines for Amusements). This chapter does not replace the need for an event organizer to obtain this guidance. The intention of this chapter is to highlight some areas for consideration when amusements, attractions and displays are incorporated in a music event rather than at a fairground or amusement park.

35.2 Amusements and Attractions

35.2.1 If you wish to include amusement activities at your event, it is important to obtain the required safety information about the activity from the operator. This is to ensure that the siting and operation of the amusement does not:
- Compromise safety with the overall risk assessment for the event;
- Block the emergency access routes; or
- Cause audience congestion problems.

35.2.2 Points to consider when incorporating any amusement as part of the overall entertainment include the following:
- Obtain advice about the particular hazards associated with the amusement or attraction from the operator and ask them for copies of their own risk assessment and safety information. Incorporate the information into your overall risk assessment for the event.
- Obtain advice from the relevant enforcement authority (state and/or local authority) about the particular amusement. Local and state authorities should have up-to-date information concerning hazards that have been reported about a particular amusement activity.
- Investigate the operator's competence by posing the following questions: Is the operator able to demonstrate compliance with legislation or codes of practice? Do they have current insurance? Does each amusement have a current certificate of examination from an inspection body? What experience have they had in operating the amusement? What safety information can they supply about the amusement?
- Information concerned with the safe operation of the amusement should also be given to other contractors working at the event who may be affected.
- Determine appropriate setting-up times, operating times and dismantling times. Amusements should be set up before the audience enters or approaches the event. Make sure that the amusement is not dismantled until all members of the audience have left or are at a safe distance. Vehicle movements are often prohibited during events and amusement operators need to be informed about this policy.

- Ensure that suitable space has been allocated for the amusement. Space is one of the most important considerations for any amusement. This does not just include space on the ground but often space above. Obstacles such as large trees, overhead cables and power lines can cause major hazards. The sides and rear of the amusement may need barriers.
- When planning the positioning of the amusement, consider emergency access routes as well as space for audience members who may queue to ride the amusement. Space may be needed for family, friends and others to comfortably watch the amusement.
- Ensure that the operation is coordinated with the music event. Crowd management problems can arise if operators are still offering rides on the amusement after the music event has ended and if members of the audience try to have one last ride before leaving. On the other hand, it may be appropriate to continue the operation of the amusement to stagger people leaving the event. Whatever the decision, careful coordination of the activities must be planned and communicated to the operator and event personnel.
- The availability of natural light may also be an important safety factor in the operation of some amusements, particularly where color-dependent safety features are used.

35.3 Bungee Jumping

35.3.1 Permits

35.3.1.1 States typically require bungee jumping operators to obtain a permit or license before they can operate in the state. The application process for these permits differs from state to state. For example, Ohio Administrative Code Section 901:9-1-23 sets out the state's permit application requirements for bungee jumping operations. The permit must include information such as a site operation manual; a site plan with detailed information about safety zones, fences, jump zones and equipment locations; and proof of insurance coverage. The state can also require an applicant to submit a "registered engineer's report confirming that the design and construction of the equipment to be used meets engineering standards acceptable to the department and confirming that all applicable local codes have been complied with."

35.3.2 Prohibited Activities

35.3.2.1 States also place specific limitations on how and where bungee jumping operations can operate. These laws commonly place restrictions on what kind of jumping is allowed, where operators can set up a jump site and what kind of equipment is or isn't allowed.
For example, the state of South Carolina Code of Laws 52-19-30 sets out the limitations on bungee jumping in the state. These prohibitions include: bungee jumping can only be done from a fixed platform; jumpers cannot use an ankle harness; jumping cannot take place over water, sand or any other surface other than an air bag; and tandem or multiple jumping is not allowed.

35.3.3 Insurance and Notification

35.3.3.1 States also require that bungee jumping operators are properly insured. The amount and kind of insurance required differs, but no bungee jumping operation is allowed to be open unless proof of insurance is first obtained. Operators also have a duty to report injuries or risk losing their permits or licenses. For example, the state of West Virginia imposes both insurance and notification requirements on all bungee jumping operators in the state. West Virginia Code 21-12-11 states that all operators must have an approved insurance policy at all times during operation. The policy must be no less than $300,000 per person and $1,000,000 in total for each site or platform the operator has. Further, West Virginia Code 21-12-8 states that an operator must notify state officials within 24 hours of an accident or fatality at any bungee jumping site.

35.3.3.2 Ensure that bungee jumping operators belong to a reputable association and meet the state's requirements.

35.3.3.3 Pre-booking is recommended for those wishing to take part in bungee jumping and this should be discussed with the bungee jumping operator. In particular, advertising before the event may need to be considered.

35.4 Inflatable Bouncing Devices

35.4.1 There are many types of inflatables that can be used for bouncing upon, including: open-sided or flat beds, open-fronted (e.g., castles), and totally enclosed.

35.4.2 Hazards include being blown over or away by the wind, splitting of the fabric, accidental spilling of users, injury to the users by themselves or other users, overcrowding, air loss due to blower disconnection, power supply failure and inadequate means of escape if there is fire. Each inflatable should be thoroughly examined annually for any deterioration by a competent person or company. Height and age restrictions are often necessary for the safe operation of these bouncing devices and such information should be made visible to the audience wishing to take part.

35.4.3 Choose an appropriate site for the device that is large enough for easy entry/exit and allows for cushioning of any hard ground under and around it.

35.4.4 Comprehensive plans should be developed that address any actions required during weather situation that can affect the operation of the device such as high winds and rain.

35.4.5 Stakes used for tie-down of the device should be long enough and pounded deep enough into the ground to sufficiently anchor the device and prevent it from moving under all reasonably expected conditions. If the device is for outdoor use, do no rent it without including a means of anchorage.

35.4.6 The recommended maximum number of children that can be safely accommodated at any one time will be influenced by the age and size of the children who are to use the inflatable.

35.4.7 At least one responsible adult should constantly supervise and manage children while it is being used. More than one adult should be used when everyone in the device cannot be seen by one adult supervising the device. Those supervising the activity in and around the device should be familiar with the limitations, safety concerns, and proper use of the device.

35.4.8 A rotation system should be designed to avoid the mixing of different ages or sizes of children.

35.4.9 Appropriate behavior of children should be enforced (e.g., Children should not climb on walls, attempt acrobatics [e.g., somersaults], or take food or drink onto the device.)

35.4.10 Make sure to protect children and others from electrical equipment, wiring, and tie-downs that may be associated with the device. Make sure to route foot traffic away from the tie-downs and anchors.

35.4.11 Extra insurance coverage may be necessary if the device is rented.

35.4.12 Adults and children should not be in bouncy devices together. Adults should only use bouncy devices made specifically for adult size and weight.

35.4.13 Inspection should include:
- Is the bouncy device securely anchored? All anchorage points should be used and, if situated on hard ground, tie-down straps should be affixed to solid points.
- Are impact-absorbing mats positioned at the open side of the bouncy castle, extending a sufficient distance forward to ensure sufficient protection? (Mats may not be necessary on soft ground.)
- Is there at least one person constantly supervising the children in and on the device? An attendant who is collecting money cannot also supervise the activity.
- Does the bouncy device seem overcrowded? If children are constantly knocking into each other, the attendant may not be following the maximum load recommendations.
- Are children of different ages/sizes mixed? If the demand is great the attendant should operate a rotation to avoid larger children crushing smaller ones.
- Are children instructed to remove sharp articles of clothing like shoes, buckles and jewelry and is the rule enforced?
- Is there evidence that the attendant is controlling the children? Horseplay should not be allowed, and children should not climb on the walls of the inflatable.

35.4.14 For inflatable devices used as promotional displays, such as rooftop inflatables, dancing inflatables, helium balloons, hot air balloons, parade balloons and more, check the standards offered by the U.S.-based Inflatable Advertising Dealer Association (http://inflatableads.org/).

35.5 Flight Simulators and Computer Games

35.5.1 Motion simulators, 4D video presentations and "Game Based Learning" are popular and expanding rapidly into many events.

35.5.2 Care should be taken choosing the environments and designated areas where these activities take place. A replica domain is often used instead of an actual domain, e.g., in paintball or laser tag a virtual forest may be constructed of padded obstacles to reduce injuries to participants. The goal of these attractions is to cause the participant to think, act, experience consequences and pursue goals as if they were real and to find the right course of action based on experimentation—making choices and experiencing the consequences.

35.5.3 Motion simulation hazards include:
- Contact with the outer simulator body during use.
- Contact with the access door when opening automatically.
- Trapping hazard between the simulator body and the surrounding floor.
- Failure of the mountings and/or structure.

35.5.4 Precautions to consider:
- Limit the simulator so it can only be operated with a key. This key should be held by the supervisor on duty and only signed out to authorized personnel;
- Restrict access to the simulator room to card access or other mechanical lockout;
- Position the simulator behind a metal barrier and a gate. Opening the gate stops the simulator's motion;
- The travel and velocity parameters are limited within the control unit;
- Participants will be given full training and safety rules for the use of the simulator;
- One participant in each group will be nominated as being responsible for the safety during each session. e.g., Regular maintenance by contractors to ensure that mountings, and fixtures are satisfactory;
- User maintenance at yearly intervals for visual checks on mountings and fixtures;
- A time lock prevents the use of the simulator outside of portered hours;
- Guards and barriers to prevent access to the moving parts of the simulator;
- Physical interlocked guards will not prevent a person from being caught behind the fencing/guarding whether deliberate or not;
- Endeavor to keep good visibility around the simulator for the operator, this will decrease the risk of accidentally containing someone in the simulator area;
- Line out a "danger area" around the safety perimeter radius in an appropriate caution color.

35.5.5 Mobile flight simulators and associated rides are often very heavy pieces of equipment that need specific ground conditions to operate safely, access requirements and ample space around the device.

35.5.6 Example of Ride Height and Weight signage:
- Minimum height is 48 inches, maximum height is 76 inches (1.93 m);
- Riders feet must be able to reach the foot plate;
- Maximum weight is 250 lbs (113.4 kg) per rider;
- Must be able to fit in seat fully restrained and keep hands through wrists straps and hold handles always;
- Pregnant women, those with high blood pressure, heart conditions, aneurysms, recent surgery or illness, anyone with back, neck or bone injuries, balance/ear problems and diabetics should not ride.

35.5.7 Basic checklist of things to consider include:
- Designated area and layout of the attraction;
- Attendants;
- A means of safely canceling, stopping and restarting the experience;

- Participant/user-required safety equipment e.g., helmets, pads;
- Maintenance and sanitation of the environment e.g., spraying or wiping down frequently handled items with a disinfectant;
- Signage posting policies regarding age, size and weight limitations, health warnings, pregnancy warnings, etc., and enforcing the policies;
- Queue line management, ingress and egress.

35.6 Mobile Amusement Rides and Attractions

35.6.1 Mobile amusement rides and attractions or individual may be incorporated at outdoor events and some larger indoor events. Standards for amusement rides are set by the ASTM International, F-24 Committee on Amusement Rides and Devices. There is also a series of plant and machinery guidance notes that detail the safety requirements.

35.7 Circuses

35.7.1 General

35.7.1.1 It is relatively unusual for a complete circus to be part of a music event. However, it is not unheard of to find circus performers demonstrating their talents (e.g., fire eating, stilt walking and juggling) in and around the venue or site itself. Make sure to brief performers on audience safety (e.g., emergency exits should always be kept clear) and advise them where and when to start their act.

35.7.1.2 The enforcement responsibility for circuses usually falls to local authorities, except for a small number of circuses operating on certain properties (e.g., state fairs), which may be the responsibility of state agencies.

35.7.1.3 Other state laws and regulations may also apply to circuses such as those required for the handling of animals, amusement rides and food service.

35.7.2 Staff

35.7.2.1 All performers should have appropriate permits and undergo training to perform in the circus.

35.7.2.2 The troupe should have a licensed nurse on staff, and as many performers as possible should be trained in first aid.

35.7.2.3 The circus should provide staff members with the protective clothing they need for their acts. This clothing and any other apparatus used during performances should be inspected before each show to make sure they have not been damaged.

35.7.3 Insurance

35.7.3.1 The circus company needs multiple types of insurance including medical for employees and liability for injuries to customers/the public.

35.7.4 Building or Tent

35.7.4.1 Each state will have rules/regulations regarding the construction and use of tents and similar structures. Please see Chapter 19, *Structures*, for more details and become familiar with all local construction and maintenance requirements for such structures.

35.7.4.2 Some of the most important issues related to tents include the maintenance of multiple, open exits for everyone inside; adequate ventilation; and, assuring that the tent materials, construction and electrical features all meet local fire and life safety code requirements.

35.7.4.3 Tents larger than a certain size usually require a permit. Make sure to check the local requirements. The permit process can help assure compliance with all other requirements.

35.7.5 Aerial Acts

35.7.5.1 Aerial acts should have safeties in place for practice and performance that are suitable for the act being performed. Such safeties may include crash pads, nets, spotters or other devices. It should be noted that for some aerial acts, the intended safety device will create a greater secondary risk than not having the device.

35.7.5.2 The European Federation of Professional Circus Schools (http://www.fedec.eu/) recommends that aerial performers have protective devices during the show (e.g., nets, foam mattresses, harnesses and/or tethers). Protective devices may not always be possible for some acts, and in these situations a plan of action should be devised ahead of time in case the performer falls.

35.7.6 Fire Acts

35.7.6.1 Special precautions must be taken if the circus uses fire in any of its acts. Many states have very specific requirements for the use of fire in an entertainment venue in front of a proximate audience. Become familiar and comply with all local live fire requirements for such situations.

35.7.6.2 Whenever there is fire involved in an act, no less than one person should be specifically made responsible for fire protection and firefighting activities. Although safety is everyone's responsibility, specific assignments must be made to provide a level of fire protection that is commensurate with the hazard.

35.7.6.3 All personnel charged with fire protection and firefighting tasks should be fully trained and equipped to perform their duties. All staff should be trained in dealing with fires and their specific role(s) should an unplanned, unfriendly fire ignite. See Chapter 4, *Fire Safety*, for more details.

35.7.6.4 All performers operating with or near the flame effect should wear fire-resistant clothing.

35.7.6.5 Appropriate methods of fire extinguishment (e.g., appropriate fire extinguisher, fire hose, water bucket) must be available when fire is used.

35.7.6.6 Fire, when used in an act, should be used at a safe distance from spectators. Many states have specific minimum distances for this.

35.7.7 Animal Acts

35.7.7.1 Animal exhibitors are required to be licensed by the U.S. Department of Agriculture. The Animal Welfare Act regulates the use of most circus animals and states that there must be "sufficient distance and/or barriers between the animals and the viewing public to assure the safety of the public and the animals. Trained handlers, leashes and stages, for example, are not substitutes for sufficient distance and/or barriers."

35.7.7.2 Many states also have requirements regarding the storage and use of "exotic" animals in an entertainment venue. Become familiar and comply with all local rule, regulations and laws pertaining to the use of live animals.

35.7.8 Safety Professionals

35.7.8.1 Safety rules vary from state to state as do rules governing the use of animals, insurance and many other aspects of a circus performance. This list is not exhaustive, and anyone interested in circus safety should consult with professionals to ensure the safest atmosphere. There are many safety consultants available for all types of circus acts, so contact a professional to guarantee a safe environment.

35.7.8.2 Circuses using incendiary devices must comply with all local and state ordinances as interpreted by the local authorities with jurisdiction over the event.

35.8 Promotional Displays

35.8.1 Companies sponsoring events may wish to advertise their product by way of a promotional display. These can range from advertising balloons and inflatables, purpose-made structures, video and virtual reality electronic games through to smaller merchandising stands.

35.8.2 It is easy to overlook the effect that some of these displays might have on the safety of the event. Display structures should be handled in the same fashion as other temporary structures used for the event (see Chapter 19, *Structures*), including structural strength and integrity, and safety procedures pursuant to the Operations Management Plan. Obtain information as to the type of equipment that will be brought on site, its method of erection and particular hazards the equipment may pose. Drawings of any special structures should also be obtained along with the methods of erection and dismantling.

35.8.3 Consider placement of promotional displays during the venue design phase to ensure that they do not obstruct emergency exit routes or hamper audience movement around the site. Inflatable balloons and displays must have appropriate space allocated to them and be suitably anchored. Banners of all types, soft goods and other materials capable of creating a sail effect must be designed for rapid lowering in the event that the wind loads exceed those in the emergency action plan.

35.8.4 Those bringing the equipment onto site must also be instructed on the site safety policy. Any advertising stands should be treated in the same way as merchandising stands. Electrical equipment must come equipped with the relevant electrical test certificates and be installed by a competent electrician (see Chapter 17, *Electrical Installations and Lighting*).

35.8.5 Some large-scale marketing events take place around major sporting events or air shows. Each activation of a marketing event should be considered a standalone, event-within-an-event. Attention should be paid to all details as if it were the main event. It is easy to underestimate the amount of effort needed to coordinate multiple event marketing activations. Be diligent in the expectations placed on staff assigned to coordinate multiple activations.

36. Acknowledgements

The *Event Safety Guide* is the product of all of the effort and expertise of people who have come before us to make events safer. Our attempt to compile and update industry best practices into one user-friendly resource would obviously have been impossible without the countless contributions of our predecessors. We gratefully acknowledge the work of all the associations referenced throughout this document.

Of course, the *Event Safety Guide* is much more than a mash-up of existing resources. It is, instead, the result of collaboration between subject matter experts in the many far-flung issues that must be addressed in order to run a safe and successful event. It would take nearly another book to list all of the contributors to the Guide, and because it is a "living document," any such acknowledgement would be outdated before it was committed to print. So, to the *Guide*'s many dedicated contributors, we offer our sincere appreciation, as well as our request that you continue to offer your ideas and experiences as issues arise.

There are a few people whose contributions to the *Event Safety Guide*, and to furthering ESA's mission of "life safety first," has been so significant that we would be remiss in not acknowledging them individually.

Event Safety Alliance Board of Directors

Jim Digby, President
Donald Cooper, Director
Harold Hansen, Treasurer
Charlie Hernandez, Director

Steven Adelman, Vice President
Steve Lemon, Director
Stuart Ross, Director
Roger Sandau, Director

The Event Safety Alliance Founding Contributors

Keith Bohn, TOMCAT USA
John Brown, Brown United Staging
James Chippendale, Doodson Insurance
    Brokerage
Benny Collins, Production Manager
John Conk, Production Manager
Anthony Davis, Anthony Davis Entertainment
    Services, Inc.
Jim Evans, Mountain Productions
Mary Lou Figley, Stageco US
Tim Franklin, Theta Consulting
Jeff Giek, Rhino
Brian Haas, Vistalogix Corporation
Hadden Hippsley, Production Manager
Kent Jorgensen, IATSE
Ken Keberle, Event Safety Specialist

Laurie Kirby, IMFCON
Dave Lester, Clair Global
Tom McClain, Projects of Interest
Debi Moen, Projection, Lights and Staging
    News, Event Safety Guide General Editor
Matt Monahan, Just a Bunch of Roadies
Scott Nacheman, Thornton Tomasetti
Tim Roberts, The Event Safety Shop
Alan Rowe, IATSE Local 728
Karl Ruling, PLASA
David Shaw, Indiana State Fair
Ron Stern, Production Manager
Mike Tierney, Pepsi Center
David VandenHeuvel, Weather Decision
    Technologies/WeatherOps

Beyond these inspiring individuals who all contributed greatly to the success of this effort, we must recognize that completing a project like this often falls to a select few who spend countless volunteer hours—often at a high personal cost—working at the mechanics of design, lay out, writing, editing, reviewing and administrative support. Without the hard work, dedication and perseverance of these few, this book would simply never have been possible.

- For their key work with the insurance industry and drumming up support for the Event Safety Alliance's mission, Scott Carroll (Take 1 Insurance), James Chippendale, and Roger Sandau;

- For his remarkable service as the online voice of the Event Safety Alliance, Jacob Worek, the ESA's new Director of Operations—social medial director, blog editor, current affairs researcher—and part of the organization's inner circle of visionaries;

- For his technical expertise, dependability and leadership skill, Steve Lemon, the one who coordinates the efforts of the writing teams and contributes so much to the content;

- For her ability to coordinate everyone's efforts in such a gentle and kind way, Melissa "Missy" Allgood, the one who always seems to know exactly how far to push to compel others to action;

- For his sage legal advice and master writing skills, Steve Adelman, the one whose faith and dedication to the mission is matched only by his talent, patience and generosity;

- For his writing, editing, technical and organization skills, Donald Cooper, the Managing Editor of the *Event Safety Guide*; the one who is often delivered moose feces and asked to turn it into rose petals—by tomorrow—and he does; and

- For his selfless, enduring response to the tragedy of Indiana in a way that will save lives in the future, Jim Digby, the leader of this overall effort who continues to teach us the importance and relevance of character, hard work, commitment, and endurance.

On a very personal note: Dr. Donald Cooper has given every moment of free time during his so called "retirement" to craft the language of this guide, and has worked tirelessly to see it come to life. We have thrown all manner of curve balls at him, which he has hit without once growing weary. Don was purposely chosen to lead the effort because of his impartiality. He is not, nor does he aspire to be, an entertainment professional. Rather, he is a safety and emergency response professional—a man who does not want to see harm come needlessly to anyone.

Dr. Cooper deserves more credit than words can express, and we as an industry will forever be indebted to him for taking this mission on when he could have been sipping boat drinks on a tropical island. I am deeply humbled by his wisdom, his participation, and his endurance. On behalf of everyone involved with the Event Safety Alliance, the event industry, and those whose lives you may save as a result of this work, we thank you Dr. Cooper.

~ Jim Digby

37. References

American College of Emergency Physicians Publications. Provision of Emergency Medical Care for Crowds. American College of Emergency Physicians Publications, 1989-90.

Australian and New Zealand Food Standards Authority. Food Standards Code. Canberra: Australian Government Publishing Service, Australian and New Zealand Food Standards Authority, 1987.

Australian National Health and Medical Research Council. Australian Guidelines for Recreational Use of Water. Canberra: Australian Government Publishing Service, National Health and Medical Research Council, 1990.

Australian Uniform Building Regulations Coordinating Council. Building Code of Australia. Australia: Australian Uniform Building Regulations Coordinating Council, 1990.

Barbera, J. A., et al. "Urban Search and Rescue." Emergency Medicine Clinics of North America May 1996.

Berlonghi, Alexander. (1990). The special event risk management manual. Dana Point, CA: Alexander Berlonghi.

Berlonghi, Alexander. (1996). Special event security management, loss prevention and emergency services. The Special Event Liability Series, Volume 2. Dana Point, CA: Alexander Berlonghi.

Berlonghi, Alexander E. "Understanding and Planning for Different Spectator Crowds." Engineering for Crowd Safety. Ed. R.A. Smith and J.F. Dickie. Elsevier Science Publications B.V., 1993.

Billie, P., et al. "Public Health at the 1984 Summer Olympics: The Los Angeles County Experience." American Journal of Public Health June 1988.

Bock, H. C., et al. Demographics of Emergency medical Care at the Indianapolis 500 Mile Race (1983 - 1990) October 1992.

Canadian Government. Aid of the Civil Power: Chapter N, Sections 274-285, in Revised Statutes of Canada. Canada: Canadian Government, 1985.

Chapman, K.R., et al. "Medical Services for Outdoor Rock Music Festivals." CMA Journal 15 April 1982: 935-938.

City of Fremantle. Concerts in Fremantle. Western Australia: City of Fremantle, 1996.

City of Keene. "Special Event Planning Checklist." New Hampshire: City of Keene.

"Controlling the Rock Festival Crowd." Security World June 1980: 40-43. http://www.crowdsafe.com

Cooper, D.C., Editor. (2005). Fundamentals of Search and Rescue. Burlington, MA: Jones & Bartlett.

Curry, Jack. Woodstock—The Summer of Our Lives. New York: Weidenfeld & Nicolson, 1989.

Defense Threat Reduction Agency. Weapons of Mass Destruction Handbook. Washington: Defense Threat Reduction Agency, 1 July 1999.

Department of the Treasury: Bureau of Alcohol, Tobacco and Firearms. "ATF Vehicle Bomb Explosion Hazard and Evacuation Distance Tables." Washington, 22 Dec. 1999. http://www.disastersrus.org/emtools/Terrorism/bomb_sem/bomatftb.htm

Donald, Ian. "Crowd Behavior at the King's Cross Underground Disaster." Easingwold Papers No. 4: Lessons Learned from Crowd-Related Disasters. Yorkshire: Emergency Planning College, 1992.

Emergency Management Australia. Australian Emergency Management Manual—Disaster Medicine. Australia: Emergency Management Australia, 1995. (Second edition due 1999.)

Emergency Management Australia. Australian Emergency Manual—Disaster Recovery. Australia: Emergency Management Australia, 1996. (Second edition due 2000.)

Emergency Management Australia. Australian Emergency Manuals Series: Part III, Volume 1, Manual 1—Emergency Catering. Australia: Emergency Management Australia, 1998.

Emergency Management Australia. Australian Emergency Manuals Series: Part III, Volume 2, Manual 1—Evacuation Planning. Australia: Emergency Management Australia, 1998.

Emergency Management Australia. Australian Emergency Manuals Series: Part III, Volume 2, Manual 2—Safe and Healthy Mass Gatherings. Australia: Emergency Management Australia, 1998.

Emergency Management Australia. Australian Emergency Manuals Series: Part III, Volume 3, Manual 1—Multiagency Incident Management. Australia: Emergency Management Australia, 1998.

Emergency Management Australia. Australian Emergency Manuals Series: Part III, Volume 3, Manual 2—Community and Personal Support Services. Australia: Emergency Management Australia, 1998.

Emergency Management Australia. Australian Emergency Manuals Series: Part IV, Manual 2—Operations Centre Management. Australia: Emergency Management Australia, 1996.

Emergency Management Australia. Australian Emergency Manuals Series: Part IV, Manual 9—Communications. Australia: Emergency Management Australia. 2nd ed. 1998.

"Emergency Medicine: Rock and Other Mass Medical." Emergency Medicine. June 1975: 116-129.

Federal Emergency Management Agency. Special Events Contingency Planning: Jobs Aid Manual. Emmitsburg, Maryland: Federal Emergency Management Agency, March 2005, Updated May 2010.

Federal Emergency Management Agency [FEMA]. (2008). IS-100.b: Introduction to the incident command system, course materials. Washington, DC: FEMA. Available online at: http://training.fema.gov/emiweb/is/is100blst.asp

Federal Emergency Management Agency [FEMA]. (2010). IS-200.b: ICS for single resources and initial action, course materials. Washington, DC: FEMA. Available online at: http://training.fema.gov/emiweb/is/is200b.asp

Federal Emergency Management Agency, Emergency Management Institute. The Emergency Planning Process: Self Instruction. Emmitsburg, Maryland: Federal Emergency Management Agency, June 1997.

Federal Emergency Management Agency, Emergency Management Institute. Tools for Emergency Planning. Emmitsburg, Maryland: Federal Emergency Management Agency, Emergency Management Institute, June 1997.

Federal Emergency Management Agency, National Fire Academy. Emergency Medical Services: Special Operations. Emmitsburg, Maryland: Federal Emergency Management Agency.

Franaszek, J. "Medical Care at Mass Gatherings." Annals of Emergency Medicine May 1986: 148-149.

Fruin, John J. "Crowd Dynamics and Auditorium management." *Auditorium News*. May 1984: 48-53.

Fruin, John J. "Causes and Prevention of Crowd Disasters." Student Activities Programming. Oct. 1981: 48-53.

Goldaber, Irving. "Is Spectator Violence Inevitable?" Auditorium News April 1979: 4-7.

Great Britain Health and Safety Commission, Home Office and the Scottish Office. "Guide to Health, Safety and Welfare at Pop Concerts and Similar Events." London: Great Britain Health and Safety Commission, Home Office and the Scottish Office, 1993.

Guide to safety at sports grounds (Fourth edition) Department of National Heritage 1997 ISBN 978 0 11 300095 1

Hanna, James A. Emergency Preparedness Guidelines for Mass, Crowd-Intensive Events. Canada: Emergency Preparedness Canada, 1995.

---. "Rock and Peace Festivals—The Field Hospital." Disaster Planning for Health Care Facilities. 3rd ed. Ottawa: Canadian Hospital Association, 1995. 247-256.

Health Department of Western Australia. "Operational Guidelines for Rave Parties, Concerts, and Large Public Events." Western Australia: Health Department of Western Australia, 1995.

Herman, Gary. Rock 'N' Roll Babylon. London: Plexus Publishing, 1982.

Hillmore, Peter. Live Aid. Parsippany, N.J.: Unicorn Publishing, 1985.

"Hillsborough: Inquiry Highlighted Differing Approach to Operational Messages." Fire. Great Britain, Aug. 1989: 7-8.

"Hillsborough: An Earlier Call Would Probably Not Have Saved Lives." Fire. Great Britain, Sept. 1989: 7.

Hopkins, Jerry. Festival. New York: Macmillan, 1970.

International Association of Venue Managers [IAVM]. (2002). Safety and security best practices planning guide for theaters and performing arts centers. Coppell, Texas: IAVM. More information available online at: www.iavm.org. Also available by email request of security@iavm.org.

International Association of Venue Managers [IAVM]. (2002). Safety and security best practices planning guide for arenas, stadiums & amphitheaters. Coppell, Texas: IAVM. More information available online at: www.iavm.org. Also available by email request of security@iavm.org.

International Association of Venue Managers [IAVM]. (2002). Safety and security best practices planning guide for convention centers/exhibit halls. Coppell, Texas: IAVM. More information available online at: www.iavm.org. Also available by email request of security@iavm.org.

International Association of Venue Managers [IAVM]. (2002). Safety and security best practices planning guide for emergency preparedness. Coppell, Texas: IAVM. More information available online at: www.iavm.org. Also available by email request of security@iavm.org.

International Association of Venue Managers [IAVM]. (2013). Severe/hazardous weather preparedness plan and guideline. Coppell, Texas: IAVM. Available online at: www.iavm.org.

International Association of Venue Managers [IAVM]. (2013). Severe weather preparedness & planning for public assembly venues and events, 2-day course. Coppell, Texas: IAVM. More information available online at: www.iavm.org.

International Association of Venue Managers [IAVM]. (2013). Academy for Venue Safety and Security [AVSS], 5-day course. Coppell, Texas: IAVM. More information available online at: www.iavm.org.

James, S.H., et al. "Medical and Toxicological Aspects of the Watkins Glen Rock Concert." Journal of Forensic Sciences, n.d. (circa 1974): 71-82.

Leonard, R.B. "Medical Support for Mass Gatherings." Emergency Medicine Clinics of North America May 1996.

Lewis, J. M. "A Protocol for the Comparative Analysis of Sports Crowd Violence." International Journal of Mass Emergencies and Disaster 1988: 221-225.

Lewis, J.M. "Theories of the Crowd: Some Cross-Cultural Perspectives." Easingwold Papers No. 4: Lessons Learned from Crowd-Related Disasters. Yorkshire: Emergency Planning College, 1992.

Lichtenstein, Irwin. "EMS at Rock Concerts." Fire Chief Magazine Nov. 1983: 44-46.

Madden, Turner. (1998). Public assembly facility law: A guide for managers of arenas, auditoriums, convention centers, performing arts centers, race tracks and stadiums. Irving, TX: IAVM.

Managing crowds safely. A guide for organisers at events and venues HSG154 (Second edition) HSE Books 2000 ISBN 978 0 7176 1834 7

Mariano, J. P. "First Aid for Live Aid." JEMS Feb. 1986.

Miami-Dade County Office of Emergency Management. Concept of Operations Plan, SuperBowl XXXIII. Florida: Miami-Dade County Office of Emergency Management. Jan. 1999.

National Center for Missing and Exploited Children® [NCMEC]. (2011). Law enforcement policy and procedures for reports of missing and exploited children: A model. Alexandria, VA: NCMEC. Retrieved on December 25, 2013, from www.missingkids.com

National Center for Missing and Exploited Children® [NCMEC].(2011). "What Should You Do If You See A Child Who Appears To Be Lost?" Alexandria, VA: NCMEC. Retrieved on December 25, 2013, from www.missingkids.com/en_US/publications/PDF15A.pdf

National Domestic Preparedness Office. "WMD Threats: Sample Guidelines Reissue." Special Bulletin. Washington: National Domestic Preparedness Office, 12 Jan 2000.

National Interagency Fire Center. "ICS Glossary." In Incident Command System National Training Curriculum. Boise, Idaho: National Interagency Fire Center, Oct. 1994.

National Interagency Fire Center. "ICS Position Descriptions and Responsibilities." In Incident Command System National Training Curriculum. Boise, Idaho: National Interagency Fire Center, Oct. 1994.

National Interagency Fire Center. "Organizing for Incidents or Events, Module 8." In Incident Command System National Training Curriculum. Boise, Idaho: National Interagency Fire Center, Oct. 1994.

National Institute for Occupational Safety and Health [NIOSH], U.S. Department of Health and Human Services [DHHS]. (1998). Criteria for a recommended standard: Occupational noise exposure. NIOSH Publication 98-126. Washington, DC: DHHS. Retrieved on January 22, 2013, from http://www.cdc.gov/niosh/docs/98-126/pdfs/98-126.pdf

National Oceanic and Atmospheric Administration [NOAA], National Weather Service [NWS]. (2012). Lightning safety: Large venues: A toolkit for lightning preparedness and related safety decision making. Silver Spring, MD: NOAA/NWS. Available online at: www.lightningsafety.noaa.gov.

National Oceanic and Atmospheric Administration [NOAA], National Weather Service [NWS]. (2013). StormReady® program and recognition. Silver Spring, MD: NOAA/NWS. Available online at: http://www.stormready.noaa.gov.

National Safety Council for the International Association of Amusement Parks and Attractions. *Fixed-Site Amusement Ride Injury Survey*, 2010 Update, Revised December 2011. Retrieved 15 November 2012 from http://www.nsc.org/news_resources/ injury_and_death_statistics/Documents/Report%202010- Sep_2011_rev%2012%205%2011.pdf

Occupational Safety and Health Administration [OSHA]. (2013). Occupational noise exposure. Retrieved on January 31, 2013, from http://www.osha.gov/SLTC/noisehearingconservation/index.html

Occupational Safety and Health Administration [OSHA]. (January 20, 1999). OSHA technical manual. Directive Number: TED 01-00-015 [TED 1-0.15A].

Ounanian, L. L. "Medical Care at the 1982 U.S. Festival." Annals of Emergency Medicine May 1986: 25-32.

Parrillo, S.J. "Medical Care at Mass Gatherings: Considerations for Physician Involvement." Prehospital and Disaster Medicine Oct.-Dec. 1995.

Parrillo, S. J. "EMS and Mass Gatherings" 16 Nov. 1999.

Pauls, J.L. Observations of Crowd Conditions at Rock Concert in Exhibition Stadium. Ottawa: National Research Council of Canada, April 1982.

Pennsylvania Emergency Management Agency. "First Responder's Guide: Terrorism Incidents." Pennsylvania Emergency Management Agency. 7 Jan. 2000.

PLASA. (2005). Introduction to modern atmospheric effects, 4th edition. NY, New York: PLASA.

PLASA. (1994). Recommended Practice for DMX512: A guide for users and installers, 2nd edition. NY, New York: PLASA.

PLASA. (2000). Camera crane operator's handbook. NY, New York: PLASA.

Public Entity Risk Institute. http://www.riskinstitute.org.

Queensland Police Service: Drug and Alcohol Co-ordination. Alcohol, Safety and Event Management: A Resource to Assist Event Managers to Conduct Safer Public Events. Queensland: Queensland Police Service, 1997.

Radiation safety of lasers used for display purposes HSG95 HSE Books 1996 ISBN 978 0 7176 0691 7

Research to develop a methodology for the assessment of risks to crowd safety in public venues CRR204 HSE Books 1999 ISBN 978 0 7176 1663 0

Rosenman, Joel, et al. Young Men With Unlimited Capital. New York: Harcourt 1974.

Ryan, S., and M. Carey. "Key Principles in Ensuring Crowd Safety in Public Venues." Engineering for Crowd Safety. Ed. R. A. Smith and J. A. Dickie. Elsevier Science Publications, 1993.

Sanders, Arthur B., et al. "An Analysis of Medical Care at Mass Gatherings." Annals of Emergency Medicine May 1986: 17-21.

Schlight, Judith, et al. "Medical Aspects of Large Outdoor Festivals." The Lancet 29 April 1972: 948-952.

Taylor, Derek. It Was Twenty Years Ago Today. New York: Simon & Schuster, 1987.

Temporary demountable structures: Guidance on design, procurement and use (Third edition) Institution of Structural Engineers, UK, 2007. ISBN 978 090129745 7

Thompson, James M., et al. "Level of Medical Care Required for Mass Gatherings." Annals of Emergency Medicine April 1991: 78-83.

Thompson, G., Ed. (1993). The focal guide to safety in live performance. New York, NY: Focal Press (now Taylor and Francis) Publishers.

Threats: Critical Infrastructure, Key Assets – from the Department of Homeland Security, Buffer Zone Planning Program

U.S. Consumer Product Safety Commission, Office of Compliance and Field Operations, Division of Defects Investigation, *Directory of State Amusement Ride Safety Officials*, July 2010.

U.S. Department of Homeland Security [DHS]. (2013). Homeland security information network [HSIN]. Available online to DHS partners at: http://www.dhs.gov/homeland-security-information-network. Items available include:

- Protective Measures Guide for the U.S. Outdoor Venues Industry, June 2011
- Protective Measures Guide for U.S. Sports Leagues, 2008
- Evacuation Planning Guide for Stadiums, Fall 2008

U.S. Department of Homeland Security [DHS]. (2013). Commercial facilities sector training and resources. Available online to at: http://www.dhs.gov/cfsectortraining. Items available include:

- Suspicious activity awareness video: Check It! - Why is it important?
- Check It! - How to check a bag? training video
- Risk Self-Assessment Tool for Stadiums and Arenas
- How to respond to an active shooter situation
- Soft target awareness course
- Protective measures course

U.S. Department of Transportation, Pipeline and Hazardous Materials Safety Administration. (2012). *Explosives (EX) approvals: Regulatory guidelines for shipping and transporting fireworks*. Washington, DC: PHMSA, retrieved on December 25, 2013, from http://phmsa.dot.gov/staticfiles/PHMSA/DownloadableFiles/Files/Hazmat/EX%20brochure.pdf

U.S. Programs, Domestic Emergencies Unit: Save the Children. (September 2007). *The unique needs of children in emergencies: A guide for the inclusion of children in emergency operations plans*. Westport, CT: Save the Children Federation, Inc. Downloaded on November 30, 2013, from http://www.idph.state.ia.us/hcci/common/pdf/children_in_emergencies_planning_guide.pdf

Wardrope, J., et al. "The Hillsborough Tragedy." British Medical Journal. Nov. 1991.

Weiner, Rex, et al. Woodstock Census. New York: Fawcett Columbine, 1979.

Wertheimer, Paul L. Crowd Management - Report of the Task Force on Crowd Control and Safety. Cincinnati: City of Cincinnati, July 1980.

Whitehead, J. "Crowd Control Can Be Critical In Emergencies." Emergency Preparedness Digest Oct.-Dec. 1989: 12-15.

Working together on firework displays: A guide to safety for firework display organisers and operators HSG123 (Third edition) HSE Books 2006 ISBN 978 0 7176 6196 1

Wyllie, R. "Setting the Scene" Easingwold Papers No. 4: Lessons Learned from Crowd-Related Disasters. Yorkshire: Emergency Planning College, 1992.

37.1 Standards, Regulations and Model Codes

Although the latest versions of the following standards are listed, some states require the use of older versions because a specific edition (year) of the standard is used in the language of the state's law or administrative rule. Event organizers must work with local authorities having jurisdiction to determine which version of a standard or code is required to be used. Also, please note that not all standards published by each of the standard setting organizations is listed herein, and all standard setting organizations are not necessarily included in this list. The standards and codes listed were selected by the editors to include those referenced in this guide and some others that may be relevant to the live event industry. This is not a comprehensive or exhaustive list.

American National Standards Institute [ANSI], Washington, DC: ANSI. (www.ansi.org)
- ANSI/ASA S12.6, Subject-fit test method for measuring the real ear attenuation of hearing protectors, 2008.
- ANSI/ISEA Z89.1. American National Standard for Industrial Head Protection, 2009.
- ANSI Z136.1. American National Standard for Safe Use of Lasers, 2007, in conjunction with the Laser Institute of America (www.lia.org).

American Pyrotechnics Association [APA], Bethesda, MD: APA. (www.americanpyro.com)
- APA Standard 87-1, Standard for the Construction and Approval for Transportation of Fireworks, Novelties, and Theatrical Pyrotechnics, 2001.

American Society of Civil Engineers [ASCE], Reston, Virginia: ASCE. (www.asce.org)
- ASCE/SEI 55-10, Tensile Membrane Structure
- SEI/ASCE 37-02, Design Loads on Structures During Construction
- SEI/ASCE 23-97, Specifications for Structural Steel Beams with Web Openings
- ASCE/SEI 19-10, Structural Applications of Steel Cables for Buildings
- SEI/ASCE 11-99, Guideline for Structural Condition Assessment of Existing Buildings
- ASCE/SEI 7-10, Minimum Design Loads for Buildings and Other Structures

ASME, New York, New York: ASME. (www.asme.org)
- ASME *International Boiler and Pressure Vessel Code*, Section VIII, 2013 Edition.

International Code Council [ICC]. Country Club Hills, Illinois: ICC. (www.iccsafe/org)
- International building code, 2012.
- International energy conservation code, 2012.
- International existing building code, 2012.
- International fire code, 2012.
- International fuel gas code, 2012.
- International mechanical code, 2012.
- International plumbing code, 2012.
- International private sewage disposal code, 2012.
- International property maintenance code, 2012.
- International wildland-urban interface code, 2012.

- ICC 300, Bleachers, folding and telescoping seating, and grandstands, 2012.
- ICC A117.1, Accessible and usable buildings and facilities, 2009.
- ICC G3, Global guideline for practical public toilet design, 2011.
- ICC G2, Guideline for acoustics, 2010.

National Fire Protection Association [NFPA], Quincy, Massachusetts: NFPA. (www.nfpa.org)
- NFPA 1: Fire code, 2012.
- NFPA 70: National electric code, 2011.
- NFPA 70B: Recommended practice for electrical equipment maintenance, 2010.
- NFPA 13: Standard for the installation of sprinkler systems, 2013.
- NFPA 14: Standard for the installation of standpipes and hose systems, 2010.
- NFPA 30: Flammable and combustible liquids code, 2012.
- NFPA 54, National fuel gas code, 2012.
- NFPA 55, Compressed gases and cryogenic fluids code, 2013.
- NFPA 58: Liquefied petroleum gas code, 2011.
- NFPA 59A, Standard for the production, storage, and handling of liquefied natural gas (LNG), 2013.
- NFPA 72: National fire alarm and signaling code, 2013.
- NFPA 80: Standard for fire doors and other opening protectives, 2013.
- NFPA 101: Life safety code®, 2012.
- NFPA 101B: Code for means of egress for buildings and structures, 2002.
- NFPA 102: Standard for grandstands, folding and telescopic seating, tents, and membrane structures, 2011.
- NFPA 140, Standard on motion picture and television production studio soundstages, approved production facilities, and production locations, 2013.
- NFPA 160: Standard for the use of flame effects before an audience, 2011.
- NFPA 289: Standard method of fire test for individual fuel packages, 2013.
- NFPA 297: Guide on principles and practices for communications systems, 1995.
- NFPA 430, Code for the storage of liquid and solid oxidizers, 2004.
- NFPA 551: Guide for the evaluation of fire risk assessments, 2013.
- NFPA 610: Guide for emergency and safety operations at motorsports venues, 2009.
- NFPA 701: Standard methods of fire tests for flame propagation of textiles and films, 2010.
- NFPA 730: Guide for premises security, 2011.
- NFPA 1123: Code for fireworks display, 2014.
- NFPA 1124: Code for the manufacture, transportation, storage, and retail sales of fireworks and pyrotechnic articles, 2013.
- NFPA 1126: Standard for the use of pyrotechnics before a proximate audience, 2011.
- NFPA 1561: Standard on emergency services incident management system, 2008.
- NFPA 1600: Standard on disaster/emergency management and business continuity programs, 2010.
- NFPA 5000: Building construction and safety code, 2012.

Occupational Safety and Health Administration [OSHA], U.S. Department of Labor, Washington, DC: OSHA. (www.osha.gov); Electronic Code of Federal Regulations (e-CFR) can be accessed at http://www.ecfr.gov.

- OSHA Publication 3071, Job hazard analysis, 2002 (revised).
- OSHA Publication 3074, Hearing conservation, 2002 (revised).
- OSHA Publication 3075, Controlling electrical hazards, 2002.
- OSHA Publication 3079, Respiratory protection, 2002 (revised).
- OSHA Publication 3088, How to plan for workplace emergencies and evacuations, 2001.
- OSHA Publication 3122, Principal emergency response and preparedness—requirements and guidelines, 2004.
- OSHA Publication 3124, Stairways and ladders, 2003.
- OSHA Publication 3143, Industrial hygiene, 1998.
- OSHA Publication 3151-12R, Personal protective equipment, 2003.
- OSHA Publication 3317, Best practices guide: Fundamentals of a workplace first-aid program, 2006.
- OSHA Publication 3335, Preparing and protecting security personnel in emergencies, 2007.
- OSHA Publication 3494, Fire service features of buildings and fire protection systems, 2012.
- OSHA 29 CFR 1904.10, Recording criteria for cases involving occupational hearing loss.
- OSHA 29 CFR 1910.95, Occupational noise exposure.
  - Appendix A, Noise exposure computation
  - Appendix B, Methods for estimating the adequacy of hearing protector attenuation
  - Appendix C, Audiometric measuring instruments
  - Appendix D, Audiometric test rooms
  - Appendix E, Acoustic calibration of audiometers
  - Appendix F, Calculations and application of age corrections to audiograms
  - Appendix G, Monitoring noise levels non-mandatory informational appendix
  - Appendix H, Availability of referenced documents
  - Appendix I, Definitions
- OSHA 29 CFR 1910.132, Personal protective equipment, general requirements.
- OSHA 29 CFR 1910.133, Personal protective equipment, eye and face protection.
- OSHA 29 CFR 1910.134, Personal protective equipment, respiratory protection.
- OSHA 29 CFR 1910.135, Personal protective equipment, head protection.
- OSHA 29 CFR 1910.136, Personal protective equipment, foot protection.
- OSHA 29 CFR 1910.137, Personal protective equipment, electrical protective devices.
- OSHA 29 CFR 1910.138, Personal protective equipment, hand protection.
- OSHA 29 CFR 1910.147, The control of hazardous energy (lockout/tagout).
- OSHA 29 CFR 1910.157, Portable fire extinguishers.
- OSHA 29 CFR 1910.158, Standpipe and hose systems.
- OSHA 29 CFR 1926, Safety and health regulations for construction.
- OSHA Technical Manual, Directive Number: TED 01-00-015 [TED 1-0.15A], 1/20/1999.

Underwriters Laboratories [UL], Ann Arbor, Michigan: UL. (www.ul.com)

- UL 1975: Fire tests for foamed plastics used for decorative purposes, Third Edition, 2006.
- ANSI/UL 711, CAN/ULC-S508: Rating and fire testing of fire extinguishers.
- ANSI/UL 8, CAN/ULC-S554: Water-based agent fire extinguishers.
- ANSI/UL 154, CAN/ULC-S503: Carbon dioxide fire extinguishers.
- ANSI/UL 299, CAN/ULC-S504: Dry chemical fire extinguishers.
- ANSI/UL 626, CAN/ULC-S507: Water fire extinguishers.
- ANSI/UL 2129, CAN/ULC-S566: Halocarbon clean agent extinguishers.

PLASA, New York, New York: PLASA. (www.plasa.org). Note that PLASA's Technical Standards Program (TSP) is ANSI accredited and highly recommended.

- ANSI E1.1, Entertainment technology: Construction and use of wire rope ladders, 2012.
- ANSI E1.2, Entertainment technology: Design, manufacture and use of aluminum trusses and towers, 2006.
- ANSI E1.3, Entertainment technology: Lighting control systems – 0-10V analog control specification, 2001, revised 2011.
- ANSI E1.4, Entertainment technology: Manual counterweight rigging systems, 2009.
- ANSI E1.5, Entertainment technology: Theatrical fog made with aqueous solutions of di- and trihydric alcohols, 2009.
- ANSI E1.6-1, Entertainment technology: Powered hoist systems, 2012.
- ANSI E1.6-3, Selection and use of chain hoists in the entertainment industry, 2012.
- ANSI E1.8, Entertainment technology: Loudspeaker enclosures intended for overhead suspension—classification, manufacture and structural testing, 2005.
- ANSI E1.9, Reporting photometric performance data for luminaries used in entertainment lighting, 2007, revised 2012.
- ANSI E1.11, Entertainment technology: USITT DMX512-A, asynchronous serial digital data transmission standard for controlling lighting equipment and accessories, 2008.
- ANSI E1.14, Entertainment technology: Recommendations for inclusions in fog equipment manuals, 2001, revised 2007.
- ANSI E1.15, Entertainment technology: Recommended practices and guidelines for the assembly and use of theatrical boom & base assemblies, 2006, revised 2011.
- ANSI E1.16, Entertainment technology: Configuration standard for metal halide ballast power cables, 2002, revised 2012.
- ANSI E1.17, Entertainment technology: Architecture for control networks (ACN), 2010.
- ANSI E1.19, Recommended practice for the use of class A ground-fault circuit interrupters (GFCIs) intended for personal protection in the entertainment industry, 2009.
- ANSI E1.20, Entertainment technology: RDM-remote device management over USITT DMX512 networks, 2010.
- ANSI E1.21, Temporary ground-supported overhead structures used to cover stage areas and support equipment in the production of outdoor entertainment events (latest edition).
- ANSI E1.22, Entertainment technology: Fire safety curtain systems, 2009.
- ANSI E1.23, Entertainment technology: Design and execution of theatrical fog effects, 2010.

- ANSI E1.24, Entertainment technology: Dimensional requirements for stage pin connectors, 2012.
- ANSI E1.25, Recommended basic conditions for measuring the photometric output of stage and studio luminaries by measuring illumination produced on a planar surface, 2012.
- ANSI E1.26, Entertainment technology: Recommended testing methods and values for shock absorption of floors used in live performance venues, 2006, revised 2012.
- ANSI E1.27-1, Entertainment technology: Standard for portable control cables for use with USITT DMX512/1990 and E1.11 (DMX512-A) products, 2006, revised 2011.
- ANSI E1.27-2, Entertainment technology: Recommended practice for permanently installed control cables for use with ANSO E1.11 (DMX512-A) and USITT DMX512/1990 products, 2009.
- ANSI E1.28, Guidance on planning followspot positions in places of public assembly, 2011.
- ANSI E1.29, Product safety standard for theatrical fog generators that create aerosols of water, aqueous solutions of glycol or glycerin, or aerosols of highly refined alkane mineral oil, 2009.
- ANSI E1.30-1, EPI 23, device identification subdevice, 2010.
- ANSI E1.30-3, EPI 25, time reference in ACN systems using SNTP and NTP, 2009.
- ANSI E1.30-4, EPI 26, device description language (DDL) extensions for DMX512 and E1.31 devices, 2010.
- ANSI E1.30-7, EPI 29, allocation of internet protocol version 4 addresses to ACN hosts, 2009.
- ANSI E1.30-10, EPI 32, identification of draft device description language modules, 2009.
- ANSI E1.31, Entertainment technology: Lightweight streaming protocol for transport of DMX512 using ACN, 2009.
- ANSI E1.32, Entertainment technology: Guide for the inspection of entertainment industry incandescent lamp luminaries, 2012.
- ANSI E1.34, Entertainment technology: Measuring and specifying the slipperiness of floors used in live performance venues, 2009.
- ANSI E1.35, Standard for lens quality measurements for pattern projecting luminaries intended for entertainment use, 2007.
- ANSI E1.36, Model procedures for permitting the use of tungsten-halogen incandescent lamps and stage and studio luminaries in vendor exhibit booths in convention and trade show exhibition halls, 2007, revised 2012.
- ANSI E1.37-1, Additional message sets for ANSI E1.20 (RDM) – part 1, dimmer message sets, 2012.
- ANSI E1.40, Recommendations for the planning of theatrical dust effects, 2011.
- ANSI E1.41, Recommendations for measuring and reporting photometric performance data for entertainment luminaries utilizing solid state light sources, 2012.

38. Glossary of Useful Terms

TERM	DEFINITION
A-Weighting	An artificial filter which is applied to measuring devices to make them more accurately mimic the way a human ear responds to sound. A-weighting of a sound measurement effectively ignores a proportion of the bass frequency because the ear is less sensitive to low register sounds.
Accident	The National Safety Council defines an "accident" as an undesired event that results in personal injury or property damage.
Agency	An agency is a division of government with a specific function, or a nongovernmental organization (e.g., private contractor, business, etc.) that offers a particular kind of assistance. In ICS, agencies are defined as jurisdictional (having statutory responsibility for incident mitigation) or assisting and/or cooperating (providing resources and/or assistance).
AHJ	Authority having jurisdiction (see below).
Air-Inflated Structure	A building where the shape of the structure is maintained by air pressurization of cells or tubes to form a barrel vault over the usable area. Occupants of such structures do not occupy the pressurized areas used to support the structure (International Fire Code, 2009).
Air-Supported Structure	A structure wherein the shape of the structure is attained by air pressure, and occupants of the structure are within the elevated pressure area (International Fire Code, 2009).
Appointed Medical Provider	A competent organization chosen by the event organizer to provide overall management of medical, ambulance and first-aid services at an event (see Chapter 5, *Medical, Ambulance and First Aid Management*).
Approved	Acceptable to the authority having jurisdiction (NFPA and ICC).
Area of Refuge	An area where persons unable to use stairways can remain temporarily to await instructions or assistance during emergency evacuation (International Fire Code, 2009). At least one state's code uses the term "Areas of Rescue Assistance" as an equivalent term.
Assembly Occupancy (Group A)	"Assembly occupancy" (Group A) is a specific classification of specific classification of building occupancy. It includes, among others, the use of a building or structure, or a portion thereof, for the gathering together of persons for purposes such as civic, social or religious functions; recreation, food or drink consumption; or awaiting transportation. More specifically, an "A-4" assembly occupancy includes arenas and skating rinks, and an "A-5" assembly occupancy includes amusement park structures, bleachers, grandstands, and stadiums (International Fire Code, 2009).
Attenuation	The ability of hearing protection or other material (such as a wall or tent lining) to absorb sound and reduce the amount of energy transmitted.

At-Will	Generally, "At-Will" simply means to do as one chooses. However, employment at-will is a common-law rule that an employment contract of indefinite duration can be terminated by either the employer or the employee at any time for any reason; also known as terminable at will. The at-will category encompasses all employees who are not protected by express employment contracts that state that they may be fired only for good cause.
Authority Having Jurisdiction (AHJ)	An organization, office, or individual responsible for enforcing the requirements of a code or standard, or for approving equipment, materials, an installation, or a procedure (NFPA).
Automatic Sprinkler System	An automatic sprinkler system, for fire protection purposes, is an integrated system of underground and overhead piping designed in accordance with fire protection engineering standards. The system includes a suitable water supply and a network of specially sized or hydraulically designed piping installed in a structure or area, generally overhead, to which automatic sprinklers are connected in a systematic pattern. The system is usually activated by hear from a fire and discharges water over the fire area (International Fire Code, 2009).
AV or A/V	Audio Visual
Banter	Chit-chat, often in spirit of humor
Bus Loop	Location where buses load and unload
Chain of Command	A series of management positions in order of authority.
Clear Text	The use of plain English in radio communications transmissions. No Ten Codes or agency-specific codes are allowed when using Clear Text.
Command	The act of directing and/or controlling resources by virtue of explicit legal, agency, or delegated authority. May also refer to the Incident Commander.
Command Center	See "Incident Command Post"
Competent Person (OSHA)	One who is capable of identifying existing and predictable hazards in the surroundings or working conditions which are unsanitary, hazardous, or dangerous to employees, and who has authorization to take prompt corrective measures to eliminate them (29 CFR 1926.32(f)). By way of training and/or experience, a competent person is knowledgeable of applicable standards, is capable of identifying workplace hazards relating to the specific operation, and has the authority to correct them. Some standards add additional specific requirements which must be met by the competent person.
DAR	Daily Action Report
dbA	Acceptable (A-weighted) noise (decibel) levels. A-weighting is used to measure hearing risk and for compliance with OSHA regulations that specify permissible noise exposures in terms of a time-weighted average sound level or daily noise dose.

Decibel (dB)	A measure of the energy in a sound wave. The human ear is incredibly adaptable and can detect quiet sounds and tolerate momentary loud noises with a huge range of energies. If these were written in normal numbers, it would span from 1 to 100,000,000,000,000 which is an unwieldy number to use in calculations or measurements. To make things simpler, sound is measured using a logarithmic scale – so instead of going from 1 to 1014 the numbers go from 1 to 140. Using a logarithmic scale means some of the everyday way we use numbers no longer works. For example, if you had an amplifier which made 100 dB of sound, and you stood it next to an identical amplifier, there wouldn't be 200 dB of sound. Instead there would be 103 dB. So as a rough rule of thumb +/- 3 dB means a doubling of halving of the energy in a sound. This is further complicated by the fact that the ear does not have a linear response to sound energy, so perceived sound is not directly proportional to energy.
Division	Divisions are used to divide an incident into geographical areas of operation. In ICS, a Division is located within the organization between the Branch and the Task Force/Strike Team. (See Group.) Divisions are identified by alphabetic characters for horizontal applications and, often, by floor numbers when used in buildings.
DRS	Daily Run Sheet
DOL	Department of Labor
Egress	The action of going out of or leaving a place.
Electric Match	An electric device that contains a small amount of pyrotechnic material that ignites when a current flows through the device (NFPA 1123, 2010).
Emergency	Any incident(s), human-caused or natural, that requires responsive action to protect life or property.
Emergency Management Coordinator/ Director	The individual within each political subdivision that has coordination responsibility for jurisdictional emergency management.
Emergency Operations Center (EOC)	The physical location at which the coordination of information and resources to support domestic incident management activities normally takes place. An EOC may be a temporary facility or may be located in a more central or permanently established facility, perhaps at a higher level of organization within a jurisdiction. EOCs may be organized by major functional disciplines (e.g., fire, law enforcement, and medical services).
Emergency Operations Plan (EOP)	The plan that each jurisdiction has and maintains for responding to appropriate hazards.
Event	A scheduled, planned, non-emergency activity.
Exit	That portion of a means of egress system that is separated from other interior spaces of a building or structure by fire-resistance-rated construction and opening protective as required to provide a protected path of egress travel between the exit access and the exit discharge (International Fire Code, 2009).

Exit Access	That portion of a means of egress system that leads from any occupied portion of a building or structure to an exit (International Fire Code, 2009).
Exit Discharge	That portion of a means of egress system between the termination of an exit and a public way (International Fire Code, 2009).
Exposure Limit Value = 87dB(A) or 140dB (C-weighted)	A measure of the energy which actually reaches the persons' ear, i.e., taking account of variations in daily or weekly routine and the attenuating effect of hearing protection. This relates to the cumulative Personal Exposure which varies as the individual carries out different duties and not to measurements of noise in the environment.
Fire Code	A set of standards established and enforced by government for fire prevention and fire and life safety. Consult the local authority having jurisdiction for details regarding the applicable fire code.
Fore Code Official	The fire chief or other designated authority charged with the administration and enforcement of the fire code, or a duly authorized representative (International Fire Code, 2012).
Fire Watch	A temporary measure intended to ensure continuous and systematic surveillance of a building or portion thereof by one or more qualified individuals for the purposes of identifying and controlling fire hazards, detecting early signs of unwanted fire, raising an alarm of fire and notifying the fire department.
Fireworks	Any composition or device for the purpose of producing a visible or an audible effect the entertainment purposes by combustion, deflagration or detonation that meets the definition of 1.4G fireworks or 1.3G fireworks as set forth herein (International Fire Code, 2012). Fireworks, 1.4G. Small fireworks devices containing restricted amounts of pyrotechnic composition designed primarily to produce visible or audible effects by combustion. Not considered explosive materials. Fireworks, 1.3G. Large fireworks devices, which are explosive materials, intended for use in fireworks displays and designed to produce audible or visible effects by combustion, deflagration, or detonation. Such 1.3G fireworks include, but are not limited to, firecrackers containing more than 130 mg (2 grains) of explosive composition, aerial shells containing more than 40 grams of pyrotechnic composition and other display pieces which exceed the limits for classification as 1.4G fireworks.
Frequency	The speed of vibration. All natural sounds are made up of a complex mix of sounds vibrating at different frequencies. More rapid vibration (high frequency) produces the bright treble notes, low frequency results in bass sounds. The proportion of each in a sound gives it its distinctive characteristics and tone. Frequency is measured in hertz (Hz), one hertz meaning one cycle of vibration per second. The normal human ear can detect from around 20Hz up to 20,000Hz or 20KHz.

Frequency Analysis	A mathematical measure of how much each "slice" across the frequency spectrum contributes to the overall sound. This process is also known as Octave Band Analysis – the whole frequency spectrum from 20Hz to 20kHz being divided into segments of 1/3 of an octave. It is by this process that the precise content of a sound is determined and suitable hearing protection is selected.
Front of House or "FOH"	Areas the attendees and public can access; can also be slang for the mix platform or control area at a concert.
Group	Groups are established to divide the incident into functional areas of operation. Groups are composed of resources assembled to perform a special function not necessarily within a single geographic division. (See Division.) In ICS, groups are located between Branches (when activated) and Resources in the Operations Section.
Hazard	An object, substance or circumstance which has the *potential* to cause harm. (George Thompson-Focul Guide to Safety in Live Performance. Also see "Risk" below.)
Helibase	The main location for parking, fueling, maintenance, and loading of a helicopter operating in support of an incident or event. This is often located at an airport or airfield.
Helispot	A designated location, usually temporary, where a helicopter can safely take off and land.
HSSE	Health, Safety, Security and Environment
ICT	Information and Communications Technology (Same as "IT")
Incident (OSHA)	OSHA defines an "incident" as an unplanned, undesired event that adversely affects completion of a task.
Incident (ICS)	An occurrence or event, natural or human-caused, that requires an emergency response to protect life or property. Incidents can, for example, include major disasters, emergencies, terrorist attacks, terrorist threats, wildland and urban fires, floods, hazardous materials spills, nuclear accidents, aircraft accidents, earthquakes, hurricanes, tornadoes, tropical storms, war-related disasters, public health and medical emergencies, and other occurrences requiring an emergency response.
Incident Action Plan (IAP)	An oral or written plan containing general objectives reflecting the overall strategy for managing an incident. It may include the identification of operational resources and assignments. It may also include attachments that provide direction and important information for management of the incident during one or more operational periods.
Incident Commander (IC):	The individual responsible for all incident activities, including the development of strategies and tactics and the ordering and the release of resources. The IC has overall authority and responsibility for conducting incident operations and is responsible for the management of all incident operations at the incident site.
Incident Command Post (ICP)	The field location at which the primary tactical-level, on-scene incident command functions are performed. The ICP may be collocated with the incident base or other incident facilities and is normally identified by a green rotating or flashing light.

Incident Command System (ICS)	A standardized on-scene emergency management construct specifically designed to provide for the adoption of an integrated organizational structure that reflects the complexity and demands of single or multiple incidents, without being hindered by jurisdictional boundaries. ICS is the combination of facilities, equipment, personnel, procedures, and communications operating within a common organizational structure, designed to aid in the management of resources during incidents. It is used for all kinds of emergencies and is applicable to small as well as large and complex incidents. ICS is used by various jurisdictions and functional agencies, both public and private, to organize field-level incident management operations.
Incident Management Team (IMT)	The Incident Commander and appropriate Command and General Staff personnel assigned to an incident.
KPI	Key Performance Indicator
Leq	The instantaneous measure of energy in a sound wave is constantly changing for everything except artificially generated tones. The mathematical process of taking an average of the sound energy over a given period results in a measure called the Leq. This gives the level of a steady tone which would have the same energy as the variable noise over the same period. In effect this means smoothing out the continually changing peaks and troughs to get a single average value. In sound measurements, this is normally given as an average over an 8-hour working day.
Lower Exposure Action Value = 80dB(A) 135dB (C-weighted)	The A-weighted the sound level (averaged over an 8-hour day) which is the threshold at which hearing protection should be made available to workers. Use is not compulsory. The equivalent level in peak noise is 135 dB(C)
Mag-n-Bag	Magnetometer and Bag Check; a method of security screening similar to that done at an airport.
Major Incident	An incident that does or is likely in the future to require the implementation of special or non-routine arrangements and resources from one or more emergency services. A major incident would typically involve the local authorities for:<ul><li>The initial treatment, rescue and transport of a large number of casualties;</li><li>The involvement either directly or indirectly of large numbers of people;</li><li>The handling of a large number of enquiries likely to be generated both from the public and the news media, usually to the police;</li><li>The need for the large scale combined resources of two or more of the emergency services;</li><li>The mobilization and organization of the emergency services and supporting organizations (e.g., local authority, to cater for the threat of death, serious injury or homelessness to a large number of people).</li></ul>

Major Disaster	As defined under the Robert T. Stafford Disaster Relief and Emergency Assistance Act (42 U.S.C. 5122), a major disaster is any natural catastrophe (including any hurricane, tornado, storm, high water, wind-driven water, tidal wave, tsunami, earthquake, volcanic eruption, landslide, mudslide, snowstorm, or drought), or, regardless of cause, any fire, flood, or explosion, in any part of the United States, which in the determination of the President causes damage of sufficient severity and magnitude to warrant major disaster assistance under this Act to supplement the efforts and available resources of States, tribes, local governments, and disaster relief organizations in alleviating the damage, loss, hardship, or suffering caused thereby.
Mass Gathering	A subset of a special event (defined below). Mass gatherings are usually found at special events that attract large numbers of spectators or participants. Both special events and mass gatherings require the kind of additional planning identified in special events. (FEMA's *Special Events Contingency Planning Job Aids Manual*, 2005, Updated 2010).
Means of Egress	a.k.a., Means of Egress System. A continuous and unobstructed path of vertical and horizontal egress travel from any occupied portion of a building or structure to a public way. (International Fire Code, 2009)
Membrane Structure	An air-inflated, air-supported, cable or frame-covered structure (usually further defined in the building code)(International Fire Code, 2009).
Minor Incident	A simple, undesired event (a) that adversely affects a task or process, (b) whose consequences can be managed through normal service delivery, and (c) which is not likely to escalate. A minor incident may or may not require the involvement of local authorities. Low-level crime, lost children and minor injuries may all fall into this category.
Mitigation	The activities designed to reduce or eliminate risks to persons or property or to lessen the actual or potential effects or consequences of an incident. Mitigation measures may be implemented prior to, during, or after an incident. Mitigation measures are often formed by lessons learned from prior incidents. Mitigation involves ongoing actions to reduce exposure to, probability of, or potential loss from hazards. Measures may include zoning and building codes, floodplain buyouts, and analysis of hazard- related data to determine where it is safe to build or locate temporary facilities. Mitigation can include efforts to educate governments, businesses, and the public on measures they can take to reduce loss and injury.
MSDS	Material Safety Data Sheets. See "SDS."

National Incident Management System (NIMS)	A system mandated by HSPD-5 that provides a consistent nationwide approach for Federal, State, local, and tribal governments; the private sector; and nongovernmental organizations to work effectively and efficiently together to prepare for, respond to, and recover from domestic incidents, regardless of cause, size, or complexity. To provide for interoperability and compatibility among Federal, State, local, and tribal capabilities, the NIMS includes a core set of concepts, principles, and terminology. HSPD-5 identifies these as the ICS; multiagency coordination systems; training; identification and management of resources (including systems for classifying types of resources); qualification and certification; and the collection, tracking, and reporting of incident information and incident resources.
NCC	National Command Center
Near Miss	Incidents where no property was damaged and no personal injury sustained, but where, given a slight shift in time or position, damage and/or injury easily could have occurred. The reporting and investigation of near misses has great potential to reveal how accidents and incidents can be prevented.
NGO	Non-Governmental Organization
Occupant Load	a.k.a., Design Occupant Load. The number of persons for which the means of egress of a building or portion thereof is designed (International Fire Code, 2009).
Panic Hardware	A door-latching assembly incorporating a device that releases the latch upon the application of a force in the direction of egress travel (International Fire Code, 2009).
Permissible Exposure Limits (PELs)	OSHA sets enforceable permissible exposure limits (PELs) to protect workers against the health effects of exposure to hazardous substances. PELs are regulatory limits on the amount or concentration of a substance in the air. They may also contain a skin designation. OSHA PELs are based on an 8-hour time weighted average (TWA) exposure. Permissible exposure limits (PELs) are addressed in specific standards for the general industry, shipyard employment, and the construction industry. (https://www.osha.gov/dsg/topics/pel/)
PPE	Personal Protective Equipment
PSA	Pedestrian Screening Area
Public Way	In terms of fire safety, a street, alley or other parcel of land open to the outside air leading to a public street. The International Fire Code (2009) also defines "public way" as having a minimum clear width and height of not less than 10 feet (3.048 m).
Pyrotechnics	Controlled exothermic chemical reactions that are timed to create the effects of heat, gas, sound, dispersion of aerosols, emission of visible electromagnetic radiation, or a combination of these effects to provide the maximum effect from the least volume (NFPA 1126, 2011).
Pyrotechnic Operator	The person who has overall responsibility fot the operation and safety of a pyrotechnic display (NFPA 1126, 2011).

Qualified Person (OSHA)	One who, by possession of a recognized degree, certificate, or professional standing, or who by extensive knowledge, training and experience, has successfully demonstrated his ability to solve or resolve problems relating to the subject matter, the work or the project (29 CFR 1926.32(m)).
Ready Box	A sturdy container for storage of fireworks devices to be reloaded at the discharge site of a display (NFPA 1123, 2014).
Risk	The probability that a hazard will cause actual harm. (George Thompson-Focul Guide to Safety in Live Performance. Also see *"Hazard"* above.)
SDS	Safety Data Sheets (formerly MSDS); manufacturer provided paperwork regarding a product's hazard identification and safe usage
SNR	Single Number Rating – an overall measure of the attenuation of hearing protection. The SNR gives an indication of how good a product may be at reducing overall volume, however it does not identify how frequencies are absorbed across the spectrum.
SOW	The division of work to be performed under a contract or subcontract in the completion of a project, typically broken out into specific tasks with deadlines. (BusinessDictionary.com)
Span of Control	The number of individuals a supervisor is responsible for, usually expressed as the ratio of supervisors to individuals. (Under the NIMS, an appropriate span of control is between 1:3 and 1:7.)
Special Effect (Pyrotechnics)	A visual or audible effect used for entertainment purposes, often produced to create an illusion (NFPA 1126, 2011).
Special Event	A non-routine activity within a community that brings together a large number of people. Emphasis is not placed on the total number of people attending but rather the impact on the community's ability to respond to a large-scale emergency or disaster or the exceptional demands that the activity places on response services. A community's special event requires additional planning, preparedness, and mitigation efforts of local emergency response and public safety agencies (FEMA's *Special Events Contingency Planning Job Aids Manual*, 2005, Updated 2010).
Standard Operating Procedure (SOP)	Complete reference document or an operations manual that provides the purpose, authorities, duration, and details for the preferred method of performing a single function or a number of interrelated functions in a uniform manner.
Steward	In this body of work, "steward" is used to identify a category of event staffing who are not security yet interact with the public and/or attendees to an event. A synonym may be "Host." Stewards may be paid staff or volunteer staff.
Storage Day Box, Type 3 (fireworks)	Portable outdoor explosive magazines for the temporary storage of high explosives while attended (for example, a "day-box"), subject to the limitations prescribed by 27 CFR 555.206 and 555.213. Other classes of explosives materials may also be stored in type 3 magazines. (27 CFR 555.203[c]).
Strike Team	A specified combination of the same kind and type of resources with common communications and a Leader.

Table-top Session	An effective training exercise, this is typically a meeting of those persons involved in the emergency planning and management aspects of the event to talk through, test and reveal weaknesses of the various emergency plans in place. On an event operations level, this can also be a meeting of event staff to talk-through various event operations scenarios.
Tactics	Deploying and directing resources on an incident to accomplish incident strategy and objectives.
Task Force	A combination of single resources assembled for a particular tactical need with common communications and a Leader.
Tent	A structure, enclosure or shelter, with or without sidewalls or drops, constructed of fabric or pliable materials supported by any manner except by air or the contents that it protects (International Fire Code, 2009).
Threat	An indication of possible violence, harm, or danger.
TMV	Theoretical market value
Unity of Command	The concept by which each person within an organization reports to one and only one designated person. The purpose of unity of command is to ensure unity of effort under one responsible commander for every objective
Upper Exposure Action Value = 85dB(A)	The A-weighted the sound level (averaged over an 8-hour day) which is the threshold at which the use of suitable hearing protection is mandatory to maintain the person. The equivalent level in peak noise is 137 dB(C)
VSA	Vehicle Screening Area
Weekly Exposure	Given that work tasks and hence noise exposure can vary dramatically from day to day, CNAW allows a weekly average to be taken for a persons' Noise Dose. This means warehouse and office days can be balanced against louder activities such as rehearsals or performances. Overall the personal weekly exposure must not exceed the Exposure Limit Value noted above.
WFX	Workforce Volunteers

39. Useful Addresses

International Association of Venue Managers - IAVM 635 Fritz Drive, Suite 100 Coppell, TX 75019-4442 972.906.7441 http://www.iavm.org	National Fire Protection Association 1 Batterymarch Park Quincy, Massachusetts 02169-7471 617.770.3000 http://www.nfpa.org
U.S. Department of Labor Occupational Safety & Health Administration - OSHA 200 Constitution Ave., NW Washington DC 20210 800.321.6742 http://www.osha.gov	Reed Construction Data - State Building Codes http://www.reedconstructiondata.com/building-codes
EPA - United States Environmental Protection Agency http://www.epa.gov	United States Government http://www.usa.gov
US Department of Health & Human Services 200 Independence Avenue SW Washington DC 20201 877.696.6775 http://www.hhs.gov	Society of Cable Telecommunications Engineers - SCTE 140 Phillips Road Exton, PA 19341 -1318 800.542.5040 http://www.scte.org/default.aspx
ANS/SCTE I - 121 2011 http://www.scte.org/documents/pdf/standards/ANSI_SCTE_121_2011.pdf	Professional Lighting And Sound Association - PLASA 630 Ninth Avenue Suite 609 New York, NY 10036 212.244.1505 https://www.plasa.org
International Alliance of Theatrical Stage Employees - IATSE 1430 Broadway 20th Floor New York, NY 10018 212.730.1770 http://www.iatse-intl.org	National Weather Service - NWS 1325 East West Highway Silver Spring, MD 20910 http://www.weather.gov
National Oceanic and Atmospheric Administration - NOAA 1401 Constitution Avenue, NW Room 5128 Washington DC 20230 http://www.noaa.gov	National Safety Council 1121 Spring Lake Drive Itasca, IL 60143-3201 800.621.7615 http://www.nsc.org/Pages/Home.aspx

Centers for Diseases Control - CDC 1600 Clifton Road Atlanta, GA 30333 800.232.4636 http://www.cdc.gov	American Council of Engineering Companies 1015 15th Street 8th Floor NW Washington, DC 20005-2605 202.347.7474 http://www.acec.org
The American Institute of Architects 1735 New York Avenue NW Washington DC 20006-5292 800.242.3837 http://www.aia.org	International Code Council - ICC 500 New Jersey Avenue, NW 6th Floor Washington DC 20001 888.422.7233 http://www.iccsafe.org/Pages/default.aspx
The Risk Management Society - RIMS 1065 Avenue of the Americas 13th Floor New York, NY 10018 212.286.9292 http://www.rims.org/Pages/Default.aspx	National Council of Structural Engineers Associations - NCSEA 645 N. Michigan Avenue, Suite 540 Chicago, IL 60611 312.649.4600 http://www.ncsea.com
National Electrical Contractors Association 3 Bethesda Metro Center, Suite 1100 Bethesda, MD 20814 301.657.3110 http://www.necanet.org	National Electrical Installation Standards http://www.neca-neis.org
The Event Safety Shop 59 Prince Street Bristol, BS1 4QH United Kingdom +44 (0) 117 904 6204 http://www.the-eventsafetyshop.co.uk/index.php	U.S. Fire Administration 16825 S. Seton Ave., Emmitsburg, MD 21727 301.447.1000 http://www.usfa.fema.gov/index.shtm
USA Football - Football's National Governing Body 45 N. Pennsylvania St., Suite 700 Indianapolis, IN 46204 http://usafootball.com	National Association of Police Organizations - NAPO 317 South Patrick Street Alexandria, Virginia 22314 703.549.0775 http://www.napo.org E-mail: info@napo.org
PLASA 630 Ninth Avenue, Suite 609 New York, NY 10036, USA 212.244.1505 www.plasa.org	American Supply Association 1200 N. Arlington Heights Rd., Suite 150 Itasca, IL 60143 630.467.0000 http://www.asa.net/Education/Safety-Resources.aspx

Appendix A – The National Incident Management System (NIMS) and Incident Command System (ICS)

This Appendix is derived from Chapter 3 of FEMA's *Special Events Contingency Planning Job Aids Manual* (2005, Updated 2010).

Introduction

The importance of pre-event planning, organization, and leadership cannot be over emphasized. Many suggest a planning team use the Incident Command System (ICS) to manage the event planning process effectively. In a large-scale event involving numerous agencies, people can become confused as to who is in charge, what role everyone plays, and what responsibilities everyone has. ICS is an excellent tool that can resolve these issues. This appendix discusses ICS, how it can be applied to special events, and the concept of Unified Command.

Unfortunately, even the best-planned special events may not run entirely smoothly. During any special event, you must be prepared to respond to one or more incidents that may occur during the event. The way these incidents are managed has a great deal to do with the ultimate success of the special event. Everyone must know his or her role and tasks, and where to seek information. This appendix also discusses the use of ICS during these situations.

National Incident Management System

The National Incident Management System (NIMS) provides a systematic, proactive approach to guide departments and agencies at all levels of government, nongovernmental organizations, and the private sector to work seamlessly to prevent, protect against, respond to, recover from, and mitigate the effects of incidents, regardless of cause, size, location, or complexity, in order to reduce the loss of life and property and harm to the environment. NIMS works hand in hand with the National Response Framework (NRF). NIMS provides the template for the management of incidents, while the NRF provides the structure and mechanisms for national-level policy for incident management.

NIMS integrates existing best practices into a consistent, nationwide, systematic approach to incident management that is applicable at all levels of government, nongovernmental organizations (NGOs), and the private sector, and across functional disciplines in an all-hazards context. Five major components make up this systems approach: Preparedness, Communications and Information Management, Resource Management, Command and Management, and Ongoing Management and Maintenance.

The components of NIMS were not designed to stand alone, but to work together in a flexible, systematic manner to provide the national framework for incident management. The Emergency Management Institute (EMI), located at the National Emergency Training Center in Emmitsburg, MD, offers a broad range of NIMS-related training. Additional information about NIMS and ICS training can be found at http://training.fema.gov.

Preparedness: Overview

NIMS provides the mechanisms for emergency management/response personnel and their affiliated organizations to work collectively by offering a consistent and common approach to preparedness.

Preparedness is achieved and maintained through a continuous cycle of planning, organizing, training, equipping, exercising, evaluating, and taking corrective action. Ongoing preparedness efforts among all those involved in emergency management and incident response activities ensure coordination during times of crisis. Moreover, preparedness facilitates efficient and effective emergency management and incident response activities.

This component describes specific measures and capabilities that emergency management/response personnel and their affiliated organizations should develop and incorporate into their overall preparedness programs to enhance the operational preparedness necessary for all-hazards emergency management and incident response activities. In developing, refining, and expanding preparedness programs and activities within their jurisdictions and/or organizations, emergency management/response personnel should leverage existing preparedness efforts and collaborative relationships to the greatest extent possible. Personal preparedness, while an important element of homeland security, is distinct from the operational preparedness of our Nation's emergency management and incident response capabilities and is beyond the scope of NIMS.

Communications and Information Management: Overview

Effective emergency management and incident response activities rely on flexible communications and information systems that provide a common operating picture to emergency management/response personnel and their affiliated organizations. Establishing and maintaining a common operating picture and ensuring accessibility and interoperability are the principal goals of the Communications and Information Management component of NIMS. Properly planned, established, and applied communications enable the dissemination of information among command and support elements and, as appropriate, cooperating agencies and organizations.

Incident communications are facilitated through the development and use of common communications plans and interoperable communications equipment, processes, standards, and architectures. During an incident, this integrated approach links the operational and support units of the various organizations to maintain communications connectivity and situational awareness. Communications and information management planning should address the incident-related policies, equipment, systems, standards, and training necessary to achieve integrated communications.

Resource Management: Overview

Emergency management and incident response activities require carefully managed resources (personnel, teams, facilities, equipment, and/or supplies) to meet incident needs. Utilization of the standardized resource management concepts such as typing, inventorying, organizing, and tracking will facilitate the dispatch, deployment, and recovery of resources before, during, and after an incident.

Resource management should be flexible and scalable in order to support any incident and be adaptable to changes. Efficient and effective deployment of resources requires that resource management concepts and principles be used in all phases of emergency management and incident response.

The resource management process can be separated into two parts: resource management as an element of preparedness and resource management during an incident. The preparedness activities (resource typing, credentialing, and inventorying) are conducted on a continual basis to help ensure that resources are ready to be mobilized when called to an incident. Resource management during an incident is a finite process, as shown in the below figure, with a distinct beginning and ending specific to the needs of the particular incident.

Command and Management: Overview

The NIMS components of Preparedness, Communications and Information Management, and Resource Management provide a framework for effective management during incident response.

Command and Management Overview: Incident Command System

The Incident Command System (ICS) is a standardized, on-scene, all-hazards incident management approach that:
- Allows for the integration of facilities, equipment, personnel, procedures, and communications operating within a common organizational structure.
- Enables a coordinated response among various jurisdictions and functional agencies, both public and private.
- Establishes common processes for planning and managing resources.

- ICS is flexible and can be used for incidents of any type, scope, and complexity. ICS allows its users to adopt an integrated organizational structure to match the complexities and demands of single or multiple incidents.

ICS is used by all levels of government—Federal, State, tribal, and local—as well as by many nongovernmental organizations and the private sector. ICS is also applicable across disciplines. It is typically structured to facilitate activities in five major functional areas: Command, Operations, Planning, Logistics, and Finance/Administration. All of the functional areas may or may not be used based on the incident needs. Intelligence/Investigations is an optional sixth functional area that is activated on a case-by-case basis.

As a system, ICS is extremely useful; not only does it provide an organizational structure for incident management, but it also guides the process for planning, building, and adapting that structure. Using ICS for every incident or planned event helps hone and maintain skills needed for the large-scale incidents.

ICS Management Principle	Description
Common Terminology	ICS establishes common terminology that allows diverse incident management and support organizations to work together across a wide variety of incident management functions and hazard scenarios. This common terminology covers the following: - Organizational Functions: Major functions and functional units with incident management responsibilities are named and defined. Terminology for the organizational elements is standard and consistent. - Resource Descriptions: Major resources—including personnel, facilities, and major equipment and supply items—that support incident management activities are given common names and are "typed" with respect to their capabilities, to help avoid confusion and to enhance interoperability. - Incident Facilities: Common terminology is used to designate the facilities in the vicinity of the incident area that will be used during the course of the incident. Incident response communications (during exercises and actual incidents) should feature plain language commands so they will be able to function in a multijurisdiction environment. Field manuals and training should be revised to reflect the plain language standard.

ICS Management Principle	Description
Modular Organization	The ICS organizational structure develops in a modular fashion based on the size and complexity of the incident, as well as the specifics of the hazard environment created by the incident. When needed, separate functional elements can be established, each of which may be further subdivided to enhance internal organizational management and external coordination. Responsibility for the establishment and expansion of the ICS modular organization ultimately rests with Incident Command, which bases the ICS organization on the requirements of the situation. As incident complexity increases, the organization expands from the top down as functional responsibilities are delegated. Concurrently with structural expansion, the number of management and supervisory positions expands to address the requirements of the incident adequately.
Management by Objectives	Management by objectives is communicated throughout the entire ICS organization and includes: • Establishing overarching incident objectives. • Developing strategies based on overarching incident objectives. • Developing and issuing assignments, plans, procedures, and protocols. • Establishing specific, measurable tactics or tasks for various incident management functional activities, and directing efforts to accomplish them, in support of defined strategies. • Documenting results to measure performance and facilitate corrective actions.
Incident Action Planning	Centralized, coordinated incident action planning should guide all response activities. An Incident Action Plan (IAP) provides a concise, coherent means of capturing and communicating the overall incident priorities, objectives, and strategies in the contexts of both operational and support activities. Every incident must have an action plan. However, not all incidents require written plans. The need for written plans and attachments is based on the requirements of the incident and the decision of the Incident Commander or Unified Command. Most initial response operations are not captured with a formal IAP. However, if an incident is likely to extend beyond one operational period, become more complex, or involve multiple jurisdictions and/or agencies, preparing a written IAP will become increasingly important to maintain effective, efficient, and safe operations.

ICS Management Principle	Description
Manageable Span of Control	Span of control is key to effective and efficient incident management. Supervisors must be able to adequately supervise and control their subordinates, as well as communicate with and manage all resources under their supervision. In ICS, the span of control of any individual with incident management supervisory responsibility should range from 3 to 7 subordinates, with 5 being optimal. During a large-scale law enforcement operation, 8 to 10 subordinates may be optimal. The type of incident, nature of the task, hazards and safety factors, and distances between personnel and resources all influence span-of-control considerations.
Incident Facilities and Locations	Various types of operational support facilities are established in the vicinity of an incident, depending on its size and complexity, to accomplish a variety of purposes. The Incident Command will direct the identification and location of facilities based on the requirements of the situation. Typical designated facilities include Incident Command Posts, Bases, Camps, Staging Areas, mass casualty triage areas, point-of-distribution sites, and others as required.
Comprehensive Resource Management	Maintaining an accurate and up-to-date picture of resource utilization is a critical component of incident management and emergency response. Resources to be identified in this way include personnel, teams, equipment, supplies, and facilities available or potentially available for assignment or allocation.
Integrated Communications	Incident communications are facilitated through the development and use of a common communications plan and interoperable communications processes and architectures. The ICS 205 form is available to assist in developing a common communications plan. This integrated approach links the operational and support units of the various agencies involved and is necessary to maintain communications connectivity and discipline and to enable common situational awareness and interaction. Preparedness planning should address the equipment, systems, and protocols necessary to achieve integrated voice and data communications.
Establishment and Transfer of Command	The command function must be clearly established from the beginning of incident operations. The agency with primary jurisdictional authority over the incident designates the individual at the scene responsible for establishing command. When command is transferred, the process must include a briefing that captures all essential information for continuing safe and effective operations.

ICS Management Principle	Description
Chain of Command and Unity of Command	• Chain of Command: Chain of command refers to the orderly line of authority within the ranks of the incident management organization. • Unity of Command: Unity of command means that all individuals have a designated supervisor to whom they report at the scene of the incident. These principles clarify reporting relationships and eliminate the confusion caused by multiple, conflicting directives. Incident managers at all levels must be able to direct the actions of all personnel under their supervision.
Unified Command	In incidents involving multiple jurisdictions, a single jurisdiction with multiagency involvement, or multiple jurisdictions with multiagency involvement, Unified Command allows agencies with different legal, geographic, and functional authorities and responsibilities to work together effectively without affecting individual agency authority, responsibility, or accountability.
Accountability	Accountability: Effective accountability of resources at all jurisdictional levels and within individual functional areas during incident operations is essential. Adherence to the following ICS principles and processes helps to ensure accountability: • Resource Check-In/Check-Out Procedures • Incident Action Planning • Unity of Command • Personal Responsibility • Span of Control • Resource Tracking
Dispatch / Deployment	Resources should respond only when requested or when dispatched by an appropriate authority through established resource management systems. Resources not requested must refrain from spontaneous deployment to avoid overburdening the recipient and compounding accountability challenges.
Information and Intelligence Management	The incident management organization must establish a process for gathering, analyzing, assessing, sharing, and managing incident-related information and intelligence.

Command and Management Overview:
Multiagency Coordination Systems

Multiagency coordination is a process that allows all levels of government and all disciplines to work together more efficiently and effectively. Multiagency coordination occurs across the different disciplines involved in incident management, across jurisdictional lines, or across levels of government. Multiagency coordination can and does occur on a regular basis whenever personnel from different agencies interact in such activities as preparedness, prevention, response, recovery, and mitigation.

Often, cooperating agencies develop a Multiagency Coordination System (MACS) to better define how they will work together and to work together more efficiently; however, multiagency coordination can take place without established protocols. MACS may be put in motion regardless of the location, personnel titles, or organizational structure.

Initially the Incident Command/Unified Command and the Liaison Officer may be able to provide all needed mulitagency coordination at the scene. However, as the incident grows in size and complexity, off-site support and coordination may be required.

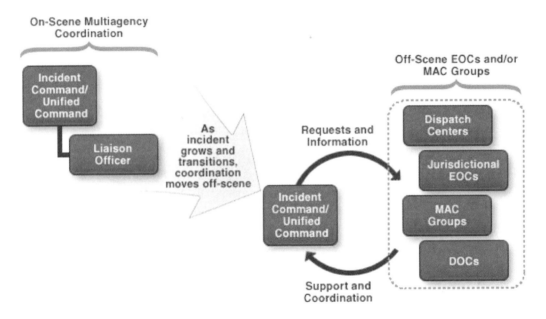

Integral elements of MACS are dispatch procedures and protocols, the incident command structure, and the coordination and support activities taking place within an activated Emergency Operations Center. Fundamentally, MACS provide support, coordination, and assistance with policy-level decisions to the ICS structure managing an incident.

Command and Management Overview: Public Information

Public Information consists of the processes, procedures, and systems to communicate timely, accurate, and accessible information on the incident's cause, size, and current situation to the public, responders, and additional stakeholders (both directly affected and indirectly affected). Public information must be coordinated and integrated across jurisdictions, agencies, and organizations; among Federal, State, tribal, and local governments; and with NGOs and the private sector.

Well-developed public information, education strategies, and communications plans help to ensure that lifesaving measures, evacuation routes, threat and alert systems, and other public safety information are coordinated and communicated to numerous audiences in a timely, consistent manner.

A Joint Information System (JIS) provides the mechanism to organize, integrate, and coordinate information to ensure timely, accurate, accessible, and consistent messaging across multiple jurisdictions and/or disciplines with nongovernmental organizations and the private sector. A JIS includes the plans, protocols, procedures, and structures used to provide public information. Federal, State, tribal, territorial, regional, or local Public Information Officers and established Joint Information Centers (JICs) are critical supporting elements of the JIS.

A Joint Information Center (JIC) is a central location that facilitates operation of the Joint Information System. The JIC is a location where personnel with public information responsibilities perform critical emergency information functions, crisis communications, and public affairs functions. JICs may be established at various levels of government or at incident sites, or can be components of Multiagency Coordination Systems. A single JIC location is preferable, but the system is flexible and adaptable enough to accommodate virtual or multiple JIC locations, as required.

Incident Command System Forms

Copies of the following Incident Command System forms can be found on the NIMS Resource Center at http://www.fema.gov/nims:

ICS 201, Incident Briefing	ICS 213, General Message
ICS 202, Incident Objectives	ICS 214, Unit Log
ICS 203, Organization Assignment List	ICS 215, Operational Planning Worksheet
ICS 204, Assignment List	ICS 215a, Incident Safety Analysis
ICS 205, Incident Radio Communications Plan	ICS 216, Radio Requirements Worksheet
ICS 206, Medical Plan	ICS 217, Radio Freq. Assignment Worksheet
ICS 207, Organizational Chart	ICS 218, Support Vehicle Inventory
ICS 209, Incident Status Summary	ICS 220, Air Operations Summary
ICS 210, Status Change Card	ICS 221, Demobilization Plan
ICS 211, Check-In List	ICS 308, Resource Order Form

Incidents Occurring During a Special Event

As discussed above, certain incidents occurring during a special event may dictate the need for a specific Incident Commander to manage that particular incident (e.g., isolated structure fire, vehicle crash, HazMat incident, structure collapse, multiple casualty incident, etc.).

When an incident occurs within a special event, immediate action must be taken to control and manage the incident. As the incident grows, the issues that must be considered will grow as well. The Incident Commander of the special event may assign command of the emergency incident to a ranking responder. This responder must take initial steps to bring order to the incident, just as in situations that require more traditional applications of ICS.

The Incident Commander of the special event may authorize the responder to implement his or her own command structure and/or call upon the resources of the event command structure. This responder must:

- Assess the situation.
- Determine whether human life is at immediate risk.
- Establish the immediate priorities and objectives.
- Determine whether there are adequate and appropriate resources on-scene or ordered.
- Establish an appropriately located on-scene Command Post (CP), if needed.
- Establish an appropriate initial command structure, if needed.
- Develop an action plan.
- Ensure that adequate safety measures are in place.
- Coordinate activity for all Command and General Staff.
- Consider whether the span of control is approaching, or will soon approach, practical limits, taking into account the safety of all personnel.
- Determine whether there are any environmental concerns that must be considered.
- Monitor work progress and coordinate with key people.
- Review and modify objectives and adjust the action plan as necessary.
- Approve requests for additional resources or for the release of resources.
- Keep the overall event Incident Commander informed of incident status.
- Authorize release of information to the news media.
- Order the demobilization of the incident, when appropriate.

Appendix B1 – Bomb Threat Checklist

BOMB THREAT CHECKLIST
Place by each telephone. Duplicate as necessary.
(Source: Special Events Contingency Planning Job Aids Manual, 2005, updated 2010, p. A-68)

Exact date and time of call: _____

Exact words of caller: _____

_____

Questions to ask
1. When is the bomb going to explode? _____
2. Where is the bomb? _____
3. What does it look like? _____
4. What kind of bomb is it? _____
5. What will cause it to explode? _____
6. Did you place the bomb? _____
7. Why? _____
8. Where are you calling from? _____
9. What is your address? _____
10. What is your name? _____

Caller's Voice (Please circle appropriate terms.)

calm	disguised	nasal	angry	broken
stutter	slow	sincere	lisp	rapid
giggling	deep	crying	squeaky	excited
stressed	accent	loud	slurred	normal

If voice is familiar, whom did it sound like? _____
Were there any background noises? _____
Remarks:_____

_____

Person receiving call: _____
Telephone number where call was received: _____
Report call immediately to: _____
(Refer to bomb incident plan.)

Appendix B2 – Bomb Threat Stand-Off

This table is derived from the Appendix of FEMA's Special Events Contingency Planning Job Aids Manual (2005, Updated 2010, p. A-69).

Threat Description	Explosive Capacity	Lethal Airblast Range	Mandatory Evacuation Distance	Desired Evacuation Distance
Pipe Bomb	5 LBS / 2.3 KG	25 FT / 8 M	70 FT / 21 M	850 FT / 259 M
Briefcase or Suitcase Bomb	50 LBS / 23 KG	40 FT / 12 M	150 FT / 46 M	1850 FT / 564 M
Compact Sedan	220 LBS / 100 KG	60 FT / 18 M	240 FT / 73 M	915 FT / 279 M
Sedan	500 LBS / 227 KG	100 FT / 30 M	320 FT / 98 M	1050 FT / 320 M
Van	1000 LBS / 454 KG	125 FT / 38 M	400 FT / 122 M	1200 FT / 366 M
Moving Van or Delivery Truck	4000 LBS / 1814 KG	200 FT / 61 M	640 FT / 195 M	1750 FT / 534 M
Semi-Trailer	40,000 LBS / 18,144 KG	450 FT / 137 M	1400 FT / 427 M	3500 FT / 1607 M

Explosive Capacity is based on maximum volume or weight of explosives (TNT equivalent) that could reasonably fit or be hidden in a suitcase or vehicle.

Lethal Airblast Range is the minimum distance personnel in the open are expected to survive blast effects. This minimum range is based on anticipation of avoiding severe lung damage or fatal impact injury from body translation.

Mandatory Evacuation Distance is the range within which all buildings must be evacuated. From this range outward to the Desired Evacuation Distance, personnel may remain inside buildings but away from windows and exterior walls. Evacuated personnel must move to the Desired Evacuation Distance.

Appendix C – Requirements for Outdoor Event Structures, Preparation Checklist

Use this list as a tool in conjunction with requirement details in *Structures* chapter.

	Action	Completion Date	Verified
1.	Structure assembly drawings		
2.	Stamped engineering calculations		
3.	Rigged component list with description		
4.	Rigging plot overlay on structure		
5.	Site layout drawing		
6.	Permits		
7.	Inspection records of components		
8.	Operations Management Plan		
9.	Local weather service resource established		
10.	Responsible individuals identified		
11.	Pre-event meeting reviewing OMP		
12.	Completion certificate of structure		

Appendix D – Requirements for Outdoor Event Structures, Key Personnel

Use this list as a tool in conjunction with requirement details in Structures chapter. Positions listed must have a designated person.

Title	Name	Phone	Main Work Area (i.e. stage, FOH, office)
Stage Manager			
Artists' Representative			
Promoter Representative			
Temporary Structure Vendor Crew Lead			
Weather Monitor			
Crowd Management Representative			
Weather Action Team Leader			